WITHDRAWN-UNL

Industrial Crystallization

Industrial Crystallization

Edited by
J. W. Mullin
University of London

Plenum Press . New York and London

Library of Congress Cataloging in Publication Data

Symposium on Industrial Crystallization, 6th, Ústí nad Labem, Czechoslovak Republic, 1975.
Industrial crystallization.

Symposium held Sept. 1-3, 1975.
Includes index.
1. Crystallization—Congresses. I. Mullin, John William. II. Title.
PT156.C57S95 1975   660.2'84298   76-10859
ISBN 0-306-30945-9

Proceedings of the Sixth Symposium on Industrial Crystallization held at Ústí nad Labem, Czechoslovakia, September 1-3, 1975

©1976 Plenum Press, New York
A Division of Plenum Publishing Corporation
227 West 17th Street, New York, N.Y. 10011

All rights reserved

No part of this book may be reproduced, stored in a retrieval system, or transmitted, in any form or by any means, electronic, mechanical, photocopying, microfilming, recording, or otherwise, without written permission from the Publisher

Printed in the United States of America

Preface

Industrial Crystallization Symposia have been organized by the Crystallization Research Group at the Czechoslovak Research Institute for Inorganic Chemistry, Usti nad Labem, since 1960. Over the years, the increasing popularity of the unit operation of crystallization has been clearly demonstrated by the steady increase in numbers of both the papers presented and the attendances at the meetings.

The 6th Symposium (1-3 September 1975) was organized jointly with the European Federation of Chemical Engineering Working Party on Crystallization, and the 44 papers presented were arranged into four sessions - A: Secondary Nucleation, B: Crystal Growth Kinetics, C: Crystal Habit Modification, D: Crystallizer Design, E: Industrial Crystallizer Operation and Case Studies. The same groupings are preserved in this edited version of the proceedings.

This is the first time that the Industrial Crystallization Symposium papers have appeared in one volume. After the 5th (1972) Symposium, authors were encouraged to submit their papers to an international journal specializing in crystallization. However, the results were not altogether satisfactory in that less than one third of the papers presented at the meeting were offered for consideration. This time, therefore, the organizing committee decided to attempt to keep the papers together by making arrangements for their publication by Plenum Press.

A distinctive feature of the Industrial Crystallization Symposia is the intimate blend of fundamental and applied research work on a truly world-wide basis. At the 1975 meeting papers were presented by authors from Bulgaria, Czechoslovakia, DDR, France, Holland, Japan, Poland, Spain, Switzerland, UK, USA, USSR, West Germany and Yugoslavia. The work reported indicates the tremendous diversity of interest in a subject whose interdependent parts range from solid-state physics to phase equilibria, from diffusional mass transfer to fluid mechanics, from strength of materials to solution structure.

PREFACE

In editing this version of the proceedings of the 6th Symposium I have received invaluable help from many sources. I should like first to thank my colleagues on the symposium organizing committee, Dr. J. Nývlt and Dr. R. Rychly (Czechoslovakia), Prof. E.J. de Jong (Holland), Prof. U. Fasoli (Italy), Dr. R.F. Strickland-Constable (UK) and Dr. R. Messing (West Germany), for their help with the preliminary refereeing procedures. I am also indebted to the symposium session chairmen, and to others who participated in the meeting, for help and advice in collating and clarifying many of the papers. And last, but by no means least, I want to thank Miss Valerie H. Potter and her colleagues at University College London for undertaking the daunting task of retyping a substantial proportion of the manuscripts.

London, 1976

J.W. MULLIN

Contents

Session A
SECONDARY NUCLEATION

Secondary Nucleation - A Review 3
   G.D. Botsaris

Attrition and Secondary Nucleation in
   Agitated Crystal Slurries 23
   P.D.B. Bujac

Collision Breeding from Solution or Melt 33
   R.F. Strickland-Constable

Secondary Nucleation of Potash Alum 41
   K. Toyokura, K. Yamazoe, J. Mogi, N. Yago,
   and Y. Aoyama

Control of Particle Size in Industrial NaCl-
   Crystallization 51
   C.M. van 't Land and B.G. Wienk

Secondary Nucleation and Crystal Growth as
   Coupled Phenomena 61
   J. Estrin and G.R. Youngquist

The Grinding Mechanism in Fluid Energy Mill and
   Its Similarity to Secondary Nucleation 67
   S. Okuda

Secondary Nucleation in Agitated Crystallizers 77
   U. Fasoli and R. Conti

Session B
CRYSTAL GROWTH KINETICS

Theory and Experiment for Crystal Growth from Solution:
    Implications for Industrial Crystallization 91
  P. Bennema

Computer Simulation of Crystal Growth from Solution 113
  T.A. Cherepanova, A.V. Shirin, and V.T. Borisov

Surface Perfection as a Factor Influencing the
    Behaviour of Growth Layers on the {100} Faces
    on Ammonium Dihydrogen Phosphate (ADP) Crystals
    in Pure Solution 123
  V.R. Phillips and J.W. Mullin

Growth Rate of Citric Acid Monohydrate Crystals in
    a Liquid Fluidized Bed 135
  C. Laguerie and H. Angelino

The Occurrence of Growth Dispersion and Its
    Consequences 145
  A.H. Janse and E.J. de Jong

Kinetics of TiO_2 Precipitation 155
  O. Söhnel and J. Mareček

Investigations on the Ostwald Ripening of Well
    Soluble Salts by Means of Radioactive Isotopes 163
  A. Winzer and H.-H. Emons

Recrystallization in Suspensions 173
  J. Skřivánek, S. Žáček, J. Hostomský, and V. Vacek

Crystal Growth Kinetics of Sodium Chloride
    from Solution 187
  R. Rodriguez-Clemente

Session C
CRYSTAL HABIT MODIFICATION

Survey of Crystal Habit Modification in Solution 203
  R. Boistelle

Some Problems of Crystal Habit Modification 215
  E.V. Khamskii

CONTENTS

Solvent Effects in the Growth of Hexamethylene
Tetramine Crystals 223
   J.R. Bourne and R.J. Davey

Effect of Isodimorphously Included Co(II), Fe(II)
and Cu(II) Ions on the Crystal Structures of
$ZnSO_4 \cdot 7H_2O$ and $MgSO_4 \cdot 7H_2O$ 239
   C. Balarew and V. Karaivanova

The Effect of Ionic Impurities on the Growth of
Ammonium Dihydrogen Phosphate Crystals 245
   R.J. Davey and J.W. Mullin

Influence of Mn(II), Cu(II) and Al(III) Ions on
Size and Habit of the Ammonium Sulphate Crystals 253
   M. Broul

The Influence of Surface-Active Agents on Gypsum
Crystallization in Phosphoric Acid Solutions 263
   J. Schroeder, W. Skudlarska, A. Szczepanik,
   E. Sikorska, and S. Zielinski

Isomorphous Series in Crystals of $MeSO_4 \cdot nH_2O$ Type 269
   S. Aslanian and C. Balarew

The Transformation of Amorphous Calcium Phosphate
into Crystalline Hydroxyapatite 277
   LG. Brečević and H. Füredi-Milhofer

The Effect of Fe(III) Ions Upon the Growth of
Potassium Dihydrogen Phosphate Monocrystals 285
   J. Karniewicz, P. Posmykiewicz, and B. Wojciechowski

Session D
CRYSTALLIZER DESIGN

Crystallizer Design and Operation 291
   J.W. Mullin

The Design of a Crystallizer 303
   K. Toyokura, F. Matsuda, Y. Uike, H. Sonoda,
   and Y. Wakabayashi

Design Criteria for DTB-Vacuum Crystallizers 311
   A.G. Toussaint and J.M.H. Fortuin

Integral Design of Crystallizers as Illustrated by the
Design of a Continuous Stirred-Tank Cooling
Crystallizer 319
   C.J. Asselbergs and E.J. de Jong

Design of Batch Crystallizers 335
   J. Nývlt

Population Balance Models for Batch Crystallization 343
   H.M. Hulburt

The Crystallization of Some Chloro-Organic Compounds
   in a Batch Crystallizer with Programmed Cooling 353
   I.V. Melikhov, E.V. Mikhin, and A.M. Peckler

A New Technique for Accurate Crystal Size Distribution
   Analysis in an MSMPR Crystallizer 363
   S. Jančić and J. Garside

Session E
CRYSTALLIZER OPERATION AND CASE STUDIES

Transient Behavior in Crystallization - Design Models
   Related to Plant Experiences. 375
   J.P. Shields

Stability and Dynamic Behaviour of Crystallizers 391
   B.G.M. de Leer, A. Koning, and E.J. de Jong

The Importance of Classification in Well-Mixed
   Crystallizers 403
   A.H. Janse and E.J. de Jong

Purification of $H_3PO_4 \cdot \frac{1}{2}H_2O$ by Crystallization 413
   Y. Aoyama and K. Toyokura

The Operation of a $NaOH \cdot 3\frac{1}{2}H_2O$ Crystallizer by Direct
   Contact Cooling 421
   T. Akiya, M. Owa, S. Kawasaki, T. Goto, and K. Toyokura

Progress in Continuous Fractional Crystallization 431
   G.J. Arkenbout, A. van Kuijk, and W.M. Smit

The Prevention of Deposition on Crystallizer Cooling
   Surfaces Using Ultrasound 437
   M.J. Ashley

Case Study of Incrustation in an Industrial
   Salt Crystallizer 449
   C.M. van 't Land and K.J.A. de Waal

Operation of a Large-Scale KCl Crystallization Plant 461
   W. Wöhlk and G. Hofmann

Index 471

Secondary Nucleation

SECONDARY NUCLEATION - A REVIEW

GREGORY D. BOTSARIS

Department of Chemical Engineering

Tufts University, Medford, Massachusetts 02155, U.S.A.

1. INTRODUCTION

1.1 Definitions and Importance

Crystallizing systems, that is supersaturated systems in which crystals are already present, can form new nuclei at conditions under which primary (spontaneous) nucleation would not occur. This phenomenon leads to the definitions which will be followed in this paper:

a) Secondary is the nucleation which occurs, irrespectively of its mechanism, only because of the presence of crystals of the material being crystallized. When no crystals are present no nucleation occurs.

b) The terms primary or spontaneous nucleation will be used to characterize any nucleation that is not secondary in nature. This includes both the classical homogeneous and the heterogeneous nucleation.

Secondary nucleation is not a newly discovered phenomenon. As early as 1906, Miers[30] reported that supersaturated solutions which were stable when left undisturbed often nucleated when a seed crystal was introduced. What is rather recent, however, is the realization that in most industrial crystallizers, secondary nucleation always accompanies the spontaneous primary nucleation, and moreover, the former is apt to be the predominant mechanism of nucleation. This realization led to a widespread and continuously increasing activity in the field. Estrin[14] has presented an excellent review of the large number of recent developments. In this paper we will also discuss these developments. The emphasis, however, will be different, the main objective being to outline the present understanding of the phenomenon and point out the desired directions for the future.

Interest and research activity in secondary nucleation is currently aiming at two directions:
a) The understanding of the <u>phenomenon</u> of secondary nucleation. A greater understanding of the mechanism is essential before nucleation rate data obtained in the various studies can be extrapolated to crystallizers of different dimensions and design.
b) The study of the implications of the present understanding of the mechanism of secondary nucleation for the crystallization <u>processes.</u> The main objective here is the development of kinetic equations, which correlate the rate of secondary nucleation with the operating conditions of a crystallizer. Heavy reliance is still placed on empirical correlations. However, these correlations are continuously modified and improved by a <u>qualitative</u> transfer of the knowledge obtained in (a). A quantitative transfer is definitely the objective of future studies.

Although a large number of investigations are pursuing simultaneously both these aims, in this review these two objectives will be treated separately, in Sections 2 and 3 respectively.

Since secondary nucleation results from the interaction of a growing crystal with a supersaturated solution, these two systems will be briefly described next.

1.2. The Growing Crystal

The layer growth model (33) predicts the existence of monomolecular steps on the surface of a growing crystal. These steps originate either from two-dimensional nuclei appearing on the surface of the crystal or most probably (at low supersaturations) from screw dislocations which act as continuous sources of steps. Due to the usually high density of dislocations, steps may originate simultaneously from numerous points on the crystal. At the same time mechanisms (for instance adsorbed impurities, series of edge dislocation, etc.) operate which can cause bunching of the monomolecular steps to larger steps. The net result of all these interactions is the appearance of a surface "roughness" consisting of <u>microscopic</u> structures and protrusions even in crystals appearing visually smooth. Of course in certain cases crystals may grow dendritically. Dendrites, however, are rather <u>macroscopic</u> irregularities which can be observed visually. The degree of roughness will depend on the supersaturation of the solution in which the crystal is growing (6).

1.3. The Supersaturated Solution

The generally accepted model of a supersaturated solution predicts the existence in the solution of ordered aggregates of various sizes, the so-called embryos; these are formed by bi-molecular reactions according to the scheme:

$$\begin{aligned} A + A &\rightleftarrows A_2 \\ A_2 + A &\rightleftarrows A_3 \\ &\vdots \\ A_{i-1} + A &\rightleftarrows A_i \end{aligned} \qquad (1)$$

where A represents a single solute molecule.

There is a steady state distribution of these embryo sizes for a given supersaturated solution and an expression for this distribution has been derived (18). If one of the embryos happens to exceed a particular size, then thermodynamics predicts that this aggregate, which now will be called nucleus, would be stable and grow in size spontaneously. The above particular size is called critical nucleus size and depends inversely on the supersaturation. The classical nucleation theory has derived expressions which allow an estimation of both the size of this critical nucleus and of the rate of its formation, i.e. the so-called nucleation rate. The form of the expression for the latter demonstrates clearly the critical nature of the spontaneous nucleation process. In supersaturated solutions in which no spontaneous nucleation is observed, the supersaturation is low enough for the nucleation rate to be extremely small-for all practical purposes zero.

2. UNDERSTANDING THE PHENOMENON OF SECONDARY NUCLEATION

A study of the secondary-nucleation investigations indicates that the question of how a crystal can generate new crystals could be conveniently split into two other questions: first, what is the source of the potential nuclei and second, how these nuclei are extracted from the source and displaced into the bulk solution to initiate the new crystals. There are multiple answers to both questions and previous works have demonstrated that more than one mechanism can simultaneously be active in both the above stages (11, 14, 15)

2.1. Sources of Secondary Nuclei

The proposed sources of the secondary nuclei can be grouped into three classes: sources on the crystal surface, sources in the supersaturated solution and sources belonging to an "intermediate" phase.

a) <u>Sources on the Crystal Surface.</u> These sources include the protruding parts of a growing crystal surface, as discussed in Section 1.2 and in general organizations of atoms which belong and are part of the crystal seed. Most prominent in this category are the needle-like dendrites which grow on crystals under certain conditions; that is, relatively high supersaturations (fast growth) and conditions in which the growth is mass transfer controlled.

The detachment of tiny dendrites by convective currents shear from NH_4Cl and NH_4Br crystals have been observed by Melia and Moffitt (28,29) in experiments done under a microscope. Clontz and McCabe (9) were able to link the occurrence of dendritic growth on a crystal seed with secondary nucleation rates. The more pronounced the dendritic morphology was the higher the nucleation rate.

There is ample evidence, however, (10,36) that secondary nucleation can occur without visible dendritic growth. Although in this case the microscopic roughness could be invoked as a source of nuclei, certain investigators have proposed the existence of atom clusters belonging to an "intermediate" state, as possible nuclei sources.

b) <u>Sources belonging to an "intermediate phase".</u> Various models have been proposed at times assuming the existence of solute clusters which belong neither to the liquid phase nor to the solid (crystal) phase (36,9). They simply exist as a semi-ordered regime on or near the crystal surface, before they are committed to and become part of the crystal.

There is no theoretical evidence for the existence of such a regime and the experimental evidence is rather weak (21). In addition, it is not necessary to invoke such solute-molecule organization as sources of nuclei. The sources on the crystal or the solution appear to be sufficient to explain all the various experimental data.

SECONDARY NUCLEATION

c) *Sources in the supersaturated solution.* The embryos existing in a supersaturated solution could become nuclei, if somehow were "activated" or the critical nucleus size was effectively lowered. This could be accomplished by various proposed mechanisms (4,11,32,40) which will be described below.

d) *Identifying the source of secondary nuclei.* An experimental method, capable of determining whether or not the nuclei originated from the crystal per se or from the surrounding supersaturated solution, has been reported in the literature (11). The study made use of the right- and left-handed optical properties of the sodium chlorate crystals (d- or l-crystals). A particular kind of crystal (i.e, d or l) was introduced as a seed in supersaturated solutions of $NaClO_3$. The product crystals were collected and analyzed to determine whether they were d or l. The findings of this work are illustrated in Figures 1 and 2.

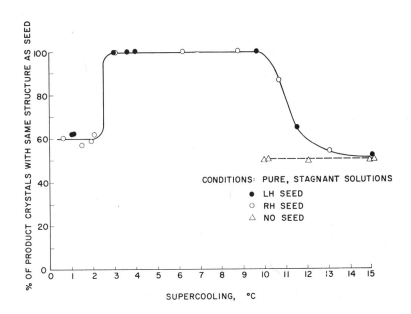

Fig. 1. Type of product crystals obtained in pure, stagnant solutions (Ref. 11)

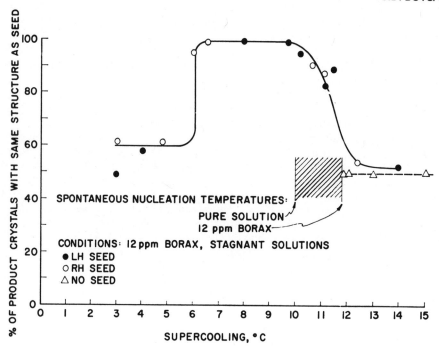

Fig. 2. Type of product crystals obtained in impure, stagnant solutions (Ref. 11)

Figure 1 shows the type of product obtained in pure solutions, while Figure 2 shows those obtained in solutions containing borax as an impurity. As shown, at high supersaturations, when spontaneous nucleation would occur even in the absence of seed, both types appear (50% of each). In the supersaturation range, however, where only secondary nucleation occurs, different types of crystals were obtained at different conditions. Over most of the supersaturation range the product crystals were all of the same type as the seed, which indicates that the nuclei were derived directly from the parent crystal. At other ranges, however, sources from the solution appear also to contribute, since the percentage of the product of the same form as the parent becomes smaller than 100%. This occurs at the lowest levels of supersaturations and in the case of impure solution (Figure 2) also at levels close to the point of spontaneous nucleation.

Mechanisms for the "removal" of nuclei from the solution will be described below.

2.2. Removal of Nuclei from Sources

Six mechanisms have been identified or proposed for the removal of nuclei from the sources, or stated in a different way, for the conversion of the "potential nuclei" to nuclei.

a) Spontaneous removal. Kliya (24) and Saratovkin (41) have demonstrated that NH_4Cl dendrites can detach themselves from the surface even in the absence of any liquid-shearing action. They attributed this to a phenomenon known as dendrite coarsening, which involves a faster growth in the tip than in the base of a dendrite. The dendrite assumes a non-uniform cross section, and its base being narrower would have a higher solubility (because of the Gibbs-Thomson effect). If these solubility effects become significant enough, it is possible to envision conditions under which the base of a dendrite might be dissolving away while the tip and the bulk crystal were still growing.

Denk and Botsaris (11) suggested dendrite coarsening as the mechanism for nuclei removal in stagnant or mildly agitated solutions.

b) Removal by fluid shearing forces. A recent paper by Sung, Estrin and Youngquist (45) has confirmed what previous investigations (11,23,36) also demonstrated experimentally: that fluid mechanical forces could effect the generation of secondary nuclei.

Lal et al. (25) and Clontz and McCabe (9) used the same system ($MgSO_4 \cdot 7H_2O$) as Sung et al. They failed, however, to observe any nuclei breeding at $\Delta T < 4°C$, while Sung et al. did observe nucleation at such supersaturations. The latter advanced a very convincing explanation for this discrepancy: particles (potential nuclei) are removed from the crystal surface in both cases. However, their size is such that they do not survive at the low supersaturations of the former investigators' experiments. In Sung's experiments the generated potential nuclei were immediately introduced into a region of high supersaturation where they could survive and grow to crystals.

This notion, of the survival of a fraction of the generated particles, which fraction depends on the supersaturation the particles are exposed to, has been expressed also by Strickland-Constable in his "survival theory" (25). This theory was originally stated to describe breeding due to collisions. It assumes that particles produced by a collision are in the same size range as the critical nucleus at that supersaturation. Particles smaller than the critical nucleus re-dissolve, while those larger than the critical size survive.

The "survival theory" assumes also that the number and the size of the particles removed from the crystals is independent of supersaturation. Sung et al., however, found that the number of particles removed from the crystal by the shearing action of the fluid does depend on the supersaturation.

c) <u>Removal by collision of a crystal with an external surface</u>. This mechanism of removal has been conclusively established by the pioneered work of two groups of investigators. Strickland-Constable and co-workers (19, 20) took particular care in eliminating initial breeding (i.e. the breeding due to the crystalline "dust" carried by the seed into the supersaturated solution) and established conditions under which the seeds could not produce secondary nuclei. Then the seeds were touched by a glass rod or were allowed to slide slowly along the bottom of the crystallizer. McCabe and co-workers (9, 22) employed a controlled impact between the seed crystals and rods made of various materials. In both types of studies a large number of secondary nuclei were produced even by a simple touch of the parent crystal surface.

An important finding of the above studies was that the contacts produced no visible defects in the parent crystal. This clearly distinguishes this phenomenon of "contact" nucleation from a (macro) attrition breeding process. In the latter process the fragmentation leaves macroscopic impressions on the crystal.

Another important feature of this method of removal is the dependence of the number of nuclei removed from the source on supersaturation. The higher the supersaturation the larger the number of nuclei. This is explained by the "survival theory" on the assumption that the supersaturation affects the percentage of particles which survive. The physical process of removal is assumed independent of the supersaturation; this contention appeared to be supported by Garabedian and Strickland-Constable's data (19). However, a recent report (31) pointed out that in the above study the seeds were not probably left inside the supersaturated solution long enough before sliding, to develop the surface growth configuration (see Section 1.2) characteristic of that supersaturation.

Actually the same report indicates that when the crystals are left to grow before sliding, the number of nuclei produced is not anymore reproducible from crystal to crystal. This was attributed to the fact that it is highly improbable for two different growing crystals to have the same exactly profile of growth irregularities on the surface, even when the experimental conditions are the same.

The practical implication of this finding is, of course, that one additional uncertainty is introduced into any attempt to express quantitatively the rate of this process of nuclei removal.

Various materials have been employed as contacting surfaces. Johnson et al. (22) contacted single crystals with a stainless steel covered by rubber or polypropylene. Other investigators studied nucleation rates in suspension crystallizers in which the agitators and propellers or the crystallizer walls or both were covered or made out of polymeric or elastomeric materials. In all cases the nucleation rates were reduced (15, 39, 42).

Attempts have been made to correlate the effectiveness of a <u>contacting</u> surface with its hardness (22). Tai et al. (46) on the other hand have reported a correlation between ease of nuclei removal and hardness of the <u>contacted</u> crystals. At this point, however, it is not clear whether or not hardness is the relevant property as opposed to another mechanical property; for instance: toughness or Young's modulus (as suggested by Evans et al., 15).

The study of the effect of the nature of a material on nuclei removal has both practical and theoretical significance. It may lead to practical means of reducing nucleation in a crystallizer. It also may clarify what is happening during the contact of the two surfaces: that of a crystal with an external surface.

Estrin (14) noticing the similarity between friction processes and the contact nucleation, suggested that investigators should look into the abundance of past works on friction.

An interesting observation made in a recent study (5) pointed out that not all nuclei produced in contact nucleation may be secondary nuclei. The study duplicated certain contact nucleation experiments using a glass dummy in place of a sodium chlorate crystal seed. At low supersaturations no nuclei were observed. At high supersaturations, however, a certain number of nuclei were produced. These nuclei, according to the definition in Section 1.1, cannot be considered as secondary nuclei. They probably resulted from a kind of "shock" or heterogeneous nucleation which in the case of crystal seeds, is operating in addition to the secondary nucleation.

This observation underscores once again the importance of incorporating in any secondary nucleation study, blank experiments and experiments with dummy seeds.

d) <u>Removal by collisions of crystals with one another</u>. Clontz and McCabe (9) in their pioneering experiments contacted stationary single crystals with another crystal of the same material and they observed the production of nuclei. Actually crystal-crystal contacts gave from two to five times as many nuclei as the crystal-metal rod contacts.

This observation demonstrates that secondary nuclei can be removed from crystals by collisions with other crystals.

e) Removal of nuclei by increasing the effective supersaturation around the crystal. The existence of secondary nuclei originating from the solution has been demonstrated experimentally (see Section 2.1d). Two models have been proposed to explain their "removal", or in other words, the conversion of "potential nuclei" in the solution to nuclei. Both are based on an increase of the "effective" local supersaturation in the liquid layer next to the crystal. The first (11), which is speculative at present, states that the liquid layer next to the crystal may differ from the bulk liquid in the structure of the solvent (water) molecules. An ordering of the water molecules near the interface may lead to secondary nucleation if the solubility of the crystallizing species in the ordered water structure was less than the solubility in the bulk solution. This would mean that the local supersaturation in that liquid layer would be much higher than the bulk supersaturation.

The other "removal" mechanism has been proven experimentally (4). It is the so called Impurity-Concentration-Gradient (ICG) mechanism which reasons as follows: in some solutions, the only reason why spontaneous nucleation does not occur in the bulk of the solution is that impurity is present. If now a crystal is introduced into such solution and the impurity is readily incorporated into the growing crystal (e.g. Pb^{++} in KCl crystals) an impurity concentration gradient may develop in the boundary layer or stagnant film next to the seed crystal. Thus the impurity concentration near the crystal surface may be reduced to such a level that nucleation becomes possible inside this liquid layer.

f) Removal of nuclei by activating existing embryos in the solution. Other speculative proposals for the "removal" of nuclei from the solution involve the activation of the embryos existing in the solution.

Mullin and Leci (32), suggested that seed crystals provide a place for the embryos to become attached weakly and to grow. The grown embryos may then be stripped away from the seeds by the fluid shearing action. By the time they are detached a number of the embryos would have become larger than the critical nucleus and would thus grow into macroscopic crystals.

Another paper noted (3) that during the removal of particles from the parent crystal by a collision mechanism, only a fraction of the particles may become nuclei: those whose size is larger than the critical nucleus size. The particles, however, whose size is smaller than the critical, will be injected into the supersaturated solution as embryos. It can be hypothesized that these embryos could affect the distribution of the existing embryos of homogeneous or heterogenous origin and the steady state of Equations (1) and may thus trigger a primary nucleation process. The author used this mechanism to explain a type of nucleation reported by Youngquist and Randolph (47).

g) <u>Evaluating the relative importance of the various removal mechanisms</u>. Evans, Margolis and Sarofin (15), studying the nucleation of ice in an agitated batch crystallizer, demonstrated that more than one mechanisms of removal can be operating simultaneously in a crystallizing system.

They designed experiments in which each of the different mechanisms of removal could be successively supressed. For instance, the crystal-crystal collisions were supressed by operating the crystallizer at very low ice concentrations. By coating successively various parts of a crystallizer (propeller, walls, etc.) with a soft rubber, the relative importance of collisions with the different surfaces in a crystallizer could be identified. The addition of inert particles of varying size and concentration permitted the study of the contribution of crystal-crystal collisions.

Figures 3 and 4 are representative of their data. In both figures the nucleation rate <u>per crystal</u>, β, is plotted against the sub-cooling ΔT.

Figure 3 shows the reduction in the nucleation rate, when all surfaces in the crystallizer were coated with RTV Silicone (lower line). The points represent nucleation rates when only different combinations of surfaces were coated. Figure 4 presents the effects of added polyethylene particles. For instance, adding 10 wt% particles with a diameter of 0.9mm to the RTV-silicone-rubber coated crystallizer, brings the nucleation rate up to the level of the rate in an uncoated crystallizer.

The authors assumed an additivity principle, stating "that the overall nucleation rate with two or more mechanisms of removal ($\dot{N}_t$) is the linear sum of the actual nucleation rate attributable to each mechanism of removal ($\dot{N}_i$)"

$$\dot{N}_t = \dot{N}_1 + \dot{N}_2 + \ldots \qquad (2)$$

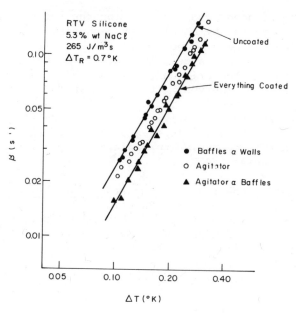

Fig. 3. Nucleation rates per crystal for an agitation power of 265 $J/m^3 s$ with different combinations of coated surfaces. (Ref. 15)

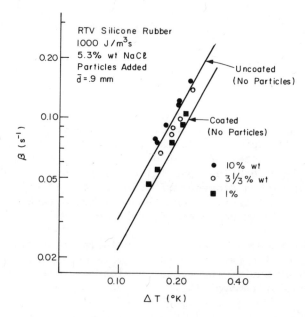

Fig. 4. Effect of added particle concentration on the nucleation rate per crystal in the coated crystallizer. (Ref. 15)

From the additivity principle follows that the nucleation rate per crystal due to two mechanisms, β_{1+2}, is related to the nucleation rates β_1 and β_2, which would prevail if either mechanism acted alone by the equation:

$$\beta_{1+2} = \beta_1 \left(\frac{\beta_1}{\beta_{1+2}}\right)^{n_1} + \beta_2 \left(\frac{\beta_2}{\beta_{1+2}}\right)^{n_2} \qquad (3)$$

where n_1 and n_2 is the power dependence of the nucleation rate in each mechanism on the crystal size.

Evans et al. utilizing equations of the form of (3), were able to express quantitatively the contribution of each mechanism of removal. For instance, in a continuous agitated crystallizer (MSMPR) operated at 10% by weight ice, crystal-crystallizer collisions were responsible for 55% and crystal-crystal collisions for 20% of the overall nucleation rate. The remaining 25% was attributed by Evans to fluid shear. However, any of the other removal mechanisms, described above, could conceivably be responsible.

In addition, it was found that the agitator was responsible for 75% of the nucleation due to crystal-crystallizer collisions while the baffles contributed the remaining 25%.

The work of Evans et al. has definitely opened a new path for the future developments in the field

2.3 Rate Controlling Stages

It can be concluded from the discussions in Sections 2.1 and 2.2 that the overall nucleation rate will be determined by an interrelation of the <u>production rate</u> of potential nuclei at the source and of the <u>rate of their removal</u> from the source.

The processes which determine the production rate of nuclei are different, depending on the source of nuclei. For instance, when the nuclei originate in the solution according to ICG mechanism (see 2.2e), the production rate is related to the rate with which the impurity is incorporated into the growing crystal. When the parent crystal is the source of nuclei, the production rate will be determined by the rate with which the protrusions on the crystal surface are regenerated.

The rate of impurity incorporation as well as the rate of protrusions regeneration are both related to the growth process of the crystal. Thus, secondary nucleation is definitely affected by the growth process. At the same time, however, the growth, being a structure sensitive process, will in turn be affected by the removal step of the secondary nucleation. In other words, there is a coupling between crystal growth and secondary nucleation.

Two extreme cases of rate controlling stage may be visualized. For instance, in collision breeding the nucleation rate may be either surface regeneration-limited or removal-limited.

Although in most of the reported experimental studies with single crystals the secondary nucleation appears to be removal-limited, instances of possible regeneration-limited nucleation have been presented. Cayey and Estrin (7) obtained results suggesting that small crystals must attain some minimum size before they are capable of producing nuclei. A possible explanation for this may be found in Strickland-Constable's suggestion (44) that very small crystals lack imperfections. Thus, the secondary nucleation in small crystals becomes surface regeneration-limited and very low.

Randolph and Sikdar (39) in their study of the effect of rubber coated impellers on nuclei breeding, have found that the difference between the effect of coated and uncoated impellers becomes very small at high agitation rates. This also could be interpreted as a case of surface regeneration-limited nucleation at intense agitation.

The above findings indicate also the importance of knowing the rate controlling step in the secondary nucleation process. If the process is surface regeneration controlled any change in the removal process (nature of crystallizer surfaces, agitation rate) may have no effect on the nucleation rate.

Bauer et al. (1) attempted to determine the conditions in a MSMPR crystallizer under which the nucleation is limited by the surface regeneration process. They used a single crystal contacted repeatedly by a McCabe type contacting device. They were hoping that by increasing the frequency of contacts they could achieve a surface regeneration-limited nucleation. Although their results were inconclusive, the method they used appears to be very promising for future experiments.

3. APPLICATION OF THE SECONDARY NUCLEATION MECHANISMS IN MODELING NUCLEATION RATES IN CRYSTALLIZERS

As our understanding of the phenomenon of secondary nucleation progresses, the following question becomes increasingly important: how can the findings and theories which were described above in Section 2 be applied to the analysis and design of crystallization operations?

The studies dealing with this question follow in general two directions: a) experimental investigations which lead to empirical correlations and b) development of theoretical models which are subsequently tested against limited number of experimental data.

3.1 Empirical Models

Once the significance of secondary nucleation in crystallizers was realized, investigators began to include in their nucleation rate expressions a parameter accounting for the presence of the crystals. In suspension crystallizers this parameter was essentially one of the moments μ_i of the crystal size distribution. Power-law models of the form

$$\frac{\dot{N}}{V} = k_N (\Delta c)^n \mu_i^m \quad (4)$$

were developed, where $\dot{N}/V$ is the nucleation rate per unit volume and Δc the supersaturation.

Larson et al. (26) correlated their experimental data using the third moment (i.e. the suspension density). Other investigators (7, 9) preferred the use of the second moment (i.e. the total surface area of the crystals). Even models involving the zeroth and first moments have been proposed (13).

Desai et al. (13) in a recent paper showed that all these models are equivalent for an unseeded (clear feed) continuous mixed suspension mixed product removal (MSMPR) crystallizer. If steady state data obtained from such a crystallizer are correlated well by one of the above models, they will be correlated equally well by any other model of the form of equation (4) involving a different moment.

Therefore, it is impossible using such data to clarify which is the correct moment for inclusion in a nucleation rate model. However, the same authors pointed out that when the crystallizer is operated with seeding, the various models are not equivalent.

Using seed experiments with potassium dichromate system they found that the third moment was the one correlating better the data.

Randolph and Cise (37), however, correlated their data of K_2SO_4, obtained also in a seeded crystallizer, by a model including the fourth moment of the crystal size distribution.

When the significance of the collision breeding was demonstrated investigators began introducing into Equation 4 also a parameter expressing the hydrodynamic conditions. Thus the following empirical models have been developed:

$$\dot{N} \alpha (RPM)^{7.8} \text{ for } NH_4SO_4 \quad (47)$$

$$\dot{N} \alpha (RPM)^{5.78} \text{ for } K_2SO_4 \quad (37)$$

$$\dot{N} \alpha (TIPS)^{3.0} \text{ for } NaCl \quad (2)$$

where RPM are the revolutions per minute of the stirrer and TIPS the impeller tip speed (ft/min).

As long as our understanding of the mechanism of secondary nucleation remains incomplete, our heavy reliance on empirical models like the above will continue. At the same time, however, the number of attempts to develop theoretical models, like those described in the next section, will increase.

3.2 Theoretical Models

The work in this area has dealt exclusively with systems in which the source of nuclei is the crystal and mechanism of removal is collisions.

In such a system the nucleation rate $\dot{N}$ can be expressed as the product of three functions:

$$\dot{N} = (\dot{E}_t)(F_1)(F_2) \quad (5)$$

where $\dot{E}_t$ = the rate of energy transfer to crystals by collision

F_1 = the number of particles generated per unit of transferred energy

and F_2 = the fraction of particles surviving to become nuclei.

Function F_1 was assumed by Strickland-Constable (19) to be independent of supersaturation ΔC. However, Sung et al. (45) have demonstrated experimentally that F_1 is a function of ΔC.

SECONDARY NUCLEATION

This appears to be more reasonable, since F_1 is expected to be governed by the surface morphology, which depends on the growth process. The growth process in turn is a function mainly of supersaturation, but also it may depend on impurities in the solution, the degree of agitation and the size of the crystal.

Function F_2 is a function of the supersaturation, the size distribution of the particles removed from the crystal, and the impurity concentration in the solution (5).

Function $\dot{E}_t$ is equal to the product of collision energy $E(r)$ and frequency of collision $\omega(r)$ of crystals in the size range r to $r+dr$ integrated over the crystal size distribution $f(r)$:

$$\dot{E}_t = \int_0^\infty \omega(r)\, E(r)\, f(r)\, dr \qquad (6)$$

deJong and co-workers (34,35) were the first to model $\dot{E}_t$ by considering all the possible collisions a crystal might undergo during its movement through a crystallizer. Evans et al. (16) extended deJong's work on modeling $\dot{E}_t$, by considering four idealized collision mechanisms in a suspension crystallizer, a) crystal-impeller, b) crystal in a turbulent flow field-crystallizer surfaces, c) crystal-crystal, as a consequence of differences in their terminal velocities and d) crystal-crystal driven by turbulent eddy motion.

The developed models indicate that the contributions of each collision mechanism to the nucleation rate are correspondingly:

a) $\alpha\ (P/V_c)\, R^{0.5} \bar{r}^{2.5}$, b) $\alpha\ (P/V_c)\, \bar{r}^{-3}$

c) $\alpha\ W_c \bar{r}^{-5.42}$, d) $\alpha\ (P/V_c) W_c \bar{r}^{-3}$

where P/V_c is the power input per unit volume of crystallizer, R the impeller diameter, $\bar{r}$ the mean crystal size and W_c the weight fraction of crystals in the suspension.

Limited experimental data obtained by Evans et al. (16) in an ice crystallizer did not contradict the above four collision models.

Models for functions F_1 and F_2 have not yet been proposed. Our knowledge of these functions is rather minimal.

In surface regeneration-limited nucleation, F_1 is expected to be coupled with $\dot{E}_t$. In removal-limited nucleation, however, F_1 can be assumed independent of the physical process of removal.

However, there is a possibility that, even in this case the surface morphology of the parent crystal will be continuously changing due to the introduction of imperfections (dislocations) by the collisions.

Function F_2 could be easily formulated if the size distribution function for the particles removed from the parent crystal were known. The only thing known, however, is that since the number of nuclei produced as a function of supersaturation is represented by an S shaped curve, the particles are probably distributed about some average size (12).

It can be concluded, therefore, that there is a definite need for development of models for F_1 and F_2.

4. OTHER ASPECTS OF NUCLEI BREEDING

A large number of investigations have appeared treating various other aspects of the phenomenon of secondary nucleation. Only certain of these investigations will be discussed here.

4.1 Crystal Abrasion

A common characteristic of all the studies of secondary nucleation discussed above, was that there is a minimum supersaturation below which no nuclei breeding was observed. This makes pure mechanical macro-attrition a probable mechanism for continuous crystallizers operating at very low steady state supersaturations.

Fasoli and Conti (17) have demonstrated that abrasion of crystals of $CuSO_4 \cdot 5H_2O$ agitated in a non-solvent liquid could produce new crystallites from the parent crystals.

4.2 Effect of Impurities

Impurities are expected to affect secondary nucleation rates through both functions F_1 and F_2 of Equation 5. Crystal growth theory predicts (33) that they affect surface morphology and thus function F_1. At the same time impurities will affect function F_2 by changing the size of critical nuclei at a given supersaturation. It is also possible that they will modify the size distribution of the particles removed from the parent crystal. Finally, as it was discussed in Section 2.2e, impurities could introduce a specific type of secondary nucleation.

Shor and Larson (43) and Liu and Botsaris (27) working with

suspension crystallizers, and Botsaris and Sutwala (5) working with single crystals, have provided a quantitative measure of the effect of impurities on nucleation rates.

5. CONCLUDING REMARKS

The objective of this review was to provide a generalized conceptual framework within which the past investigations could be viewed in perspective, unanswered questions could be exposed and fruitful areas for further work could be pointed out. This framework could, hopefully, become the common basis for future discussions among the investigators in this exciting field of secondary nucleation.

6. REFERENCES

1. Bauer, L.G., M.A. Larson and V.J. Dallons, Chem. Eng. Sci., 29 1253 (1974).
2. Bennett, R.C., H. Fiedelman and A.D. Randolph, C.E.P., 69, 86-93 (1973).
3. Botsaris, G.D., paper presented at the 5th Symposium on Industrial Crystallization, CHISA, 1972, Praha, Czechoslovakia, September 1972.
4. Botsaris, G.D., Denk, E.G., and J. Chua, AIChE Symposium Series, 68, (No. 121), 21 (1972).
5. Botsaris, G.D., G. Sutwala and G. Molina-de-Longoria, paper presented at 77th Annual AIChE Meeting, Pittsburgh, June 3, 1974.
6. Burton, W.K., N. Cabrera and F.C. Frank, Phil. Trans. Roy. Soc., A243, 299 (1951).
7. Cayey, N.W., and J. Estrin, Ind. Eng. Chem.-Fundamentals, 6, 13 (1967).
8. Cise, M.D., and A.D. Randolph, AIChE Symposium Series, 68 (121), 42 (1972).
9. Clontz, N.A., and W.L. McCabe, Chem. Eng. Progress, Symposium Series, 67 (No. 110), 6 (1971).
10. Denk, E.G., Jr., "Fundamental Studies in Secondary Nucleation", Ph.D. thesis, Dept. Chem. Eng. Tufts Univ. Medford, Mass. (1971)
11. Denk, E. G., and G. D. Botsaris, J. Crystal Growth, 13/14, 493 (1972).
12. Denk, E.G., and G. D. Botsaris, J. Crystal Growth, 15, 57 (1972).
13. Desai, R. M., J.W. Rachow and D. C. Timm, AIChE J. 20, 43 (1974).
14. Estrin, J., "Secondary Nucleation" to be published in "Preparation and Properties of Solid State Materials", Marcel Dekker, New York.
15. Evans, T.W., G. Margolis, A.F. Sarofim, AIChE J., 20, 950 (1974).
16. Evans, T.W., A.F. Sarofim and G. Margolis, AIChE J. 20, 959 (1974).

17. Fasoli U., and R. Conti, Kristal und Technik, 8, 931 (1973).
18. Frenkel, J., "Kinetic Theory of Liquids", p. 384, Dover Pub.
19. Garabedian, H., and R.F. Strickland-Constable, J. Crystal Growth, 13/14, 506 (1972).
20. Garabedian, H., and R.F. Strickland-Constable, J. Crystal Growth, 22, 188 (1974).
21. Garten, V.A., and R.B. Head, J. Crystal Growth, 6, 349 (1970).
22. Johnson, R.T., Rouseau, R.W., and W.L. McCabe, AIChE Symposium Series, 68 (121), 31 (1972).
23. Kane, S.G., T.W. Evans, P.L.T. Brian and A.F. Sarofim, AIChE J., 20, 855 (1974).
24. Kliya, M.O., Soviet Physics-Crystallography, 1, 456 (1956).
25. Lal, D.P., R.E.A. Mason, and R.F. Strickland-Constable, J. Crystal Growth, 5, 1 (1969).
26. Larson, M.A., Timm, D.C., and P.R. Wolff, AIChE J. 14, 488 (1968).
27. Liu, Yih-An, and G.D. Botsaris, AIChE J., 19, 510 (1973).
28. Melia, T.P., and W.P. Moffitt, J. Colloid Sci., 19, 433 (1964).
29. Melia, T.P., Moffit, W.P., Ind. Eng. Chem. Fundamentals 3, 313 (1964).
30. Miers, H.C., Isaac, F.J., J. Chem. Soc. 89, 413 (1906).
31. Molina-de-Longoria, G.O., MS thesis, Dept. of Chemical Eng. Tufts University, 1975.
32. Mullin J.W., and C.L. Leci, AIChE Symposium Series, 68 (121), 8 (1972).
33. Ohara, M., and R.C.Reid, "Modeling Crystal Growth Rates from Solution", Prentice-Hall, Englewood Cliffs, N.J. (1973).
34. Ottens, E.P.K., Janse, A.H., and E.J. deJong, J. Crystal Growth 13/14 500 (1972).
35. Ottens, E.P.K., and E.J. deJong, I&EC Fundamentals, 12, 179 (1973).
36. Powers, H.E.C., Nature, 178, 139 (1956). Chemistry and Industry 14, 627 (1962). Industrial Chemist. p. 351 (July, 1963).
37. Randolph, A.D., and M.D. Cise, AIChE J., 18, 798 (1972).
38. Randolph, A.D., and M.A. Larson, AIChE J., 8, 639 (1962).
39. Randolph, A.D., and S.K. Sikdar, AIChE J., 20, 410 (1974).
40. Rogacheva, E.D., A.V. Belyustin, T.K. Vyatkina and N.M. Khrenova, Soviet Physics - Crystallography, 17, 378 (1972).
41. Saratovkin, D.D., "Dendrite Crystallization," Consultants Bureau Inc., New York, p. 64-65 (1959).
42. Shah, B.C., W.L. McCabe and R.W. Rousseau, AIChE J., 19, 194 (1973).
43. Shor, S.M., and M.A. Larson, Chem. Eng. Progr. Symposium Ser. 67 (110), 32 (1971).
44. Strickland-Constable, R.F. AIChE Symp.Ser. 68 (No.121), 1 (1972).
45. Sung, C.Y., J.Estrin and G.R. Youngquist, AIChE J., 19, 957 (1973)
46. Tai, C.Y., W.L.McCabe and R.W.Rousseau, paper presented at 77th Annual AIChE Meeting, Pittsburgh, June 3, (1974)
47. Youngquist, G.R., and A.D.Randolph, AIChE J. 18, 421 (1972).

ATTRITION AND SECONDARY NUCLEATION IN AGITATED CRYSTAL SLURRIES

P.D.B. Bujac

ICI Corporate Laboratory

Runcorn, Cheshire, England

INTRODUCTION

The objective of our study is to determine methods of scaling-up nucleation rates, measured in the laboratory, for the design and optimisation of full-scale crystallisers. It is usually accepted that the dominant form of nucleation occurring in continuous industrial crystallisers is secondary nucleation. Recent research has shown that this nucleation is caused by some form of attrition mechanism, (1, 2). This mechanism is either the detachment, by impact or fluid-shear, of clusters from a parent crystal or the fracture of the parent crystal surface, by some form of comminution, and the production of discrete crystal fragments. The major factors influencing this mechanism are the system super-saturation and the crystalliser hydrodynamics. The supersaturation will control both the parent crystal surface and the survival and development of the clusters or fragments into product crystals. The crystalliser hydrodynamics will control the collision energies and collision frequencies of the parent seeds.

To understand the influence of the supersaturation and hydrodynamics we have measured the birth rates of fragments formed by attrition in various agitated vessels. We have also investigated their possible development into macro-sized particles.

The system chosen for this study was pentaerythritol (PE) - water. PE is a medium tonnage chemical and the experimental work used commercial material. It has a bi-pyramidal habit, convenient solubility and a wide metastable limit. Previous measurements had shown that the growth rate was low at room temperature but reasonable at 40°C and above. It was useful to study an organic system; much of the published work on secondary nucleation is for inorganic crystals. 0.7% aqueous nitric acid was used as the solvent for the growth and size analysis measurements to provide a conducting solution for a Coulter Counter.

ATTRITION STUDIES

We found that when a slurry of PE crystals was suspended in an agitated vessel, in a slightly supersaturated ($\sim 5^{\circ}$C sub-cooling) solution at room temperature, crystal fragments (<30 µm) were rapidly produced. The fragments were remarkably stable and there was no significant change in the numbers and size distribution over a period of a few hours. This stability enabled us to measure the fragment birth rates for different crystalliser hydrodynamics by size analysis of the total fragments over a period of time.

A typical attrition run was as follows. 80g of sieved commercial PE seeds (355-500 µm) were gently washed with 8% PE solution to remove dust and placed in a 1 l. vessel. The vessel was fitted with four vertical 10 mm baffles and a simple 52 mm diameter paddle. 800 ml of the PE solution were added and the agitation started. All particles were completely suspended. After two minutes the agitation was stopped and when the seed crystals had settled a 2 ml sample of the liquor was taken. This was diluted with 150 ml of filtered (0.6 µm filter) 8% PE nitric acid solution. The resulting suspension was size analysed using a model TA Coulter Counter fitted with either a 100 or 170 µm aperture tube. The size data were in the form of cumulative numbers greater than ~ 2 µm and the volumes of particles per channel. These raw data were subsequently converted to give the numbers distribution for the full size range. The agitation was restarted and the sampling procedure repeated at various time intervals. The sample dilution was adjusted as necessary to minimise coincidence errors during counting. In some runs the seed crystals were removed at the end (~ 3 hours), dried and resieved. It was found that the general shape of the size distribution was unaltered and the mean size had only changed by some 5%. No major breakage of PE seed crystals thus occurred under these conditions.

In all runs the pattern of fragment production, the attrition rate, was similar (see Fig. 1). We found that the rates of generation of particles were constant for approximately 20 minutes but then they fell with time. This was particularly noticeable for the small particles (<5 µm). Although this reduction in rate can be explained by 'initial breeding' (ie the removal of crystallites from a previously dried crystal surface) the data from the crystalliser runs (see later) suggest that these initial

ATTRITION AND SECONDARY NUCLEATION

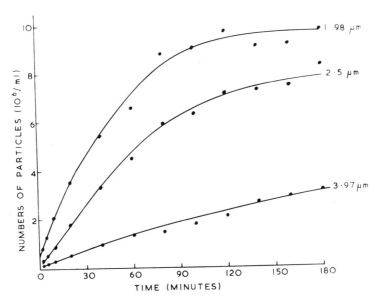

Fig. 1   The Production of Crystal Fragments

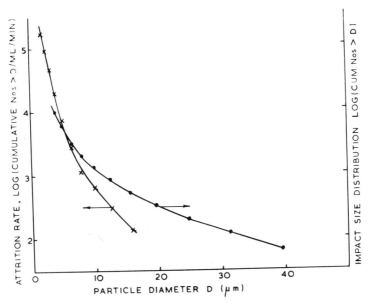

Fig. 2   Size Distribution of Attrition Particles

rates are true attrition rates. The general form of the attrition function is shown in Fig. 2. This indicates the exponential dependence of numbers on size and the ease with which fragments can be produced in agitated PE systems.

Attrition rates were measured in a 1 l. vessel for different hold-ups (H). The results (Fig. 3) show that the production rate of crystal fragments can be given by

$$\dot{m}_a (D) \alpha H^m$$

where m ∼2 for small particles (∼3 µm) but m tends to 1 for particles >10 µm. This varying dependence suggests two mechanisms of attrition. The small particles are generated by two body (crystal - crystal) collisions and the larger particles by impacts with the impeller and possibly with the baffles. A size analysis was obtained for the crushing of large PE crystals. This analysis (Fig. 2) shows that this form of attrition produces predominantly larger crystals and supports the above theory.

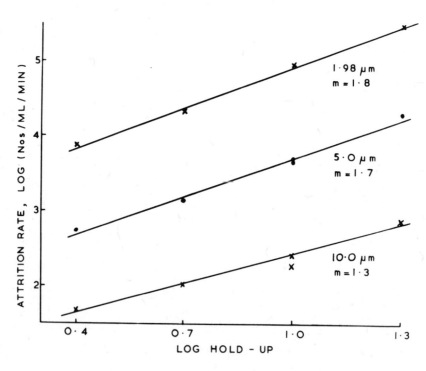

Fig. 3   Attrition vs Crystal Hold-Up

The effect of parent crystal size was examined. The results indicate that, for a given crystal hold-up, the rate is proportional to D for the small fragments. Data for large fragments are limited but they suggest a higher power dependence on D. As the number of crystals present is proportional to $1/D^3$ then obviously the seed crystal mass is critical in determining collision energies. Small particles will tend to follow the fluid streamlines and hence tend to avoid collisions. The previously noted stability of agitated suspensions of fragments implies that these small particles do not undergo further attrition.

Attrition rates were also measured in a series of agitated vessels (1 l., ·5 l. and 20 l.) to determine scale-up parameters. Different impellers and speeds were used but in all cases complete suspension occurred without aeration. The data so far obtained suggest that attrition rates can be correlated approximately with power input (Fig. 4).

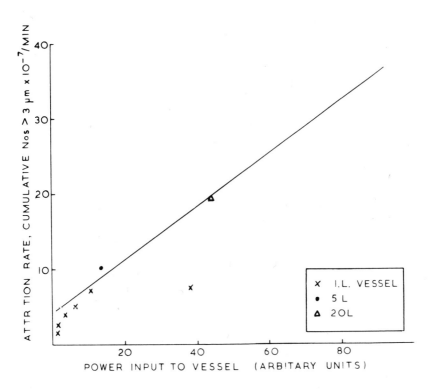

Fig. 4   Attrition Rate vs Power Input

CRYSTALLISER STUDIES

A mixed suspension mixed product removal (MSMPR) crystalliser was operated at 40°C to provide information on the numbers and size distribution of small particles in a continuous crystalliser where birth and growth occur simultaneously. A 5 l. agitated vessel, used in the attrition studies, was fitted with a recirculation loop and temperature control. The operation and analysis of these laboratory units are well established; eg see (3) for details. Samples were taken from the crystalliser at steady state conditions. One sample was filtered, dried and sieved to give the effective nucleation rate and growth rate of product crystals. A further sample was diluted with the filtered PE solution, wet screened to remove large particles and analysed with the Coulter Counter.

A typical size analysis is shown in Fig. 5. The numbers of small particles are appreciably higher than the numbers predicted by analysis of product material. The effective nucleation rate is approximately two orders of magnitude less than the small particle birth rate. These data are consistent with published data, eg K_2SO_4 results (4, 5) and confirm an inhibited growth mechanism. Attrition data for the same vessel, but corrected for hold-up and mean crystal size, are also shown in Fig. 5. These attrition data agree closely with the crystallisation results and suggest that similar attrition mechanisms occur in both vessels. The data also confirm that the lower size channels in a crystalliser are populated predominantly by birth rather than by growth.

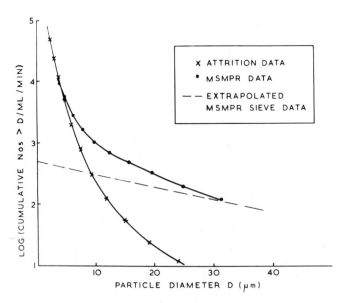

Fig. 5   Size Distribution from MSMPR Crystalliser

SMALL PARTICLE GROWTH STUDIES

The behaviour of attrition particles in different supersaturations has been studied by monitoring their size distributions directly using the Coulter Counter. A sample of particles, produced by attrition in an agitated vessel, was added to a supersaturated PE solution at $\sim 40^\circ C$. The numbers and size distribution were measured at frequent intervals for up to four hours. Over a wide range of supersaturations (σ = 5 - 30%) no significant change (<0.01 µm/min growth rate) was noted. Although stable these particles are not capable of growth at all reasonable supersaturations likely to be found in crystallisers. Further checks on the particle stability show that they rapidly dissolve in undersaturated solutions. To measure the supersaturations at which the particles might grow, a supersaturated suspension was slowly cooled. No growth occurred until at a relative supersaturation of >35% crash nucleation occurred and many large particles were formed. The mass of crystals was such that a detailed size analysis was impossible but the growth rate was estimated at >10 µm/min.

To determine whether the crystals produced at the crash point were derived from the original fragments, a dilute suspension of attrition particles were prepared at room temperature and sized using the Coulter Counter. Extrapolation of the data below 3 µm gave an estimate of the total present as $2 - 6 \times 10^5$/ml. Weighed drops of the suspension were placed on microscope slides and allowed to evaporate slowly. The final number of crystals were counted by eye; these corresponded to 5×10^5/ml. This agreement between the two methods suggest that extrapolation of the size distribution is valid (ie the distribution is continuous) and all fragments are capable of growing at sufficiently high supersaturations.

DISCUSSION

The measurements of crystal attrition indicate that two mechanisms can occur in a PE crystal suspension. These are a chipping mechanism which produces predominantly small fragments and an impact mechanism in which larger fragments are born. The chipping may be due to low energy crystal-crystal collisions. The impacting can be explained by comparatively high energy collisions between crystals and impellers or baffles. Both mechanisms correlate reasonably with the power input. The overall rate may be expressed as,

$$\dot{m}_a (D) = (P/V) (aH + bH^2)$$

The existence of two separate mechanisms may limit the scale-up of nucleation data as the rate may not necessarily be linearly proportional to the total collision energy.

The experimental work in the MSMPR crystalliser shows the mechanisms of particle generation occurring in a growing system are similar to those occurring in an attrition system. The probability of these particles growing into product crystals is however low and the effective nucleation rate is only 1% of the 'true' nucleation rate. Attempts have been made to describe the behaviour of small fragments by use of the Gibbs - Thomson relationship which predicts an enhanced solubility for small particles. This enhanced solubility will influence both the survival (6) and the growth rate (7, 8). The PE small particle growth studies do not support these theories. The particles are stable over a wide range of supersaturations and appear capable of growing only when some surface nucleation occurs. It has also been suggested (9) that the abnormal low growth rate of small crystals is due to their perfect structure as small crystals might rapidly rearrange themselves to eliminate defects. Although it is unlikely that the small particles observed in our work (2-30 μm) are perfect, it is assumed that the surface structure of the fragment determines the growth rate and only particles with an active growth site will grow. The activity and distribution of growth sites between the fragments resulting from an attrition event will depend both on the supersaturation present at the event and the supersaturation which the fragments encounter. The large difference between effective and 'true' nucleation rates suggests that the probability of a fragment having an active site is low. It is also likely that this probability will increase with surface area and that large fragments will be relatively more active.

Our experiments are continuing to determine more information on the growth of effective nuclei in an attempt to bridge the gap between the true secondary nucleation rate and the effective rate.

ACKNOWLEDGEMENTS

The author appreciates the many discussions he has had with his colleagues in the ICI Corporate Laboratory. He would like to thank W S H Sunley and R K Brock who were responsible for the experimental work.

NOMENCLATURE

a	= constant
b	= constant
D	= equivalent spherical diameter of particle
H	= crystal hold-up (g crystal/100 ml of liquor)
$\dot{m}_a(D)$	= rate of production of fragments of size (D) by attrition
m	= exponent in attrition rate equation
P	= power input
V	= volume of crystalliser
σ	= relative supersaturation

REFERENCES

1. Clontz N A and McCabe W L, Chem. Eng. Progr. Symp. Ser., No. 110, **67**, (1971) 6.

2. Strickland-Constable R F, "Kinetics and Mechanism of Crystallisation", Academic Press, London (1968).

3. Ramshaw C and Parker I B, Trans. Instn. Chem. Engrs., **51**, (1973), 72.

4. Randolph A D and Cise M D, AIChEJ. **18**, (1972), 798.

5. Randolph A D and Sikdar S K, AIChEJ. **20**, (1974), 410.

6. Garabedian H and Strickland-Constable R F, J. Crystal Growth, **13/14**, (1972), 506.

7. Wey J S and Terwilliger J P, AIChEJ. **20**, (1974), 1219.

8. Rosen H N, AIChEJ, **20**, (1974), 388.

9. Garabedian H and Strickland-Constable R F, J. Crystal Growth **12**, (1972), 53.

COLLISION BREEDING FROM SOLUTION OR MELT

R.F. STRICKLAND-CONSTABLE

Imperial College, London SW7

Present address: Combe Wood, Brasted, Westerham, Kent

SUMMARY

Clontz & McCabe {2} have found in one typical experiment with $MgSO_4 \cdot 7H_2O$ that secondary contact nucleation resulted from the expenditure of 2.2 ergs of work per nucleus.

Calculations are given in this paper to examine some of the consequences which would result from the assumption that the nuclei are formed by fracture of the parent crystal, and that they are of the order of size given by the Gibbs-Thomson equation. The calculations suggest that the minimum work needed to form a compact crystalline nucleus is of the order of 10^{-6} ergs: and the work to form a nucleus containing the same number of molecules, but spread out in a monomolecular layer, will still be only about 10^{-5} ergs. It is clear therefore that almost all the measured work must be used up in processes not directly connected with the actual fracture of the crystal. Suggestions are made as to the nature of this work. Some comments are offered on the correct form of the Gibbs-Thomson equation, and also on the theory which assumes formation of the nucleus in the liquid phase.

INTRODUCTION

Secondary nucleation from the fluid phase in a stirred suspension of growing crystals occurs very readily as a result of collisions between crystals or between crystals and the wall: this is known as collision breeding (Strickland-Constable {1}, p 117) or as contact nucleation (Clontz & McCabe {2}). The present paper discusses some aspects of this process, whose mechanism is still little understood. Accurate data on the physical properties required in the calculations are in many cases unknown; it is believed however that the rough estimates used will be sufficiently accurate for the purposes in hand.

Some theories that have been advanced to account for collision breeding assume either

A. that the new crystal results from actual fracture of the parent crystal, so that the new nucleus is actually part of the parent crystal, or

B. that the nuclei are produced in the liquid phase by some form of nucleation catalysed by the collision process.

The discussion is principally concerned with mechanism A.

SURVIVAL THEORY

The rate of collision breeding often increases markedly with supersaturation. To explain this observation in the case of theory A, it has been suggested (Ref. {1}, p 119) that the particles produced in a crystal collision frequently have a spread of sizes which fall on either side of the critical size given by the Gibbs-Thomson (also known as the Ostwald-Freundlich) equation

$$h = \frac{2\sigma_L V}{\mu - \mu^*} \qquad (1)$$

where h is an average radius,
σ_L is an average surface free energy of the small crystal in contact with the solution,
V is the molar volume of the crystal.

$\mu - \mu^*$ is the difference in chemical potential (ergs/mole) between the solute in the given solution, μ, and that in a saturated solution, μ^*, at the same temperature. It is also, of course, the difference in chemical potential between the critical sized crystal and a very large crystal†. Any crystal fragments smaller than the

---

† A more precise description of the critical nucleus is obtained from the set of equations:

$$h_i = \frac{2\sigma_i V}{\mu - \mu^*}$$

σ_i is the surface free energy of the ith face, and h_i is its perpendicular distance from a fixed point within the crystal. Since accurate values for the σ_i are seldom known, this formula, although much more elegant than (1) is of restricted practical use.

critical size will re-dissolve. The higher the supersaturation, the smaller the critical size, and the larger the proportion of surviving nuclei. The fact that there appears to be a limiting supersaturation above which no further increase in the number of nuclei occurs would indicate a lower limit to the size of particles produced. This has been referred to as the Survival Theory {1,3}.

GIBBS-THOMSON EQUATION APPLIED TO SOLUTIONS & MELTS

To use equation (1) values must be found for $\mu-\mu^*$. For solutions it is sometimes assumed (Ref. {2, 9}) that

$$\mu - \mu^* = RT \ln\left[\frac{x}{x^*}\right] \sim RT \left[\frac{x - x^*}{x^*}\right] \quad (2)$$

where x is a mole fraction. Now, in general

$$d\mu = \bar{V} dP - \bar{S} dT + (\partial\mu/\partial x)_{PT} dx \quad (3)$$

Defining the activity coefficient by means of the equation

$$d\mu = RT d\{\ln(\gamma \cdot x)\}, \text{ at constant T and P,}$$

we obtain for a small change

$$\mu - \mu^* = RT \left[\frac{x - x^*}{x^*} + \frac{\gamma - \gamma^*}{\gamma^*}\right] \quad (4)$$

Equation (2) is therefore in error by the amount $RT(\frac{\gamma - \gamma^*}{\gamma^*})$. Admittedly, neglect of this term may well introduce an error that is less than that due to the error in σ_L.

On the other hand, another method of measuring $\mu - \mu^*$ which is thermodynamically exact is described below.

If a solution saturated at T_{SAT} is cooled to the experimental temperature T, where $\Delta T = T_{SAT} - T$ is a relatively small quantity, then

$$\mu - \mu^* = (\bar{S}_L - S^*) \Delta T = \frac{(\bar{H}_L - H^*)}{T_{SAT}} \Delta T \quad (5)$$

μ, $\bar{S}_L$, $\bar{H}_L$ are the chemical potential, partial molar entropy and partial molar enthalpy of the supersaturated solution at temperature T. The * refers to the solid at the same temperature T. $\bar{H}_L - H^*$ is the differential heat of solution at saturation: this can be estimated if heats of solution are known at a series of concentrations: these are probably easier to measure than activity coefficients. This equation is thermodynamically exact, and has the further advantage that supersaturation is more often produced by cooling a supersaturated solution than by evaporation at constant temperature.

For melts $\overline{H}_L - H_S$ is simply the heat of melting, and in cases where this is known equation (5) provides an exact measure of $\mu - \mu^*$.

COLLISION BREEDING OF $MgSO_4 \cdot 7H_2O$ (Epsomite)

Free Energy Required

Clontz & McCabe {2} used a steel hammer to strike a single crystal of this salt, held in a flowing aqueous solution supersaturated by a ΔT of $2.2°C$: typically 92 nuclei (i.e. new crystals) were produced for an energy expenditure of 2.2 ergs (0.22 µJ) per nucleus. By equation (5) and using a heat of fusion of 1.8×10^{11} ergs/mole (1.8×10^4 J/mole) † we obtain

$$\mu - \mu^* = \frac{2.2 \times 1.8 \times 10^{11}}{308} = 1.28 \times 10^9 \text{ ergs/mole}$$

(128.5 J/mole). Whence by equation (1) and taking $\sigma_L = 42.4$ ergs/cm^2 (4.24 µJ/cm^2) as estimated in Appendix I

$$h = \frac{2 \times 42.4 \times 147}{1.28 \times 10^9} = 9.7 \times 10^{-6} \text{ cm} .$$

Let us assume that the critical sized crystal is a cube of side $2h = 19.4 \times 10^{-6}$ cm, and a total surface area of 22.5×10^{-10} cm^2.

To estimate the minimum thermodynamic work needed to form a nucleus by fracture of a big crystal, the surface area of the nucleus must be multiplied by twice the solid/vapour surface free energy, σ_v. (The solid/liquid surface free energy, σ_L, would apply only if the fractured surface were in constant equilibrium with the liquid, which is highly unlikely in the circumstances envisaged here.)

---

† Perry {4} p 247, gives a value of 1.34×10^{11} ergs/mole for the heat evolved on solution of Epsomite at high dilution: we have increased this to 1.8×10^{11} to allow for the fact that in the present case the solution is saturated.

If the value of σ_v is taken as 140 ergs/cm² (14 µJ) (see Appendix II) the minimum work to form one nucleus will be $2 \times 140 \times 2.25 \times 10^{-9} = 6.3 \times 10^{-7}$ ergs (6.3×10^{-8} µJ). The energy of 2.2 ergs/nucleus observed by Clontz & McCabe is therefore some 10^6 times the minimum thermodynamic work as calculated above.

In discussing the nature of the energy observed by Clontz & McCabe it must be remembered that they only considered the energy expended after actual contact had been obtained. When a crystal collides with a flat surface the chances are that a corner of the crystal will first strike the target face, resulting in a point contact. (Admittedly Clontz & McCabe sought to establish parallelism of the faces, but it may be doubtful how accurately this could be achieved on the molecular scale: furthermore, Johnson et al {6} found the faces to be rough on the microscopic scale, suggesting contact at a few points only even if the faces were parallel.) At the point of solid contact the high forces set up may understandably cause fracture of the colliding corner. But in the vicinity of the point of contact substantial hydrodynamic forces will operate over the surfaces, giving rise to plastic and elastic deformation over a volume of crystal which may be large compared with the volume of the separated nucleus. Additional work will also be needed to expel the thin liquid layer between the colliding surfaces which will occur as a result of the solid deformation. The deformation of bodies colliding in a liquid medium is a complex subject which has been discussed by (inter alia) Christensen {5}, and Gohar & Cameron {7}.

Energy may also be absorbed by deformations in the structure of the nucleus. The small fragment broken off in the collision will certainly be in a considerably disordered state, with many dislocations and mismatch surfaces: in fact, it may be nearer to an amorphous glassy condition than to a crystal. It may be of interest therefore to discuss the free energy involved in the somewhat extreme case where the nucleus is first of all separated from the parent crystal in the form of a monomolecular layer, which subsequently rearranges itself into a compact crystal.

The area of 1 gram mole spread out into a monomolecular layer
$= (V^{2/3})(\text{Avogadro's number})^{1/3} = (147)^{2/3}(6 \times 10^{23})^{1/3}$
$= 2.34 \times 10^9$ cm². The monolayer area of the nucleus will therefore

$$= \frac{(1.94 \times 10^{-5})^3}{147} \times 2.34 \times 10^9 = 1.2 \times 10^{-7} \text{ cm}^2 .$$

The surface free energy of the monolayer nucleus will

$$= 2 \times 140 \times 1.2 \times 10^{-7} = 3.3 \times 10^{-5} \text{ ergs} \quad (3.3 \times 10^{-6} \text{ µJ}) .$$

This figure is still a very long way short of the energy observed by Clontz & McCabe {2}, and it is still necessary to seek other explanations of the observed work, such as the compressive and hydrodynamic effects which we have attempted to analyse above.

Summary of Free Energy Calculations

Work observed in the experiments of Clontz & McCabe {2} = 2.2 ergs/nucleus

Calculated minimum work to separate a cubic nucleus = 6.3×10^{-7} ergs/nucleus

Calculated minimum work to separate nucleus as a monolayer = 3.3×10^{-5} ergs/nucleus

Gibbs-Thomson radius = 10^{-5} cm

It is of course realised that the fracture of crystals is an exceedingly complex subject, which for adequate treatment would demand far more space than is available in the present paper.

MECHANISM B: NUCLEATION IN THE LIQUID PHASE

All of the above arguments are based on the assumption that collision nucleation depends on a fracture mechanism (A), involving the survival theory as a necessary corollary. The writer knows as yet of no crucial experiment which is capable of deciding definitely between the fracture mechanism A, and the liquid phase mechanism B.

Admittedly the data of Garabedian & Strickland-Constable {9} appears to lend positive support to the survival theory: the work has not yet however been repeated on any system other than Epsomite, and by no other workers, which of course limits the value of the evidence.

APPENDIX I

Value of σ_L (Solid/Liquid Surface Free Energy)

This has been estimated by a method based on that used by Hollomon & Turnbull {8} in the case of solids in contact with their melts. In the case of Epsomite we propose to take a latent heat of melting close to the heat of solution at saturation. The use of Hollomon & Turnbull's method in the case of a solution rests on the following argument. The dissolution of anhydrous inorganic salts usually results in the evolution of heat, since the exothermic heat of solvation exceeds the endothermic heat of melting; whereas the solution of hydrated salts is often endothermic owing to a low heat of reaction between solid and liquid. In the case of $MgSO_4 \cdot 7H_2O$ we are considering a very fully hydrated salt dissolving in a saturated solution. The heats of solution and melting may be expected to be very similar: we will take the heat of melting as 2×10^{11} ergs/mole - slightly greater than the assumed heat of solution at saturation of 1.8×10^{11}. We furthermore assume that the surface free energy of the solid in contact with a saturated solution will be close to that of the solid in contact with its (hypothetical) melt.

Then according to the method of Hollomon & Turnbull

$$\sigma_L = \frac{\text{Heat of melting}}{2 \times A}$$

where A is the area of 1 mole spread out to a thickness of 1 molecule. For details of the method of calculation see Ref {1}, p 97.

Then $A = (\text{Avogadro's Number})^{1/3} V^{2/3} = 2.34 \times 10^9$.

$$\sigma_L = \frac{2 \times 10^{11}}{2 \times 2.34 \times 10^9} = 42.4 \ .$$

This result agrees with that given in Ref. {6}.

APPENDIX II

Value of σ_V (Solid/Vapour Surface Free Energy)

Little useful information is available concerning the value of σ_V. For ice we have $\sigma_V = \sigma_L$ + surface tension of liquid H_2O ~ 30 + 70 = 100 ergs/cm. Therefore for ice

$$\sigma_V = \frac{100}{30} \times \sigma_L = 3.3 \ \sigma_L$$

In the absence of other evidence we can take σ_v for Epsomite as
$3.3 \times \sigma_L = 3.3 \times 42.4 = 140$ ergs/cm^2.

NOMENCLATURE

h	"average radius" of critical nucleus	cm
σ_L	solid/liquid surface free energy	ergs/cm^2
σ_v	solid/vapour surface free energy	ergs/cm^2
V	molar volume of solid	cm^3/mole
μ	chemical potential	ergs/mole
$\overline{V}_L$	partial molar volume	cm^3/mole
$\overline{S}_L$	partial molar entropy	ergs/mole.K
$\overline{H}_L$	partial molar enthalpy	ergs/mole

μ, $\overline{V}_L$, $\overline{S}_L$, $\overline{H}_L$ refer to supersaturated solution at temperature T.

* superscript refers to a saturated solution, or an infinitely large crystal at the same temperature

T_s	saturation temperature of above solution	K
A	monolayer molar area	cm^2/mole

REFERENCES

1. Strickland-Constable, R.F., *Kinetics & Mechanism of Crystallization*, Academic Press, London (1968).
2. Clontz, N.A. & McCabe, W.L., A.I.Ch.E.Symp.Series No. 110, 67, 6 (1971).
3. Strickland-Constable, R.F., A.I.Ch.E.Symp.Series No. 121, 68, 1 (1972).
4. Perry, J.H. (Ed.), *Chemical Engineer's Handbook* 3rd Edition, McGraw-Hill, New York (1950).
5. Christensen, H., J.Lubrication Technology 92, 145 (1970).
6. Johnson, R.T., Rousseau, R.W. & McCabe, W.L., A.I.Ch.E. Symp.Series No. 121, 68, 40 (1972).
7. Gohar, R. & Cameron, A., Proc.Roy.Soc. A291, 520 (1966).
8. Hollomon, J.H. & Turnbull, D., Prog.Metal Physics 4, 333 (1953).
9. Garabedian, H. & Strickland-Constable, R.F., J.Crystal Growth 12, 53 (1972).

SECONDARY NUCLEATION OF POTASH ALUM

K. TOYOKURA, K. YAMAZOE, J. MOGI, N. YAGO, Y. AOYAMA

Waseda University

Nishiookubo 4Chome, Shinjukuku, Tokyo, Japan

INTRODUCTION

This study is aimed at obtaining a correlation between secondary nucleation rates and operating conditions, and their application to industrial crystallization. The secondary nucleation rates of potassium aluminium sulphate $12H_2O$ (potash alum) are observed in supersaturated solution passed through a multi-crystal bed and correlated against the solution supersaturation and Reynolds number. Mean crystal growth rates of seed crystals, fluidized by the solution, are also studied. Design theories for a classified bed crystallizer are proposed from the secondary nucleation and crystal growth rates, and the operating conditions of an industrial crystallizer are discussed.

EXPERIMENTAL

A schematic diagram of the experimental apparatus is shown in Fig.1. Supersaturated solution was fed into the crystallizer (5) located in a constant temperature bath (7) whose temperature was kept the same as that of the outlet solution from the heat exchanger. Seed crystals were put in the crystallizer to observe their effect on nucleation. Solution leaving the fluidized bed in the crystallizer flowed into the dissolution tank (8) which was heated 8°C higher than the saturation temperature of the solution.

After steady state conditions had been achieved, the overflow solution was diverted to the glass sampling cell (6) set in the bath (7). The cell (280 ml) was covered to prevent evaporation. When the cell was filled with solution, the flow was returned to

Tank (8) and the state of the solution in the cell was watched through a microscope. After several minutes, very small fines appeared uniformly over the bottom surface of the cell and they then grew larger to almost the same size. No further new crystals appeared.

When seed crystals were not added to the crystallizer, no new crystals appeared, even after two hours at the same supersaturation and flow rate. Therefore, the fine crystals in the cell were concluded as being secondary nucleii produced by the seed crystals.

The number and size change of nucleated crystals were observed in this study. Reagent grade chemicals were used. The size range of seed crystals was 0.61 to 1.7 mm. The depth of the bed in the crystallizer was 1.45 to 10.5 cm and the decrease of solution supersaturation passing through the bed was less than 4×10^{-4} mol/ℓ (H_2O).

RESULTS

Crystals resting in the cell grew as shown in Fig.2. Most of the data, except those for high supersaturation, plot as straight lines which, when extrapolated, coincide at $\ell=0$ and time θ, the time when the sampled solution passed through the bed. Growth rates of crystals smaller than about 75 μm were not measured, but if the growth rates from the slopes of the lines in Fig.2 were applied to fine crystals, the nuclei in the supersaturated solution could be considered as being born just at the time that the solution was passing through the bed in the crystallizer. From these results, the waiting time for secondary nucleation was concluded to be zero for these experiments.

The fluidized crystals had size distributions of 0.7 to 1.25 mm and 0.9 to 1.7 mm. Growth rates, G, were obtained at a superficial velocity of 2.4 cm/s and correlated with the supersaturation, ΔC, by

$$G = 5 \times 10^{-3} \Delta C^{5/3} \tag{1}$$

These data are almost the same as the mean crystal growth rates calculated from the data of Mullin and Garside (1). Therefore, growth rates of potash alum crystals may be considered approximately independent of crystal size for industrial purposes (discussed later) when the fluidized seed crystals have a size distribution and the mean crystal size is used for calculation purposes.

The secondary nucleation rate per unit volume of the supersaturated solution, N, was calculated from the observed number of crystals in the cell. The surface area of seed crystal per unit

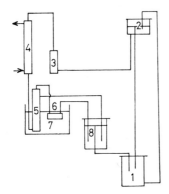
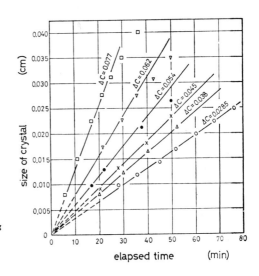

1: Tank 2: Head tank
3: Flow meter 4: Heat exchanger
5: Crystallizer (30 mm I.D., 250 mm height) including 50 mm depth of ass bead bed as strainer
6: Constant temperature bath
7: Sampling cell 8: Dissolution tank

Fig. 1 Schematic diagram of experimental apparatus

Fig. 2 Size change of crystal in the cell against the elapsed time

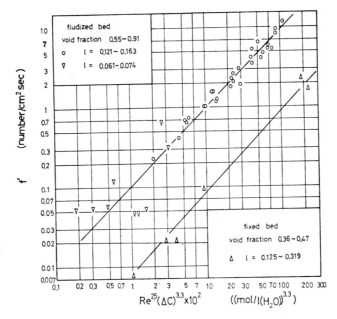

Fig. 3 Correlation between secondary nucleation rate and $Re^{2.5}(\Delta C)^{3.3}$

volume of solution in the crystallizer, a, was calculated from the void fraction of the fluidized bed. The time, θ', required for passing through the bed, was calculated by (height of bed) × (void fraction) ÷ (superficial velocity). Therefore, the secondary nucleation rate per unit area of seed crystal surface, f', was calculated by

$$f' = N/a\theta' \tag{2}$$

When f' was plotted against ΔC for a constant superficial velocity, eq.3 was obtained:

$$f' \propto \Delta C^{3.3} \tag{3}$$

When f' was plotted against Reynolds number for a constant supersaturation, eq.4 was obtained:

$$f' \propto Re^{2.5} \tag{4}$$

Plotting the data on the basis of eqs 3 and 4 (Fig.3) gave the following correlations:

fluidized bed: $\quad f'_{flu} = 10 \times \Delta C^{3.3} Re^{2.5} \tag{5}$

fixed bed: $\quad f'_{fix} = 0.85 \times \Delta C^{3.3} Re^{2.5} \tag{6}$

DESIGN OF A CYLINDRICAL CLASSIFIED BED CRYSTALLIZER FOR POTASH ALUM, USING THE CRYSTAL GROWTH RATE REPORTED BY MULLIN AND GARSIDE (1)

Design Equations

The following conditions are assumed:
(a) Piston flow of solution up through the crystallizer and temperature constant throughout.
(b) Fluidized crystals completely classified by the hydrodynamics along the axis of the crystallizer.
   These assumptions are the same as the previous report (2)
(c) Crystals generated according to the nucleation equation:

$$f' = \alpha Re^{\beta} \Delta C^{\gamma} \tag{7}$$

where α, β, and γ are constant.
(d) Crystals grow to product size according to the equations reported by Mullin and Garside (1):

SECONDARY NUCLEATION OF POTASH ALUM

$$G' = K_o \ell^b \Delta C^n \tag{8}$$

K_o, b, and n are constants. G' and ℓ are the growth rate and crystal size. From these assumptions, a height Z and cross-section area S of a crystallizer become (2):

$$Z = 3Pk_v \ell_2^{1-b} \text{ (C.F.C.)}/y_1^3 \, K_o \Delta C^n k_a \tag{9}$$

$$\text{(C.F.C.)} = \int_1^y \frac{\emptyset^n \, y^{3-b} \, dy}{(1-\varepsilon_2 y^{-1/3})\left[1+(\emptyset-1)(y^3-1)/y_1^3-1)\right]^n} \tag{9'}$$

$$S = P(1-y_1^{-3}) \, \emptyset / \, fM\Delta C_1 (\emptyset - 1) \tag{10}$$

where f = superficial velocity of solution on solvent basis, M = molecular weight, P = production rate, k_v & k_a shape factors, y = dimensionless crystal size (ℓ/ℓ_1), $\emptyset$ = dimensionless supersaturation ($\Delta C_1/\Delta C_2$), ε = voidage, Subscripts 1 and 2 are top of the crystallizer.

Design Procedure

Crystallization rate, product size (crystal size at the bottom) and voidage at top and bottom are fixed. The design calculations are performed by the following steps:

1. Number of crystals required per unit time:

$$F' = P / \rho_c k_v \ell_1^3 \tag{11}$$

2. Superficial velocity, u, assuming that the terminal velocity, u_t, follows Allen's Law:

$$u_t = I\ell, \quad \varepsilon = K(u/u_t)^{1/3}, \quad u = \varepsilon_1^3 I \ell_1 / k^3 \tag{12}$$

where I, k and K are constants.

3. Crystal size ℓ_2 at top of bed is obtained from equation 12 using ε_2 instead of ε_1.

4. f is obtained from

$$f = C'u \tag{13}$$

where C' is the concentration.

5. Z and S are calculated from the results of steps 2 and 4 by eqs 9 and 10 as function of $\emptyset$ and ΔC_1

For an assumed supersaturation ΔC_1 the secondary nucleation rate per unit cross-sectional area of the designed crystallizer, f'_{all} is calculated from

$$f'_{all} = \int_0^Z f' \, a' \, dz = \int_0^Z a' \alpha \, Re^\beta \, \Delta C^\gamma \, dZ \qquad (14)$$

where, a' is the crystal surface area per unit volume of crystallizer. When the design is satisfactory, f' should be equal to f'_o ($=F'/S$) calculated by eqs 10 and 11. f'_{all} amd f'_o are related to ΔC_1 by

$$f'_{all} \propto \Delta C_1^{\gamma - n+1} \qquad (15)$$

$$f'_o \propto \Delta C_1 \qquad (16)$$

When f'_{all} and f'_o are plotted against ΔC_1 on a log-log basis, they yield two straight lines, and the required supersaturation ΔC_1^*, is the point of intersection.

A continuous classified bed crystallizer for 30 ton/day of 0.9 crystals of potash alum is designed as an example and the results are:
  Inlet supersaturation: 1.03×10^{-2} kmol/kg H_2O
  Seed generation rate: 3.8×10^8 m^{-2} h^{-1}
  Diameter of crystallizer: 2.6 m.

DESIGN OF A CYLINDRICAL CLASSIFIED BED CRYSTALLIZER FOR POTASH ALUM USING THE MEAN CRYSTAL GROWTH RATE

When the mean crystal growth rate obtained in this study is used for the design instead of that proposed by Mullin and Garside (1), i.e. if the crystal growth rate is assumed independent of crystal size, eqs 9 and 9' become:

$$Z = 3Pk_v \ell_2 (C.F.C.)_{mean, n} / K'_o \Delta C^n k_a \qquad (17)$$

$$(C.F.C.)_{mean, n} = \int_1^y {}_1 \phi_y^n y^3 dy / (1 - \varepsilon_2 y^{1/3}) \{1 + (\emptyset - 1)(y_1^3 - 1)\}^n \qquad (17)'$$

The cross-sectional area of the crystallizer is calculated by eq 10 and the calculation method is the same as in the previous section for the same design conditions as in the previous examples the results are:
  Inlet supersaturation: 0.98×10^{-2} kmol/kg H_2O
  Seed generation rate: 3.65×10^8 m^{-2} h^{-1}
  Height of crystallizer: 2.6 m
  Diameter of crystallizer: 2.7 m

In these calculations, the mean crystal growth rate is the growth rate corresponding to the mean crystal size in the crystallizer. The results from eqs 10 and 17 are almost the same as those from eqs 9 and 10. Therefore, the mean crystal growth rate is considered to be applicable for the design of industrial crystallizer instead

SECONDARY NUCLEATION OF POTASH ALUM

of the more accurate growth rate correlation. The following design equations are therefore proposed using the mean crystal growth rate.

DESIGN OF A CONE CLASSIFIED BED CRYSTALLIZER

The model of the cone classified bed is the same as the cylindrical type except that the void fraction is constant through the bed. Therefore, the design equations become

$$Z = \frac{3Pk_V I(\varepsilon/1.5)^3 \ell^2 C^1 \; (C.F.C.)_{\varepsilon=const.n}}{Fk_a(1-\varepsilon)MK_o' \Delta C_1^n} \quad (18)$$

$$(C.F.C.)_{=const,n} = \int_1^{y_1} \phi^n y^4 dy / \{1+(\phi-1)(y^3-1)/(y_1^3-1)\}^n \quad (18)'$$

$$S = (P_1-P_2)/MC' \; (\Delta C_1 - \Delta C_2) I (\varepsilon/1.05)^3 \ell_2 \quad (19)$$

When ϕ in eq. 18' is replaced by $\bar{\phi}$, defined by

$$\bar{\phi} = (\phi-1)(y^3-1)/(y_1^3-1) + 1 \quad (20)$$

and y_1 is replaced by y, eqs 18 and 19 may be used to calculate the shape of the crystallizer walls using the dimensionless crystal size. The overall secondary nucleation rate becomes

$$F'_{all} = \int_1^{y_1} \alpha \left[\frac{(\varepsilon/1.05)^3 I \ell_2^2 y^2}{\varepsilon \nu} \right]^\beta \left[\frac{\Delta C_1}{\bar{\phi}} \left(1 + \frac{(\phi-1)(y^3-1)}{(y_1^3-1)} \right) \right]^2$$

$$\times \frac{3\rho_c P \phi^n y^2 dy}{MK_o \Delta C^n \ell_2^2 \{1+(\phi-1)(y^3-1)/(y_1^3-1)\}^n} \quad (21)$$

Since F'_{all} in eq. 21 should be equal to F'_o in eq. 11, ΔC_1^* is decided from the intersection point on the plots of F'_{all} and F'_o against ΔC_1, as in the cylindrical crystallizer.

DISCUSSION

Test data for an industrial cone crystallizer designed to produce potash alum are shown in Fig.4. The relations between product crystal size, crystallization rate and operational conditions are estimated from laboratory tests by the proposed equations as follows:
1. S_1, S_2, Z for the given crystallizer, are known
2. $P.\phi$ are set
3. $F.\varepsilon$ are assumed
4. ℓ_1 is calculated from $F.S_1$, ε by the fluidized property.

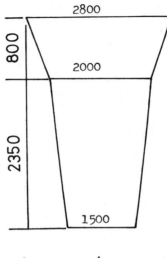

l_1 (mm) ×10⁻²	ϕ (-)	P (ton/mon)
2.2	4.2	600
2.4	5.1	580
2.8	14.0	524
3.2	21.3	454
3.4	18.9	367
4.1	29.3	260

Fig. 4 Dimensions (mm) and operational data of the cone-type industrial crystallizer

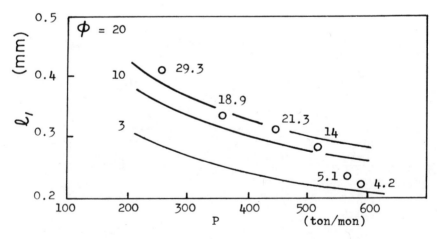

Fig. 5 Correlation between l_1 and P

5. F' is calculated from P, ℓ_1 by a material balance of crystal phases
6. ΔC_1 is calculated from P, F, $\emptyset$ by a material balance of solution phases
7. $F'_{a\ell\ell}$ is calculated from ℓ_1, ΔC_1, F, ε, the shape of the crystallizer by the proposed theory in this study
8. The operational flow rate of solution F^* (replaced by ℓ_1) is decided from comparison between F' and $F'_{a\ell\ell}$ as ΔC_1^* for the cylindrical crystallizer.

Calculations are plotted in Fig.5 using data from Fig.4. The number beside the points are values of $\emptyset$ from the test data, and they agree very well. The industrial crystallizer is about a hundred thousand times the size of the test equipment. A hundred thousand times scale up may therefore be expected using the new design theory.

REFERENCES

1. Mullin, J.W. and Garside, J., Trans. Instn.Chem. Engrs, 45 (1967) 291.
2. Toyokura, K., Krystall und Technik, 8 (5), (1973) 567.

CONTROL OF PARTICLE SIZE IN INDUSTRIAL NaCl-CRYSTALLIZATION

C.M. VAN 'T LAND* AND B.G. WIENK**

Akzo Engineering b.v., Arnhem, the Netherlands*

Akzo Zout Chemie Research, Hengelo (O), the Netherlands**

INTRODUCTION

Because the solubility of NaCl increases only slightly with temperature, pure salt is produced industrially by evaporative crystallisation, generally in multiple-effect plants. One of the installations at the Akzo Company's salt plant is shown in Fig.1. The plant has the following characteristics: (a) parallel brine feed, (b) salt discharged in series and (c) forced circulation.

The capacities of the various plants, which operate in four or five effects, vary from 150,000 to 1,000,000 metric tons per year. The diameters of the vapour separators vary from 3 to 8 m. Three main types of crystalliser are in common use.

Type 1: Forced circulation crystalliser with external screw pump and heat exchanger. See Fig.2.
Type 2: Forced circulation crystalliser (either direction feasible) with internal screw pump and heat exchanger (calandria). See Fig.3.
Type 3: Growth type crystalliser, also called an Oslo crystalliser. Clear brine circulates while the salt crystals are suspended in a fluidised bed. See Fig.4.

The criterion for choosing a crystalliser from the given types depends upon the desired crystal size. If there are no special requirements, type 1 is chosen. Salt with 50% by weight larger than about 450 μm will be produced.

If a somewhat coarser salt is desired, type 2 is chosen. Salt with 50% by weight larger than about 650 μm will be produced.

The type-3 crystalliser produces the so-called granular salt; these particles have a size between 1 and 2 mm.

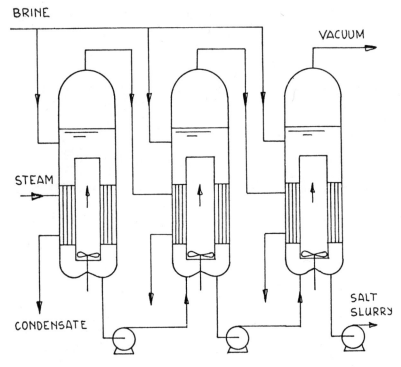

Fig. 1. Salt plant

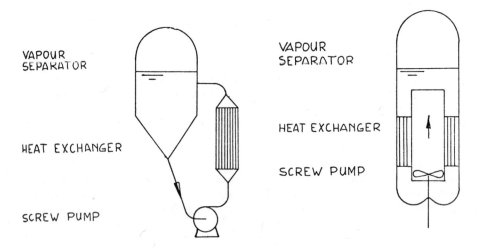

Fig. 2. Type-1 crystalliser   Fig. 3. Type-2 crystalliser

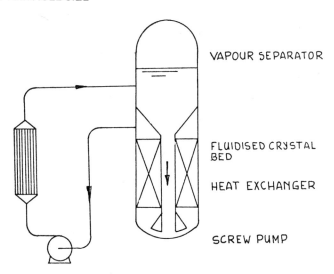

Fig. 4. Oslo crystalliser

REVIEW OF PARTICLE-SIZE STUDIES

Comparison of Type-1 and -2 Crystallisers

Table 1 reflects the actual situation. The conditions of operation are given together with the average particle size on a 50% by weight basis. The type-2 crystalliser produces a coarser salt than type-1. The reason is that the slurry velocities of type-1 exceed those of type-2. This is an important parameter. Rather high velocities are chosen to promote good heat transfer and liquid/gas mass transfer in the vapour separation.

Experiments with a Type-1 Crystalliser

The salt crystals circulating in the individual crystallisers of the multiple-effect evaporation plant are more or less of the same size. This applies to in-series and parallel salt discharges as well as to crystallisers having temperatures between 50 and 140°C. Therefore, temperature cannot be used to control particle size. Different slurry velocities were not used, but the effects of a number of independent variables were studied. The crystal size distribution (CSD) was measured as the dependent variable. The results are summarised in Table 2.

Dosing with milled salt leads to a significant reduction of the average particle size. This is the only method for particle size control which was actually installed in a salt plant equipped with type-1 crystallisers and designed by Akzo Engineering.

Much work on this topic has been carried out by Mrs M.A. van Damme-van Weele and W.M. de Wolff (1,2).

The presence of vast amounts of fine particles (~20 µm) in industrial salt crystallisers has been observed by various investigators. This was the reason for van Damme-van Weele to start an experiment in which she had a salt slurry circulating in a type-1 crystalliser for twenty-four hours without producing salt. The temperature was kept at 120°C and a growth inhibitor to prevent re-crystallisation had been added. Many small chips were obtained. The production of chips in numbers per hour exceeded the production of salt crystals under normal conditions by a factor of about 1,700! We can anticipate the production of chips under normal conditions as well.

Table 1   Comparison of Two Types of Crystalliser

		Type 1	Type 2
Crystallisation temperature,	°C	50 - 140	100 - 140
Residence time (crystals and brine),	h	1 - 2	1
Slurry density, W_i	% by wt	20 - 25	20 - 25
Circulation velocity, U_c	m/s	3.5	1
Velocity in the heater tubes, U_p	m/s	2	1.25
Screw pump tip velocity U_t	m/s	20	10
Weight average particle size, d_{50}	µm	450	650

Table 2   Experiments with a Type 1 Crystalliser

Variable	Variation	Effect (µm)
Slurry density	10 to 15% by wt	370 to 430
	10 to 30% by wt	450 to 440
Production	0.6 to 1 (a)	450 to 450
Tip velocity screw pump (b)	15 to 20 m/s	430 to 430
Dosing ferrocyanide	~5 ppm by wt in salt	440 to 400
Dosing salt cubes 2- 5 µm	70X (c)	450 to 450
Dosing salt cubes 5- 20 µm	6X (c)	460 to 440
Dosing milled salt 20-150 µm	4X (c)	440 to 285

(a) Ratio is given. (b) Velocity in the heater tubes is 2 m/s in both cases. (c) This number indicates the number of particles dosed per unit of time compared with the number of crystals leaving the crystalliser.

CONTROL OF PARTICLE SIZE

Table 3   Particle-size Studies in a Type-2 Crystalliser

Number	1	2	3	4	5	6	7	8
W_i, % by wt	20	20	20	20	10	10	10	10
U_p, m/s	1	1	0.7	0.7	1	1	0.7	0.7
u or d	u	d	u	d	u	d	u	d
d_{50}, μm	650	700	700	750	600	650	650	700

u and d stand for upward and downward circulation through the draft tube. Other independent variables tested: production (8 to 18 t/h), tip velocity screw pump (8 to 16 m/s), dosing some ppm by wt of ferrocyanide. In all 3 cases no effect was observed.

Experiments with a Type-2 Crystalliser

The results for an industrial crystalliser operating at 120°C (first effect) and receiving a clear liquid are summarised in Table 3.

Neither milled salt nor small salt cubes were added. The tip velocity of the screw pump was varied, while the circulation rate in m³/h was kept constant.

Conclusion

As a result of the work described above crystallisers able to produce salt with an average crystal size (50% by weight) in the range of 200 - 750 μm can now be built.

There is no relationship between the average crystal size and the production costs. Furthermore, we concluded that for a better understanding of what really happens in our industrial crystallisers, a study of the nucleation mechanism should be made.

The method of plotting sieve analyses as so-called population density curves has recently been introduced (3). The following section reviews the results of applying this technique to sieve analyses of industrial salt crystallisers receiving a clear liquid feed.

POPULATION DENSITY CURVES

Population density curves provide an insight into what is actually happening in a crystalliser. Ideally, the function $\log N = f(L)$ can be represented as a straight line, but in practice deviations occur. The deviations lead to conclusions concerning the characteristics of the system considered and ways to modify the CSD may manifest themselves. An 'ideal' straight line is obtained if:

The feed is clear
The contents are mixed
The state is steady
The operation is continuous
There is neither breakage nor conglomeration
The crystal growth is independent of L

Canning (4) deals with the translation of deviations to the actual characteristics.

Sampling and Sieving

It is desirable to know the CSD of the salt circulating in a crystalliser. This knowledge can only be obtained by sampling and the sample must be representative. The sample must be processed in various steps:

Cooling the sample
Separating the salt and mother liquor
Drying the crystals
Classification into many size ranges
Weighing the crystals in each size range

Except during the last step, precautions must be taken to prevent the formation of new particles, the growth of crystals and, vice versa, the disappearance of particles and the partial dissolution of crystals. Much attention has been paid to this subject.

Cooling was carried out in the presence of ferrocyanide ions, their function being the instantaneous 'freezing' of the dynamic equilibrium growth/dissolution.

Separation of salt and mother liquor was carried out on a Buchner funnel. The air sucked through the salt cake had been drawn through saturated brine to prevent water evaporation.

Drying of the salt crystals took place via brine, ethanol/water and di-ethyl ether washings on the Buchner funnel. The functions of the various washings are: 1. brine replaces mother

CONTROL OF PARTICLE SIZE

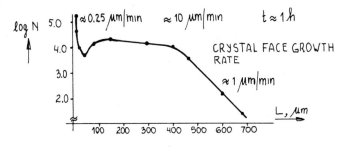

Fig. 5   Population density plot.

liquor, 2. ethanol/water replaces brine, preventing crystallisation and dissolution, 3. 100% ethanol and di-ethyl ether promote quick drying.

Classification was achieved by sieving. The large sieves had square holes of 710, 600, 500, 425, 355, 250, 125 and 63 μm. A set of micro-precision sieves with circular holes of 90, 75, 60, 45, 30, 20 and 10 μm was also used. Wet sieving was done in toluene in a trough in which ultrasonic vibrations were generated (40 Hz). The very small salt particles were also examined under a microscope.

A characteristic plot is shown in Fig.5 representing a sample drawn from the third crystalliser in a multiple-effect evaporation plant. This crystalliser (type-1 with an elutriation leg) received a clear feed (parallel salt discharge) and produced about 20 ton/h of salt under continuous, steady state operation. The crystallisation temperature was about 75°C and the slurry density 18% by weight.

Thus, the 'ideal' requirements mentioned above are met. Certainly the elutriation leg involves some classification of its contents, but this is not an objection against applying the method. Moreover, the work was repeated for crystallisers not receiving clear feeds (salt discharge in series). Qualitatively the graphs are similar to the one presented in Fig.5.

The classification and rounding-off of particles larger than 400 μm causes the downward curvature. The upward curvature for particles smaller than 60 μm is a more complicated problem. It has been noticed by other workers in the field as well. Bennett (5) and Canning (4) made similar observations. The steep slope between 0 and 60 um can be related to one (or a combination) of the following phenomena:

low rate of growth of small crystals
high death rate of small particles.

Also remarkable is the local minimum at about 60 μm. To what physical phenomenon can this minimum be attributed? These issues will be dealt with later.

Microscopic Studies

The various fractions smaller than 125 μm were observed and photographed. It appeared that the particles < 10 μm were, with few exceptions, irregularly shaped fragments. Particles ~ 100 μm were regular cubic crystals with some fragments. When going from 10 to 100 μm, the picture shifted gradually from fragments to cubes. As to the particles of 10 μm, four possibilities may be distinguished:

Irregular-shaped small particles heal to cubes, small cubes grow.
Only small cubes grow, chips re-dissolve or disappear via the outlet.
Both types re-dissolve (cubes of 40-50 μm survive only).
Only chips grow, cubes re-dissolve or disappear via the outlet.

The occurrence of undersaturation in the heater tubes was proved.

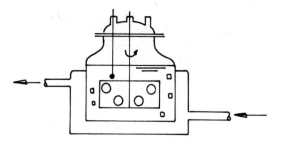

Fig. 6   Laboratory crystalliser

NUCLEATION MECHANISM

To obtain an insight in what is actually happening inside an industrial crystalliser, a series of simple growth experiments was carried out. The apparatus used is shown in Fig.6.

A well-stirred crystalliser (contents 1 litre) is thermostatically controlled at $72 \pm 0.1^\circ C$. At this temperature the filtrated brine in the vessel is brought carefully to saturation by means of large crystals of pure salt. Just before the addition (of various kinds) of seed crystals, the brine is cooled down $0.2^\circ C$. After the seeds have been added, supersaturation is created slowly in the brine by cooling. The maximum supersaturation reached depends on the temperature decrease.

For $\Delta\theta = 5^\circ C$, a $\sigma_{max} = 0.5\%$ and for $\Delta\theta = 1^\circ C$, a $\sigma_{max} = 0.1\%$ can be calculated. Gentle stirring takes place to suspend the crystals. In all experiments, except for one, nucleation occurred suddenly, 10 - 20 minutes after the start of the experiment, many perfect cubic particles (10 - 50 μm) were produced within a few seconds. When this occurs, the experiment is stopped by 'freezing' the situation with a ferrocyanide addition. After the solid

CONTROL OF PARTICLE SIZE

Table 4 O: no effect, -: not present. x A 50% by weight salt slurry from an industrial crystalliser was milled. The 350-40μm fractions were separated by wet sieving.

Experiment No.	Experimental Conditions			Seed Crystals			Results		
	Temperature decrease, °C	Cooling time, minutes	σ_{max}, %	SOURCE	Size, μm	Amount, mg	Nucleation	Crystal growth observed, μm	Chip growth observed
1	4	14	0.4	-	-	0	+	-	-
2	5	20	0.5	cubes from No. 1	10-50	600	+	100	-
3	5	20	0.5	cubes, sieve fraction industrial crystalliser	70-125	300	+	30	-
4	4	12	0.4	chips, dry milled salt	<20	200	+	-	0
5	4	14	0.4	chips, dry milled salt	20-40	200	+	-	0
6	3	12	0.3	damaged crystals from wet milling *	>350	1300	+	20	-
7	1	5	0.1	chips, wet milled salt *	<40	300	0	-	0
8	2	14	0.2	chips, wet milled salt *	<40	300	+	-	0
9	2	8	0.2	damaged crystals * and chips, wet milled salt *	>350 <40	1300 300	+	20	0
10	3	12	0.3	repetition of No. 9	As No.9	As No.9	+	20-30	0

material has settled, the crystals are collected. Samples, together with some crystal samples taken during the experiment, are studied microscopically. A survey of the experimental conditions, the seed crystal used and the results obtained is given in Table 4.

Conclusions

Cubic salt crystals, new nuclei and crystals damaged by milling can grow. Small salt chips (<40 μm) do not grow.

Chips do not grow to form cubes (they do not heal).

Growth of damaged crystals starts on the undamaged faces and from there spreads and gradually covers the faces created by milling.

DISCUSSION

During industrial salt crystallisation, many chips having a size smaller than 100 μm are formed from the large crystals by abrasion. Up until now it was taken for granted that the fine chips re-dissolve in the heater tubes. Asselbergs and de Jong have thrown doubt upon this assumption (8). Now it appears that these small chips do not grow either. They seem to act like inert material in the crystalliser being formed by abrasion and being removed through the outlet.

The production of new crystals probably proceeds via secondary nucleation mainly, giving small cubic particles (7). These small cubes grow at a high rate. More vigorous mechanical conditions lead to higher nucleation rates. High rates of diffusion-controlled growth of small crystals have been calculated by Matz (9), e.g. for 3/30/3000 µm crystals, 16/3.3/0.23 m/s respectively at σ = 0.36% and 52°C). The phenomenon of rapid growth of small particles is known to chemical engineers as well. This is expressed by $Sh = K \cdot d_p/D = 2$ (Frossling equation). Only little depletion of the adjacent stagnant layer is required to reach an appreciable size.

The steep slope between 0 and 60 µm and the local minimum at 60 µm (Fig.5) can be explained by the assumption of a superposition of two phenomena: formation and disappearance of inert chips and rapid growth of cubes. Margolis et al. (6) found for ice crystallisation with gentle stirring an upward curvature between 0 and about 100 µm and a downward curvature for larger crystal sizes.

NOMENCLATURE

CSD	crystal size distribution	-
d_{50}	weight average particle size	µm
d_p	particle size	µm
D	diffusion coefficient	m²/s
K	mass transfer coefficient	m/s
L	crystal size	µm
N	population density	number/µm litre
Sh	Sherwood number	-
t	residence time	h
U_c	circulation velocity	m/s
U_p	slurry velocity in the tubes	m/s
U_t	pump tip velocity	m/s
W_i	slurry density	% by weight
θ	temperature	°C
σ	supersaturation	%

REFERENCES

1. Van Damme-Van Weele, M.A., unpublished results (1961-2).
2. De Wolff, W.M., unpublished results (1967-8).
3. Larson, M.A. and Randolph, A.D., Chem. Eng. Prog. Symp. Ser., No.95 (1969) 1.
4. Canning, T.F., Chem. Eng. Prog., 66, 7, (1970), 80.
5. Bennett, R.C., Fiedelman, H. and Randolph, A.D., Chem. Eng. Prog. 69, 7, (1973), 86.
6. Margolis, G., Sherwood, T.K., Tibaut-Brian, P.L. and Sarofim, A.F., Ind. and Engng. Chem. 10, 3, (1971), 439.
7. Clontz, N.A. and McCabe, W.L., Chem. Eng. Prog. Symp. Ser., No.110 (1971) 6.
8. Asselbergs, C.J. and De Jong, E.J., 5th Symposium on Industrial Crystallisation, Prague, (1972).
9. Matz, G., Chemie-Ing.-Techn. 42, 18, (1970), 1134.

SECONDARY NUCLEATION AND CRYSTAL GROWTH AS COUPLED PHENOMENA

J. ESTRIN AND G.R. YOUNGQUIST

Department of Chemical Engineering
Clarkson College of Technology
Potsdam, New York 13676

There is reason to consider the mechanisms of growth of crystals of industrial importance and secondary nucleation as coupled. Neither mechanism, as it occurs in a suspension crystallizer, is understood with confidence. Since growth and nucleation occur at the crystal-solution interface concurrently, the coupling hypothesis leads to efficiency in designing experimental studies. This approach has been well demonstrated recently (Tai, McCabe, Rousseau) in a study which included growth rate determinations and contact nucleation behaviour for three systems. The dependence of growth rate upon supersaturation and of number of nuclei upon supersaturation exhibited the same functional form for each system. Recently growth rates of crystals of magnesium sulphate heptahydrate were determined in the apparatus in which secondary nucleation due to a fluid shear impulse was studied (Sung; Youngquist, Estrin, Jagannathan, Sung). The growth rate was measured just prior to application of the jet of the solution upon the crystal surface under observation. Here we briefly introduce this experimental method for measuring secondary nucleation characteristics and the correlating scheme used in treating the data. We then discuss the possible coupling of growth and nucleation mechanisms in light of the data in hand and some newly reported experimental observations.

EXPERIMENTS

The crystallizer consisted of a vertical Pyrex glass cylinder with a horizontal baffle which divided the vessel into two regions. These were separately jacketed so that different temperatures could be maintained in each. There were ports which provided for flow from the upper to the lower compartment. The seed crystal was positioned in the upper compartment, just above one port. Complete details of this apparatus are available in Sung's thesis. After the solution and seed (Mg $SO_4 \cdot 7H_2O$) were cooled to the desired temperature and level of supersaturation, ΔT_C, the growth rate of the seed crystal ((110) face) was measured using an optical reader with internal scale and rider. Solution was drawn into the submerged barrel of a syringe whose plunger was then activated by a weight to provide a submerged jet of solution of known quantity and velocity, which impinged upon the centre of the growing face. Simultaneously the solution from the vicinity of the syringe tip and seed flowed to the lower compartment maintained at lower temperature or higher supersaturation, ΔT_B. The nuclei developed here and were counted as they ultimately rested upon the flat transparent bottom.

Somewhat different arrangements were used by Sung, Estrin, and Youngquist for the magnesium sulphate system and by Estrin, Wang and Youngquist for the ice-brine system. All arrangements provided for the generation and development of nuclei by the application of a finite impulse of fluid shear and subsequent determination of numbers by a visual count. In an effort to learn about the growth mechanisms at play, the growing face of magnesium sulphate crystals were observed under low power magnification and with continuously increasing supersaturation in the bulk solution. These crystals were oriented and held in a position similar to that used in the secondary nucleation experiments referred to above.

RESULTS

Most of the results implied in the above experiments have been reported earlier (Youngquist, Estrin, Jagannathan, Sung). Not all of these are in print and some are new, so we shall list briefly results from these experiments and treatment of associated data which we consider of greatest importance:

The nuclei generated and developed are very sensitive to the supersaturation in both bulk solution in which the seed is immersed and solution at higher supersaturation in which nuclei develop; and to the velocity (or shear) of the jet causing nucleation.

The principles underlying the development and characterization of the nuclei are entirely consistent with the survival theory.

The notion of growth or regeneration limited nucleation (Evans, Sarofim, Margolis) is applicable in the interpretation of these secondary nucleation results. No change in number of nuclei was observed when the period of jet application was varied from 1.1 to

SECONDARY NUCLEATION AND CRYSTAL GROWTH

2.75 seconds (Mg $SO_4 \cdot 7H_2O$-water system), and similar qualitative behaviour was noted for the experiments using the ice-brine system. These results indicate that regeneration of the nuclei source controls the nucleation rate.

For a given, constant fluid mechanical condition and temperature of operation, the number of magnesium sulphate nuclei counted were correlated with ΔT_c according to the following relationships obtained by adaptation of classical nucleation theory:

$$N_{TOT} = NS \sum_{i=i_1}^{i_2} \exp(-Ai^{2/3} + B_1 \Delta T_c i)$$

i_1 and i_2 are defined as follows:

$$i_1 = (2A/3B_1 \Delta T_B)^3 \quad \text{and} \quad 1 = NS \exp(-Ai_2^{2/3} + B_1 \Delta T_c i_2)$$

The value of the parameter B_1 was considered known:

$$B_1 = v\Delta H/kTT_s$$

and A and NS are parameters evaluated by a fit of the data – nuclei counted, N_{TOT}, vs. ΔT_c and ΔT_B. The data fit the functional form of these relations with respect to supersaturation very well.

For the magnesium sulphate solution, the measured growth rates showed second order dependence upon supersaturation. The same rates and dependencies were observed at two temperature levels (40.6 and 47.4°C). These growth rates were obtained using rigidly held crystals growing in stagnant solution; the supersaturation entering into the correlation refers to the bulk of solution.

Qualitative changes in the surface of a growing crystal of magnesium sulphate were observed to be gradual rather than abrupt, when continuously increasing supersaturation, except that lines, which are best described as ripples, appeared at a supersaturation of 2.7 to 3°C. These changed from curvy ripples to broken lines with small increases in supersaturation. Whether or not these lines or ripples are mobile has not been determined up to this time.

DISCUSSION

We wish to discuss the extent of inherent coupling between growth and secondary nucleation. Following the leads offered by Botsaris and Denk we recognize that the sources of nuclei are either from the solid crystal *per se*, from solution *per se*, or from the solution-solid interface. The direct approach of Denk and Botsaris reveals convincingly that, at least for the enantiomorphic sodium chlorate system, contact nucleation gives rise to nuclei whose source is predominantly the solid phase. This is supported by the direct observations of Melia and Moffitt (for the dendritic ammonium chloride and bromide systems) and the effects of microattrition upon the product ammonium sulphate crystals in the experiments of Youngquist and Randolph. It appears that so long as the observed nucleation rate is removal limited, in the sense of Evans, Sarofim and Margolis, the fundamental linking of growth and nucleation is not complete; the rate of nucleation is determined by direct contacts with foreign objects or other crystals and the mechanisms implied by survival theory.

Of more interest to us here, are those nuclei which do not originate directly from the solid by attrition. Less direct evidence that nucleation occurs in solution or at the solution-solid interface is available. Most convincing that such nucleation occurs in unstirred solutions and at low supersaturations (for the sodium chlorate-water system) is the work of Denk and Botsaris (a) which showed that new crystals produced from a parent seed need not be of the same enantiomorphic form. It appears then that the nuclei did not originate as part of the parent crystal.

A serious difficulty in modelling on the basis of nucleation occurring during the growth process is the proper accounting for mass transfer effects. It is indeed surprising that the growth rates, determined <u>in situ</u> prior to the application of the jet in the experiments of Sung <u>et al</u>. showed second order dependence upon the supersaturation. Mass transfer effects were definitely present: the solution was unstirred; growth rates at two-temperature levels were the same; and the rates were about ten fold lower than those measured by Liu, Tseui and Youngquist who eliminated effects due to mass transfer. Sikdar and Randolph showed 1.7 order dependence for growth rates inferred from population balance analysis of stirred suspension crystallizer product size distributions. Because of the micron sizes of the crystals in these experiments, one would expect mass transfer effects to be significant. It is also difficult to resolve that nucleation events determine growth without having supersaturation involved exponentially.

We consider a description of the growth process which permits coupling with secondary nucleation and admits the presence of heat and mass transfer. The growth process is based upon an equilibrium interface, as described by Jackson. Using thermodynamic arguments, Jackson shows that an energy barrier must be overcome in order to deposit a complete layer of monomers upon the surface. The existence and extent of the barrier is given by the parameter α, equal to $\Delta S \xi /k$. ΔS is the molecular entropy of the phase change and ξ, $0<\xi<1$, depends on the interface orientation; $\xi \to 1$ for the most closely packed faces. The suggestion here is that the growth properties are represented by the value of α. For α less than 2, deposition upon the surface occurs easily, and the surface is less stable to uniform growth and is a potential source for dendrites and the nuclei observed. The value of α is about 2 for the magnesium sulphate-water system which permits rapid inherent growth.

Mass transfer effects existed significantly in Sung's experiments. The supersaturation ΔT_C existed "at the surface" only during application of the jet; the developing region at supersaturation ΔT_B is necessary for nuclei development as previously described. In a stirred suspension crystallizer the process is visualized as follows: turbulent eddies approach the parent to raise the level of supersaturation at the interface from the undisturbed value "ΔT_C" to a value more nearly that of the bulk solution, ΔT_B. For the magnesium sulphate-water system, the difference between ΔT_C and ΔT_B in a stirred crystallizer is much less than existed in Sung's experiments. But the fluid impulse intensity may be significantly greater and compensates for lower supersaturation. This is plausible since Sung has demonstrated the very sensitive dependence of the number of nuclei obtained upon jet intensity and supersaturation. Exactly what specie is transferred from the surface to the nearby region for development is not known.

For the ice-brine system, α for the basal plane is about 2. One expects $\xi \ll 1$ for the prism faces and higher index faces so that deposition upon these readily occurs and growth is anisotropic. From the preliminary ice-brine experiments noted earlier, the number of nuclei obtained does not appear to be sensitive to fluid shear variation, but nuclei are formed readily. Because of the high propensity of ice to form dendrites (α<2, highly anisotropic growth) one may visualize that dendrites are sheared from the polycrystalline face exposed in that experiment. This is in contrast to the observation of Garabedian and Strickland-Constable (a) who generated no nuclei in a shear field. Their system included one crystal which grew anisotropically but apparently not dendritically. Their experiment apparently did not include localized variations of shear and temperature. That fluid shear impulses may cause secondary nuclei for ice-brine in stirred suspension crystallizers has been suggested in the interpretation of the experimental results of Evans *et al*.

SYMBOLS USED

A   corresponds to $(36\pi)^{\frac{1}{3}} v^{\frac{2}{3}} \gamma / kT$ from classical theory

i   number of monomers in embryo; i_1 critical size of embryo relative to ΔT_B; i_2 size of embryo which number one in Boltzmann's distribution

NS   fitting parameter; corresponds to total number of all species at the crystal solution interface

T   absolute temperature; T_S saturation temperature

v   volume per monomer
ΔH enthalpy change per unit volume due to phase transformation
γ interfacial energy of the solution crystal interface

ACKNOWLEDGEMENT

We wish to acknowledge the assistance of C. Kumar and R. Jagannathan for exploratory experiments and treatment of the data.

REFERENCES

1. Botsaris, G.D., Denk, E.G. in "Annual Reviews of Industrial and Engineering Chemistry – 1970", Chapter 17; ACS (1972).
2. Denk, E.G., Botsaris, G.D., Journal of Crystal Growth 15, (1972) 57.
3. (a) Denk, E.G., Botsaris, G.D., J. Crystal Growth 13/14, (1972), 493.
4. Doremus, R.H., Journal of Physical Chemistry 62 (1958) 1068.
5. Estrin, J., Wang, M.L., Youngquist, G.R., AIChE J. 21, (1975), 392.
6. Evans, T.W., Sarofim, A.F., Margolis, G., AIChE J. 20, (1974), 950.
7. Garabedian, H., Strickland-Constable, R.F., J. of Crystal Growth 13/14, (1972) 506.
8. (a) Garabedian, H., Strickland-Constable, R.F., J. of Crystal Growth 22, (1974) 1.
9. Larson, M.A., Mullin, J.W., J. of Crystal Growth 20, (1973) 183.
10. Liu, C.Y., Tseui, H.S., Youngquist, G.R., Chemical Engineering Progress Symposium Series No.110, 67, (1971) 43.
11. Jackson, K.A. in "Liquid Metals and Solidification" American Society for Metals, Metal Park, Ohio (1958).
12. Melia, T.P., Moffitt, W.P., "Journal of Colloid Science 19, (1964), 433.
13. Sikdar, S.K., Randolph, A.D. "Secondary Nucleation of Magnesium Sulphate Heptahydrate in a Seeded Mixed-Magna Crystallizer", 77th Annual Meeting, AIChE, Pittsburgh, Pa., June 1974.
14. Sung, C.Y., "Secondary Nucleation of Magnesium Sulphate by Fluid Shear", Ph.D. thesis, Clarkson Collete of Technology, May, 1974.
15. Sung, C.Y., Estrin, J., Youngquist, G.R., AIChE J. 19, (1973), 957.
16. Tai, C.Y., McCabe, W.L., Rousseau, R.W., AIChE J. 21, (1975), 351.
17. Youngquist, G.R., Randolph. A.D., AIChE J. 18, (1972), 421.
18. Youngquist, G.R., Estrin, J., Jagannathan, R., Sung, C.Y., "Secondary Nucleation by Fluid Shear", 67th Annual Meeting, AIChE, Washington, D.C., December, 1974.

THE GRINDING MECHANISM IN A FLUID ENERGY MILL AND ITS SIMILARITY

TO SECONDARY NUCLEATION

SATOSHI OKUDA

Department of Chemical Engineering

Doshisha University, 602 Kyoto, Japan

INTRODUCTION

Fluid energy mills are used because of their advantages in fine grinding. The materials fed into a jet stream are accelerated to a relatively high velocity from 10 to 200 m/s and although these values are much higher than those encountered in the breeding process of crystals, there are some similarities between these two operations. In both tiny fragments are produced by the attrition of particles when they collide with each other, with the mixer or the wall of the apparatus.

Fasoli and Conti (1) have already studied the mechanism of secondary nucleation and crystal breakage in a mixed suspension crystallizer, and they also point out (2) that some interesting similarities can be observed between the formation of abrasion nuclei by each breeding crystal and the grinding performance of jet mill. The author has previously reported (3,4) on fine grinding in a nozzle, especially in a convergent-divergent (De Laval) nozzle. This paper is concerned with the behaviour of particles accelerated in such a nozzle and the inter-collision of particles for several shapes of target.

GRINDING PERFORMANCE IN THE DE LAVAL NOZZLE

Fig. 1a shows the De Laval nozzle used for grinding. The feed materials are forced into high pressure air, and the mixture of solid particles and air is passed through the nozzle. Fig. 1b shows the De Laval nozzle with secondary air conduits. Fig. 1c shows the nozzle in which feed materials are introduced into the pressure

reducing portion of the diffuser. Fig. 2-1 shows the size
distribution curves of the product obtained with secondary air
conduit type (b) nozzle. Regardless of the air pressure and the
mixing ratio (concentration), three peaks in the size distribution
curves can be seen at about 20, 200 and 450µm size. As the
pressure of main air becomes higher, the peak in the region of
200µm moves towards the smaller size and the peak in the region
of 20µm becomes higher. Comparing Figs 2-2 and 2-1, the effect
of secondary air on the pattern of the size distribution curve
can be clearly noted. Fig. 2-2 shows typical size distribution
curves for a De Laval type (a) nozzle. The peak in the region of
200µm cannot be seen in these curves. Instead of that, two peaks

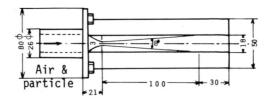

(a) De Laval type grinding nozzle.

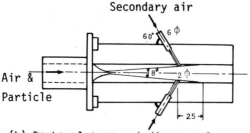

(b) De Laval type grinding nozzle
with secondary air conduits

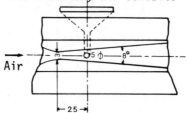

(c) De Laval type nozzle with a feed
gate at reduced pressure portion.

Fig.1 Various types of grinding nozzle

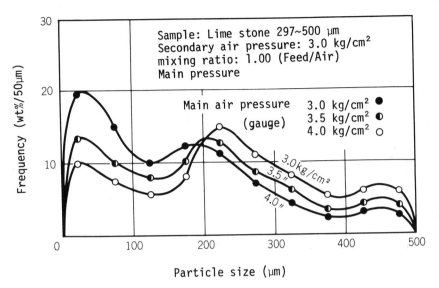

Fig. 2-1. Size distribution of the product by (b) nozzle.

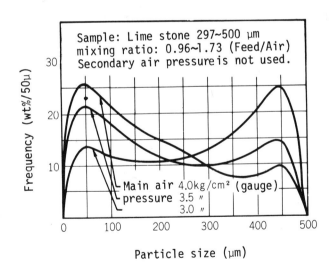

Fig. 2-2. Size distribution of the products by (a) nozzle.

are observed. The distribution curves with two peaks suggest that grinding proceeds by a rubbing action and attrition of the surface. From these experiments, it is clear that the secondary air is used to grind the feed material to the middle size of about 200µm, and these middle size particles are produced by splitting due to the collision of particles accelerated by the secondary air.

The grinding effect of an ordinary jet mill (e.g., Jet-O-Mizer, Micronizer, etc.) may be brought about by the same action as that of secondary air in this study. And this is also apparent from microscopic observation, i.e., the fine particles (<44µm) produced by the type (c) nozzle have an angular shape, but those by type (a) nozzle are rounded.

EFFECT OF CONCENTRATION ON THE GRINDING PERFORMANCE (INSIDE NOZZLE)

The frequent interparticle collisions inside the nozzle were observed by double-flash photographs taken near the nozzle throat. The mixing ratio (concentration of solid particles) is an important factor for the collision of particles and grinding. Generally, the smaller the mixing ratio, the more effective is the production of fine particles. The relation between the fine product (<104µm % to total feed material) and the mixing ratio is shown in Fig. 3.

According to Smoluchowski's expression the frequency of collision in a shearing fluid is proportional to the square of the concentration, C_s^2. On the other hand, the kinetic energy is proportional to $C_s^{-0.5}(m^{-0.5})$ by the calculation of Tanaka (5). Therefore, in the region where the mixing ratio is small (m < 1.0), the grinding product expressed by the product of collision frequency and kinetic energy (v^2) becomes nearly constant. But when the mixing ratio becomes large (m > 1.0), the number of collision in unit time decreases to $C_s^{1/3}$, since the average distance δ between the particles may be expressed by the concentration C_s as $δ = C_s^{-1/3}$ (1). The decrease of kinetic energy (v^2) may be expressed in accordance with Tanaka's calculation (5) as $V^2 \propto m^{-2/3}$. Therefore the grinding product <104µm (G%) may be represented, as shown in Fig. 3, by

$$G = a/m^{1/3} \tag{1}$$

where a is the coefficient which corresponds to m = 1 concerned with G, and m is the mixing ratio = M_s/M_a (M_s is the mass velocity of solid and M_a is that of air). Therefore, the mill capacity W (kg/s) may be given by

$$W = M_s G/100 = M_s a/100 \, m^{1/3} \tag{2}$$

$$= \frac{a}{100} M_s^{2/3} M_a^{1/3} \tag{3}$$

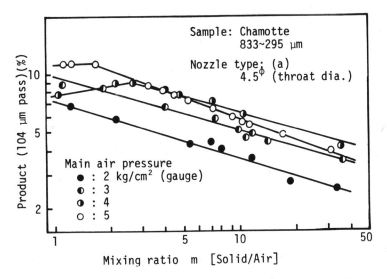

Fig. 3 Relation between product and mixing ratio

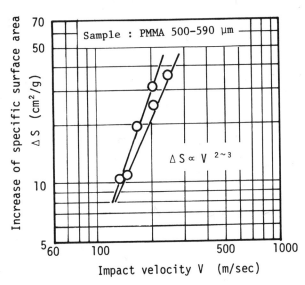

Fig. 4 Relation between ΔS and V

On the other hand, the reduction of feed size G' (%) (= 1-R, where R is the residue of feed size) has the same relation between m:

$$G' = a'/m^{1/3} \qquad (4)$$

where a' is the coefficient which corresponds to m = 1 concerned with G'. From eqs 1 and 4 we get

$$\frac{a}{a'} = \frac{G}{G'} \qquad (5)$$

which is independent of the mixing ratio. This relation means that the ratio G/G' for m = 1 may be applicable to G/G' in other cases where the mixing ratio is changed. That is, the reduction of R is proportional to the increase of G (fine product %). This means that surface grinding is dominating. From eqs 1 and 4

$$G' - G = (a' - a)/m^{1/3} \qquad (6)$$

is deduced. This equation shows that an increase of m corresponds to a decrease of (G'-G), i.e., the grinding is dominated by surface grinding. Conversely, a decrease of m corresponds to an increase of (G'-G) and the grinding is dominated by breakage grinding.

EFFECT OF CONCENTRATION ON GRINDING PERFORMANCE (OUTSIDE NOZZLE)

The behaviour of flying particles outside the nozzle and their collision with one another or with the target (wall) was investigated. Some particles in the jet stream move as a single particle, but generally the particles gather together and flow as a group like a cloud. Therefore, it is considered that they move as groups which have a certain porosity, and each particle in the group is repeating the collision and splitting.

Grinding by collision with the target may be affected mainly with impact velocity and particle concentration. The particles are accelerated in the De Laval nozzle above 100 m/s and blown into the surrounding atmosphere. The small particles (590-840μm) were decelerated outside the nozzle by the air resistance, but large particles (840-1190μm) were not decelerated markedly at a distance of 80 mm. The shape of target has considerable effects on the velocity before and after the collision.

It is also confirmed that there is an optimum angle of target with the direction of flying particles. An angle of 40 to 50° with the flow direction may be considered optimum for grinding. The increase of specific surface area is nearly proportional to the square of the impact velocity and the size of grinding product

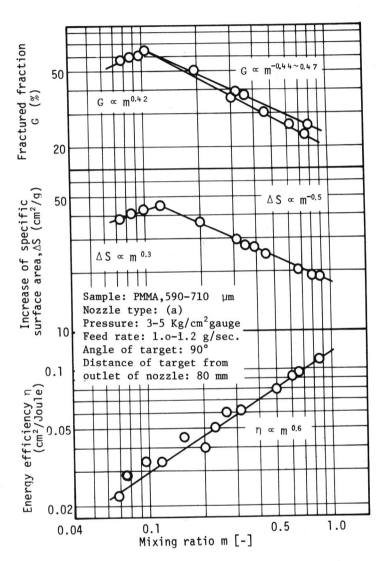

Fig. 5. G, ΔS, η vs m (Target collision)

is increased with an increase of concentration. There is also an optimum length L from the outlet of the nozzle to the target in relation with the nozzle diameter.

Within the limits of experiment (mixing ratio m = 0.1-1.0, particle diameter D_s = 250-1100μm, air pressure P = 3.0-5.0 kg/cm² gauge), the following relation for the impact velocity V (m/s) was confirmed for a 90° flat target:

$$V \propto m^{-0.11} \cdot D_s^{-0.69} \cdot P^{0.46} \quad (7)$$

Since the increase of surface area (ΔS) by collision has the relation

$$\Delta S \propto V^{2\sim3} \quad (8)$$

then, the next relation can be recognized:

$$\Delta S \propto m^{-0.22\sim-0.33} \cdot D_s^{-1.38\sim-2.07} \cdot P^{0.92\sim1.38} \quad (9)$$

Fig. 4 shows the relation of impact velocity and the increase of surface area for P = 3.0-5.0 kg/cm² gauge, mixing ratio = 0.062-0.098, target angle = 90°, distance of target from nozzle outlet L = 80 mm. It is reasonable that ΔS should be proportional to V^2, since the kinetic energy has been changed effectively to the increase of surface area of the ground particles. In some cases, ΔS is proportional to V^3. This is due to the velocity effect in grinding. (In our experiments (6) on the high sliding speed wear (up to 100 m/s) of talc, the specific wear rate is proportional to the cube of the sliding speed.) From the calculation by Tanaka (5) the decrease of V^2 with the increase of mixing ratio (m = 0.1-1.0) may be defined by

$$V^2 \propto m^{-0.5} \quad (10)$$

Therefore, the grinding fraction G (%) may be expressed (for m = 0.1-1.0) by

$$G \propto m^{-0.44\sim-0.47} \quad (11)$$

The small deviation from $m^{-0.5}$ is probably due to the velocity effect.

ENERGY EFFICIENCY OF A JET MILL

To define how the fluid energy is consumed in grinding in the fluid energy mill, the next assumption was considered.

(1) The grinding rate is expressed by the rate of increase of

GRINDING MECHANISM IN FLUID ENERGY MILL

An example of a double flash photograph of flying particles before a target: flash time interval 30μs, nozzle type (a), gauge pressure 5 kg/cm², sample PS 590-710μm, mixing ratio 0.15-0.22, target angle 70°, distance from nozzle outlet 100 mm, magnification 0.86X.

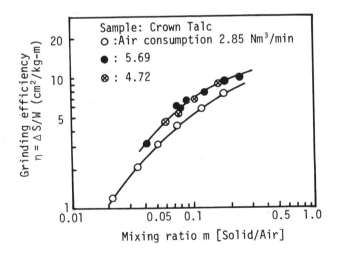

Fig. 6 Grinding efficiency versus mixing ratio

surface area, ΔS (cm^2/s).
(2) The fluid energy is expressed by the isothermal expansion energy from compressed high pressure to atmospheric pressure, and the work per unit time P is calculated.
(3) The grinding efficiency η is expressed by

$$\eta = \Delta S/P \quad (cm^2/kg \cdot m) \tag{12}$$

The energy efficiency of a fluid energy mill is very closely related to the concentration. It would be increased with increase of concentration but it would be saturated at some optimum concentration and it would be decreased at very high concentration. Using the preceding assumption, Fig. 6 shows the relation between η and mixing ratio m of the actual jet mill which was developed by the author and co-workers. From this figure it may be seen that the air pressure is independent of the η vs. m relation, and η is proportional to m up to a mixing ratio m = 0.1, i.e.,

$$\eta \propto m^{1.0} \tag{13}$$

but the slope is decreased as the mixing ratio becomes larger (m = 0.1-1.0) as mentioned above in the impact test (Fig. 5). It is expected that the slope becomes negative for m > 1.

CONCLUSIONS

The author has tried to clarify the grinding mechanism of flying particles by collision of each other and impact with a target. The grinding mechanism was classified into splitting and rubbing action of particles, and rubbing is the most effective for fine grinding. The effects of several operating factors, such as concentration, pressure and particle size were examined. This work may be a useful reference for the study of secondary nucleation.

The author wishes to express his gratitude for the assistance of his colleagues, Mr. J. Nakayama, Mr. M. Matsumoto, Mr. H. Imamura and Mr. W.S. Choi.

REFERENCES

1. Fasoli, U. and Conti, R., Kristall und Technik, 8 (1973) 931.
2. Conti, R. and Fasoli, U., RICERCA ICP, Dec. (1973) 19.
3. Okuda, S. et al., J. Soc. Materials Sci., Japan, 14 (144) (1965) 730.
4. Okuda, S., Proc. Particle Technology Seminar (Madras-India), (1971) 176.
5. Tanaka, T., Ind.Eng.Chem.Proc.Des.Dev., 12 (2) (1973) 213.
6. Okuda, S., J.Res.Assoc. Powder Tech. Japan, 12 (1) (1975) 10.

SECONDARY NUCLEATION IN AGITATED CRYSTALLIZERS

UGO FASOLI and ROMUALDO CONTI

Instituto di Chimica Industriale del Politecnico di

Torino, Italy

INTRODUCTION

The causes of secondary nucleation may be divided into two main groups. In the first are the physical-chemical causes such as cluster formation (1), the presence of impurities (2) and the catalytic behaviour of seed crystals. In the second group are the mechanical causes such as the detachment of nuclei by collisions of the crystals with themselves, the stirrer, the baffles, the walls of the crystallizer, etc. The possibility of detachment of nuclei without any mechanical action in the presence of a highly supersaturated solution has been demonstrated by photographic methods (3),(4).

COLLISION NUCLEATION

Most authors agree that the specific nucleation rate r, i.e. the number of nuclei generated by a seed crystal per unit time is a function of the power P dissipated in the crystallizer per unit volume, of the equivalent diameter D_s of the generating crystals and of the number of collisions. More exactly it has been found (5) that the dependence on P is linear while D_s enters into the relationship with an exponent between 3 and 5 (ice crystals).

A theoretical study of the influence of the above-mentioned parameters together with the size D_a of the generated nuclei and the weight concentration c_s of the generating crystals has been presented in a previous report (6).

If k is the energy needed to detach a fragment of unit area

and α_a is a surface shape factor of nucleation, the energy spent per unit time by a seeded crystal to breed new nuclei is:

$$r \, k \, \alpha_a \, D_a^2 \qquad [1]$$

This energy is equal to the power spent per unit volume P multiplied by the volume of suspension engaged by the crystal and by an efficiency coefficient, taking account that the energy spent for nucleation is only a part of the total energy spent by the crystal (the other part is dissipated by friction losses, elastic behaviour, and so on). Experimental data have shown that the volume engaged is proportional to the volume of the crystals and therefore it is possible to write:

$$r \, k \, \alpha_a \, D_a^2 = \eta_1 \, N \, \alpha_s \, D_s^3 \, P \qquad [2]$$

where η_1 is the efficiency coefficient which is a function of the ratio between the energy spent to breed new nuclei and the kinetic energy of the crystal dissipated in the collision, α_s is a volumetric shape factor and N is a dimensionless number whose meaning will be explained later.

A theoretical analysis of the problem (6) has demonstrated that it is also possible to write

$$r \, k \, \alpha_a \, D_a^2 = \eta_1 \, n \, m_s \, u \, \frac{\alpha_s \, D_s^2 \, P}{2} \qquad [3]$$

where u is the mobility of the crystal, put equal to the ratio between its velocity v and the force F applied to it, n is the number of collisions of the crystal in the time unit (and therefore equal to $1/z_s$ where z_s is the time interval between two subsequent collisions) and m_s is the mass of the crystal.

By comparing expressions 2 and 3 we can get:

$$N = \frac{1}{2} \frac{m_s v}{F z_s} \qquad [4]$$

which shows the meaning of the dimensionless number N since $m_s v$ is the momentum of the crystal and $F z_s$ the impulse of the force applied to it: N is the ratio between the theoretical maximum energy proportional to $m_s v$, and the energy dissipated in the surrounding fluid poing to the shear stress, proportional to $F z_s$.

SECONDARY NUCLEATION IN AGITATED CRYSTALLIZERS

COEFFICIENT k

The coefficient k in equations 1-3 seems to be the key factor in the explanation of the phenomenon of collision nucleation. The value of k depends on the superficial energy of the crystal and hence on the physical properties of the crystalline material. When the superficial energy or the fragility of the material increase, the value of k should decrease.

But the problem is more complex since it is necessary to consider the influence of the supersaturation too. There is no general agreement on the way supersaturation affects the rate of collision nucleation. Even if other phenomena occur, supersaturation probably affects the value of coefficient k by changing the regularity of the crystal surfaces. It is well known that fast crystal growth at very high supersaturations can cause the detachment of fragments even in a quiescent liquid: so there is a minimum value of k corresponding to the lowest supersaturation at which a spontaneous detachment of nuclei is possible, i.e. self-nucleation.

Moreover it is possible that there is a maximum value of k and of the mechanical resistance for crystals grown at very low supersaturation. The crystal might get this maximum mechanical resistance by abrasion, but in this case its shape would be rather different.

OTHER PARAMETERS

To avoid any misinterpretation of the experimental results a careful determination of the values of the parameters used in the model has been necessary. In fact the common ways for calculation, as for instance Rushton's method for power consumption of impellers and Einstein's method for mobility, led to errors that cannot be ignored. So the following methods have been used.

<u>Power spent in the volume unit P.</u> The power supplied to the suspension was measured with a torque-meter (Vibro-meter sa., Fribourg, Switzerland). The total power was calculated by multiplying the torque by the agitator rotational speed. P was calculated by dividing the total power by the volume of the suspension.

<u>Mobility u.</u> The mobility was determined by measuring the terminal falling velocity of some crystals in a glass cylinder and dividing this velocity by the weight of the crystals minus the buoyant force. This value was corrected to take into account the influence of viscosity and the hindering action of the surrounding crystals (in the crystallizer).

<u>Viscosity μ.</u> Owing to the high solid-phase concentration the

viscosity of the suspension was rather different from that of the liquid. So the fluidity of the suspension $1/\mu$ was calculated by Einstein's expression (7) and the value obtained used to get, by multiplication, the real value of the mobility.

<u>Clearance between two crystals δ.</u> If the crystals are uniformly distributed in the whole suspension, their average clearance is

$$\delta = D_s \left[\left(\frac{\rho_c \, g \, \frac{\pi}{6}}{\rho c_s} \right)^{\frac{1}{3}} - 1 \right] \qquad [5]$$

where ρ_c is the density of the crystals and g the gravitational acceleration. The values of D_s and c_s were determined during the runs, so it was possible to know the value of δ at any time.

<u>Velocity of the crystals in the suspension v.</u> Since $P \, u \, \alpha_s \, D_s^3 \, m_s$ is the energy supplied to each crystal by the stirrer, it is possible to write

$$v = Pu \, \alpha_s \, D_s^3 \qquad [6]$$

<u>Number of collisions of a crystal in the time unit n.</u> It is obviously:

$$n = \frac{v}{\delta} = \frac{1}{z_s} \qquad [7]$$

<u>Efficiency η_1.</u> The value of η_1 is normally calculated in size-reduction studies through the superficial energy of the material and the amount of energy spent in the compression to break a single grain. In this work, however, the right value of η_1 has not been calculated, since, as will be demonstrated, the development of the model does not require it.

DEVELOPMENT OF THE MODEL

Introducing expressions 6 and 7 in expression 3 we obtain:

$$r \, k \, \alpha_a \, D_a^2 = \eta_1 \frac{P^{3/2} u^{3/2} D_s^{9/2} m_s \alpha_s^{3/2}}{2\delta} \qquad [8]$$

and collecting all the experimentally evaluable quantities on the right hand side, we can write

$$\frac{k\,\alpha_a}{\eta_1 \alpha_a^{3/2}} = \frac{P^{3/2} u^{3/2} D_s^{9/2} m_s}{2\,\delta\,r\,D_a^2} \qquad [9]$$

which allows the calculation of coefficient k multiplied by $\alpha_a/\eta_1 \alpha_s^{3/2}$. Since we can think that this last expression should be reasonably constant during a run, expression 9 allows us to know the behaviour of k in the run.

To get an experimental value of k to compare roughly with the superficial energy of the crystal it is possible to use the values of η_1 that can be found in literature (η_1 varies between 0.001 and 0.01 (8), and to suppose that the new face bred by abrasion is not very different from D_a^2 ; so $\alpha_a = 1$.

EXPERIMENTAL

Three kinds of experimental runs were accomplished:
Type A with copper sulphate varying all the parameters in expression 9 to verify its validity.
Type B with nickel-ammonium sulphate to evaluate the influence of supersaturation on the coefficient k.
Type C with copper, nickel- ammonium and magnesium sulphates to put into the correlation secondary nucleation and mechanical strength of the materials.

Apparatus and Technique

The apparatus is shown in Fig.1. It consisted of a two-litre glass tank with plexiglass top and baffles (four) and a four flat-blade stainless steel stirrer. Three stirrer speeds were used: 500, 700 and 900 rpm, obtained exactly by means of a gear-working speed reducer.

To avoid any dissolution or growth of the generated nuclei the tests were made in an 'inert' liquid in which the low solubilities of the solids were not affected by temperature, since temperature gradients are always possible in a stirred vessel.

As in the previous work (6) a 50% solution of methanol in water was used for the tests with copper sulphate (tests A) and also with nickel-ammonium sulphate (tests B). Since this liquid did not seem to work very well with magnesium sulphate a third group of tests (C) with this salt was accomplished in acetone.

The granulometric analysis of the seed crystals and of the generated nuclei was made taking photographs of representative

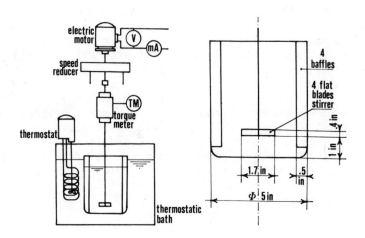

Fig. 1   Experimental apparatus

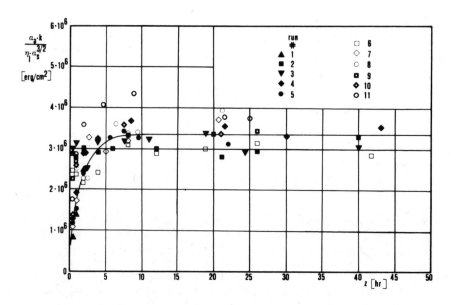

Fig. 2   Dependence of the quantity A (eq. 10) on time

SECONDARY NUCLEATION IN AGITATED CRYSTALLIZERS

samples and determining the size distribution with a model TGZ - 3 Zeiss Counter.

The number of generated nuclei was determined by filtration and weighing. Knowing their weight size distribution, their number was easily determined. The same was done for the seeded crystals and the value of the nucleation rate r was then calculated.

RESULTS

Type A runs. The experimental data are reported in Table 1 and Fig.2 which shows the dependence of the quantity A

$$A = (\alpha_a k/\eta_1 \alpha_s^{3/2}) \qquad [10]$$

on the time. After an initial period the value of quantity A appears to be constant in all the runs and to change very slowly just like coefficient k is expected to do.

Type B runs. Since in the type A runs the initial values of r are rather scattered, some tests were made with nickel-ammonium sulphate crystals of equal dimensions (D_S = 1.06 mm). The crystals were grown from 0.9 mm at different supersaturations : S = 1.08, 1.12 and 1.16.

The results are shown in Fig.3. They indicate that the scattering observed in the type A runs was due to the different supersaturations used to grow the crystals and confirm once more that secondary nucleation is largely affected by supersaturation by means of the remarkable variations caused in the value of k.

Type C runs. The experimental results are reported in Table 2 and Fig.4. They seem to show a connexion between secondary nucleation and mechanical strength of the crystalline material. It is almost impossible to quantify this result since the hardness of faces, corners and vertexes of the crystals used have not yet been well measured. However, the results are, at least qualitatively, in accordance with the higher fragility shown by magnesium sulphate in comparison with the other two salts.

CONCLUSIONS

The possibility that crystal-crystal collision is the controlling mechanism of secondary nucleation has been examined. With the proposed model the velocity of the crystals and the rate and energy of the collisions has been calculated. The rate of generation of nuclei and their surface has been determined experimentally and hence it has been possible to calculate the resistance of the

Table 1

run #		1	2	3	4	5	6	7	8	9	10	11
	rpm	500	700	900	700	700	700	700	700	700	700	700
D_s	cm	.120	.120	.120	.120	.120	.146	.146	.146	.173	.173	.173
P	$\frac{erg}{sec \cdot cm^3}$	1200	5340	9380	5800	5940	5340	5650	5940	5340	5640	5940
c_s	$\frac{g}{cm^3}$	.2655	.2655	.2655	.177	.0885	.2655	.177	.0885	.2655	.177	.0885
δ	cm	.078	.078	.078	.107	.166	.095	.130	.201	.112	.154	.238
u	$\frac{cm}{dyne \cdot sec}$	17.13	17.13	17.13	18.53	20.15	14.07	15.23	16.56	11.89	12.87	13.99
v	$\frac{cm}{sec}$	5.96	12.57	16.66	13.63	14.38	15.29	16.36	17.50	18.10	19.35	20.71
n	$\frac{1}{sec}$	76.51	161.19	213.63	127.61	86.84	161.66	126.07	86.83	161.18	125.98	86.83
D_a	cm	.00058	.00058	.00048	.00058	.00058	.00055	.00055	.00055	.00050	.00050	.00050

Table 2

run #		12	13	14	15	16	17
		$CuSO_4 \cdot 5H_2O$		$Ni(NH_4)_2(SO_4)_2 \cdot 6H_2O$		$MgSO_4 \cdot 7H_2O$	
	rpm	700	700	700	700	700	700
D_s	cm	.135	.135	.127	.119	.140	.140
P	$\frac{erg}{sec \cdot cm^3}$	5770	5770	5050	5050	5130	5130
c_s	$\frac{g}{cm^3}$	.0885	.0885	.0885	.0885	.0885	.0885
δ	cm	.186	.186	.158	.148	.161	.161
u	$\frac{cm}{dyne \cdot sec}$	51.17	51.28	72.20	77.13	48.36	48.36
v	$\frac{cm}{sec}$	26.95	26.98	27.32	25.68	26.09	26.09
n	$\frac{1}{sec}$	144.89	145.05	172.91	173.51	162.05	162.05
D_a	cm	.00055	.00055	.00050	.00050	.00020	.00020

SECONDARY NUCLEATION IN AGITATED CRYSTALLIZERS

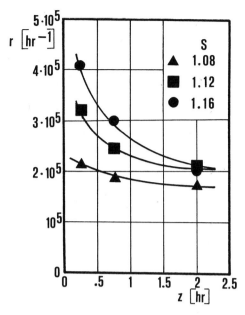

Fig. 3

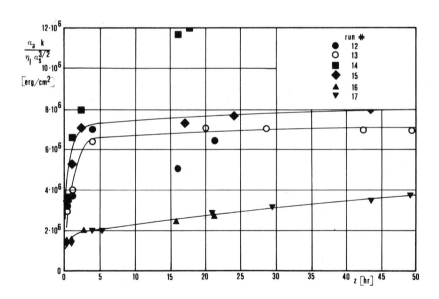

Fig. 4

crystals to abrasion (disregarding shape coefficients and breakage efficiency) which can be compared with the surface energy of the crystals.

Nevertheless resistance to abrasion is not only a function of the surface energy, it is also affected by other parameters among which the most important is the structure of the external layers.

The experimental runs have demonstrated that the resistance to abrasion of the crystals increases to a constant final value when, due to abrasion, they become more and more rounded (type A runs, Fig.2). If this final value is compared with the superficial energy of copper sulphate it is possible to get roughly the magnitude of the breakage efficiency (0.001-0.01) that agrees reasonably well with values found in literature. Naturally to get a full coincidence it would be necessary to know some other parameters, e.g. shape factors. Type B runs (Fig.3) have demonstrated that secondary nucleation is also affected by supersaturation and suggest an explanation of this phenomenon: supersaturation changes the crystal growth rate and therefore the structure of their external layers and their resistance to abrasion.

Type C runs are a first step to compiling a list of abrasion resistances of some important crystalline materials.

On the basis of the work described in this report, it should soon be possible to gain a better understanding of the phenomenon of secondary nucleation.

REFERENCES

1. Mullin, J.W. and Leci, C.L., Chem.Eng.Progress, Symposium Series, 68 (121) (1972) 8.
2. Denk, E.G. and Botsaris, G.D., J.Crystal Growth, 13/14 (1972) 493.
3. Kliya, M.O., Soviet Phys.-Cryst., 1 (1956) 456.
4. Saratovkin, D.D., Dendritic Crystallization, Consultants Bureau, New York (1959) 64.
5. Evans, T.W., Margolis, G, Sarofim, A.F., AIChE J., 20 (1974) 950, 959.
6. Fasoli, U.,and Conti, R., Kristall und Technik, 8 (1973) 931.
7. Einstein, A., Ann.Physic., 19 (1906) 289.
8. Rumpf, H., 1° Europaischen Symposium Zerkleinern, VDI - Verlag - Gmbh - Dusseldorf, (1962) 1.

NOMENCLATURE

c_s	weight concentration of the generating crystals
D_a	average diameter of the nuclei
D_s	average diameter of the generating crystals
F	force acting on the crystal
g	gravity acceleration
k	energy needed to detach a nucleus of unit area
m_s	mass of a generating crystal
n	collision rate per crystal
N	ratio between the energy dissipated in a collision and the energy dissipated by the crystal in the surrounding fluid.
P	power spent in the volume unit
r	nucleation rate per crystal
S	supersaturation
u	mobility
v	velocity
z	time
z_s	time between two subsequent collisions
α_a	shape coefficient of nucleation
α_s	volumic shape coefficient
δ	clearance between two crystals
η_1	efficiency coefficient
μ	viscosity
ρ_c	density of the crystalline material

Crystal Growth
Kinetics

THEORY AND EXPERIMENT FOR CRYSTAL GROWTH FROM SOLUTION:

IMPLICATIONS FOR INDUSTRIAL CRYSTALLIZATION

P. BENNEMA

Delft University of Technology

Delft, The Netherlands

INTRODUCTION

In a previous paper (1) it was shown that a unified crystal growth theory does not exist, but a set of complementary theories has been developed, each dealing with a certain aspect of the crystal growth process. Special growth theories for growth from solution have not been developed: theories applicable to solution growth are modified general theories.

As a starting point for this paper we will introduce a scheme of existing crystal growth theories or models equivalent to the scheme presented before (1) (see Fig. 1). In this survey attention will be focussed on recent developments. Results from theories will be compared with experimental data. Implications for population balance models, as now being widely used in industrial crystallization, will be discussed.

FIVE THEORIES OF CRYSTAL GROWTH

Theory I: the morphological theory of Hartman and Perdok (H-P)

The essentials of the H-P theory have been discussed previously (1,2) and will not be treated here. On the basis of the H-P theory crystallographic faces occurring on crystal growth forms (habits) can be predicted. Application of this theory may be relevant for industrial crystallization because

(i) the habits of crystals can be explained. These determine to a large extent such properties of bulk industrial crystals such as ease of filtration, caking tendencies during storage, etc.

(ii) deviations from the ideal habits may be traced back to the role of impurities (3) or the role of the solvent (4). Cases of crystal habits varying within one run or from one run to another may be traced back to variable concentrations of impurity.

(iii) in fundamental studies on crystal growth with technological implications it is essential to understand the crystallographic structure of the crystallographic plane under consideration, since these properties determine to a high extent the crystal growth kinetics. For example, on the {100} prism faces of ADP the elongated ellipsoidal growth fronts with the long axis in the b direction and the short axis in the c direction were explained in a preliminary way by the fact that the {001} faces could be divided into separate zones of positive and negative ions giving an anisotropy in the surface diffusion process (5).

Another example is $MgSO_4 \cdot 7H_2O$ (6). Preliminary investigations have shown that the ions in a slice of an end-face are connected to each other with weak bonds. This may explain that above a certain supersaturation this face grows in an irregular way. This in turn may give rise to a kind of secondary nucleation due to the fact that peaks developing on the end faces may break off. Due to a combination of factors which probably partly go back to the crystal structure, leading to a certain morphology and growth kinetics, instabilities occur in a continuous crystallizer in which $MgSO_4 \cdot 7H_2O$ crystals crystallize. This was observed and simulated by Ottens (7).

Theories II and III: statistical surface models and computer simulation of crystal surfaces and crystal growth

As mentioned before (1) an important set of models for the theory of crystal growth is the set of crystal-fluid interface models which may be called lattice gas models. In these models the whole interface is divided into blocks of the same size and shape, each block being either in a solid or a fluid state. Models, where fluid cells in the solid phase (and vice versa) are ruled out, are the most relevant. Introduction of this Solid on Solid (SOS) condition makes the solid-fluid interface identical to the surface of a Kossel crystal. In the last twenty years important theoretical work on SOS models has been carried out (8,9,10,11). Except for the one-layer BCF model (8), the models mentioned suffer from the use of the so-called zeroth or Bragg-Williams approximation, which obscures to an extent the physical significance of the results. Recently, therefore, SOS surface models have been simulated by Monte Carlo techniques using fast digital computers (12,13,14). On the basis of these models the growth process has been simulated so that two-dimensional nucleation and other processes could be studied (13,14,15).

There is now a tendency in the field of crystal growth theories to simulate crystal surfaces and crystal growth with Monte Carlo simulations and to compare the results with existing and newly developed (analytical) models in which (coupled) differential equations are solved, again using a computer. This development leads to a stronger physical, and especially statistical mechanical, foundation of crystal surfaces and the crystal growth process (16,17). Moreover, simulation of crystal growth leads to models, which come closer to the real complicated growth process. Therefore it is to be expected that by using computer simulation experiments real experimental observations can be explained.

For experimentalists working in the field of industrial crystallization the statistical mechanical and Monte Carlo models mentioned above may seem to be complicated and perhaps not relevant to their problems. However, it is an advantage of the use of computers, that the essential points resulting from the theories can now be easily visualized.

In Fig. 2 a surface of a Kossel-like crystal with a step is presented. This surface is characterized by the dimensionless factor α (= $4\phi/2kT$) first introduced by Jackson (9). ϕ is the bond energy between neighbouring units, k the Boltzman constant and T the absolute temperature. The factor 4 is added since each solid block is surrounded by four other blocks within the (horizontal) slice of a Kossel crystal. Therefore α characterizes the bond situation of a crystallographic plane relative to the temperature. ϕ can be generalized to include solid-fluid and fluid-fluid bonds. If the temperature increases or ϕ decreases so that α decreases, it follows both from the statistical models and computer simulations that the surface becomes rougher.

Now it is possible to simulate crystal growth by making the chances for condensation exp $\Delta\mu/kT$ higher than the chances for evaporation (dissolution), where $\Delta\mu = \mu_f - \mu_s$ is the difference in chemical potential of growth units in the fluid and the solid states. $\Delta\mu/kT$ is roughly equal to the relative supersaturation. In Fig. 3 (dimensionless) (normalized) rate versus supersaturation curves as obtained from the computer are presented. The data were obtained from a home-made special purpose computer, which was especially designed to simulate crystal growth (18).

It can be seen from Fig. 3 that for values of $\alpha > 3.5$ there is a 'two-dimensional' nucleation barrier for growth, since for low $\beta = \Delta\mu/kT$ values the growth rate is zero. It was indeed found that for $\beta = \Delta\mu/kT < 0.5$ and $\alpha > 3.5$ the 'measured' curves could be fitted with a two-dimensional nucleation curve where the nucleus above nucleus also called birth and spread model was used. This

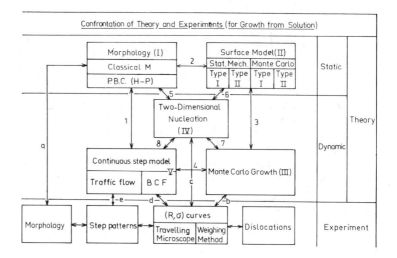

Fig. 1 Confrontation of theory and experiments. In the Surface Models II two types were distinguished: Type I (Solid on Solid models) and Type II models, where the Solid on Solid condition is not introduced. The relations between the five types of theories and the way to check these theories are discussed in ref. 1.

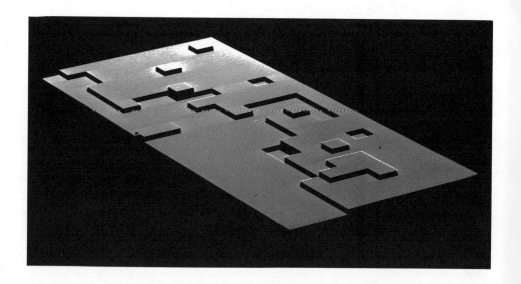

Fig. 2 Computer-drawn surface with a step. Here $\alpha = 4$.

means that on the whole surface nuclei are formed and on top of
these new nuclei are formed, etc. It turned out that the events
occurring on the surface are much less regular than would be
obtained from analytical growth models. Even so, the nucleation
models give a good description of the (R,β) curves obtained from
computer simulation experiments (see Fig. 3).

Work is now in progress to analyse in more detail what happens
on the surface (19). Recently the shape of the rate versus β
curves for the high supersaturation range has been explained on
the basis of analytical pair approximation models (20). The
transition from linear growth kinetics to a two-dimensional
nucleation growth kinetics for $\alpha \simeq 3.1$ can be explained by assuming
that for $\alpha \simeq 3.1$ a transition point for surface roughening exists.
Such a value follows indeed from the one layer model of BCF based
on the exact solution of Onsager and to a certain extent from a
recent multilayer model simulated on a computer (17). Monte Carlo
simulations were generalized for anisotropic surfaces (21) (where
$\phi_x \neq \phi_y$ contrary to the Kossel crystal, where $\phi_x \neq \phi_y$; extensive calculations with a special purpose computer are in progress) and to the
(001) face of a fcc crystal (22). The use of computer simulations
in crystal growth is very recent. However, already in 1966 computer
simulations were carried out by Chernov (23) - see also the work of
Binsbergen (15).

It is appropriate now to mention the influence of steps on the
growth and the properties of steps. As discussed above, it follows
from computer simulation experiments that for sufficiently high α
and low β values crystals do not grow. However, if a step is
available growth occurs at low β. This is in agreement with the
computer simulation experiments shown in Fig. 4. The reason is
that except for the absolute zero (T = 0) steps are rough (see
Fig. 3). This results from statistical calculations (8,17,24) and
computer simulation studies (see also ref. 25).

Theories IV: two-dimensional nucleation models

Concerning the two-dimensional nucleation theories IV (see
Fig. 1) we have already mentioned above that these theories are
used in conjunction with computer simulation studies. A survey
of two-dimensional nucleation theories is given by Ohara and Reid
(26). The shape of the critical nucleus was studied by BCF (8) and
these shapes are calculated in ref. 27.

Theories V: continuous step models

Since crystals grow from solution with facets it may be assumed
that these facets have reasonably high α values. Such faces can
only grow at low supersaturations if steps are permanently available.
A source of steps is the centre of a growth spiral as proposed by

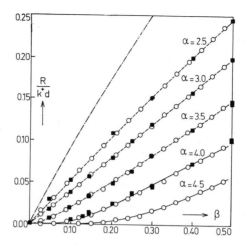

Fig. 3 Dimensionless growth rate versus $\beta(= \Delta\mu/kT)$ for $0 < \beta < 0.5$. Results of Gilmer and Bennema[11,12], O special purpose computer results[14], dashed line = empirical relation[14], solid lines = nucleation formula resulting from a birth and spread model.

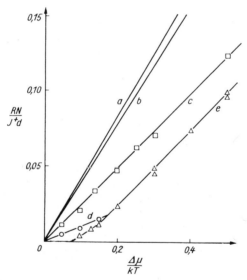

Fig. 4 Computer simulation curves for a (001) face of a Kossel crystal. (a) $\alpha = 3$, with steps, (b) $\alpha = 3$, without steps, (c) $\alpha = 4$, with steps — □ —, (d) $\alpha = 4$ with a fivetimes lower step density than (c) — O —, (e) $\alpha = 4$ without steps — ▲ —. For values of $\beta = 0.2$ a compound mechanism does occur (see curve e). (See ref. 44)

Frank (8). Since steps originating from a growth spiral are permanently available the transport of growth units to the surface, surface diffusion and the integration of growth units in the kinks of the steps become the processes determining the crystal growth mechanism. Therefore BCF (8) developed a model in which four relaxation times could be distinguished, namely τ_{desolv}, τ_{deads}, τ_{sdiff} and τ_{kink} (28,29,30). Here τ_{desolv} is the relaxation time for entering the adsorption layer (for this a desolvation process is required), τ_{deads} corresponds to the deadsorption process for going from the surface to the solution, τ_{sdiff} to surface diffusion and τ_{kink} to the process of entering the kinks of the step from the surface. From the BCF theory the well known equation

$$R = C \sigma^2/\sigma_1 \tanh (\sigma_1/\sigma)$$

results. For $\sigma \ll \sigma_1$ a parabolic law is obtained and for $\sigma \gg \sigma_1$ a linear law (see Fig. 5). C and σ_1 are characteristic constants, C gives the slope of the $R(\sigma)$ curve for $\sigma \gg \sigma_1$ and σ_1 characterizes the curvature for $\sigma < \sigma_1$. In this formula the resistance for kink integration is neglected. If this is taken into consideration the parabolic depression obtained is steeper than that given by the formula above.

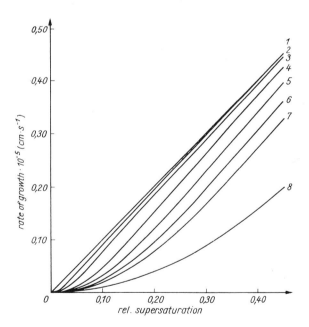

Fig. 5 BCF curves according to the tanh expression above:
Curve 1, $\sigma_1 = 10^{-3}$; curve 2, $\sigma_1 = 5.10^{-2}$;
curve 3, $\sigma_1 = 10^{-1}$; curve 4, $\sigma_1 = 0.2$;
curve 5, $\sigma_1 = 0.3$; curve 6, $\sigma_1 = 0.4$;
curve 7, $\sigma_1 = 0.5$; curve 8, $\sigma_1 = 1.0$.

The BCF model was generalized to include the role of volume diffusion, i.e. transport from growth units to the surface, by Gilmer, Ghez and Cabrera (31). The volume diffusion model of Chernov (32) can be considered as a special case of the GGC model. The implications of the GGC model for the practice of crystal growth have been worked out recently (33,34). It was shown that when the step integration resistance is neglected $R(\sigma)$ curves could be described in a satisfactory way by a BCF curve. The constant σ_1 determining the curvature of the BCF curve does not change very much; the slope C, however, may change considerably (see Fig. 6a-c). It was found that the role of volume diffusion probably never can be ruled out even under very favourable stirring conditions. (See ref. 35.)

The BCF theory is based among others on the idea that far from the centre the steps can be considered as an infinite train of equidistant steps. Gilmer (30) developed a theory for small deviations from equidistant step-trains and this theory was compared with steptrains having a large deviation from their equidistant positions (the equidistant steptrain being used as a reference). In these simulations the so-called Schwoebel effect (36) has also been introduced. Here the relaxation time for entering the steps from the right side is not equal to the relaxation time for entering from the left side. As already shown by Schwoebel, this gives rise to instabilities in trains of steps leading to 'head on collisions' and the formation of macrosteps which may reduce the effectiveness of the step propagation mechanism. This was also found by the computer simulation experiments. However, recent computer experiments have shown that it takes a long time before such head on collisions do occur even with a strong Schwoebel effect (37,38). We also do not believe that the Schwoebel effect is a general mechanism so that the occurrence of complicated step patterns on crystal surfaces in most cases probably cannot be explained on the basis of a Schwoebel effect. Complicated step patterns occurring on crystal surfaces probably must be attributed to fluctuations in the volume diffusion field. We note that recently Monte Carlo simulations of step propagation of a train of steps with four steps was compared with the simulations resulting from solving coupled differential equations; the results of the two types of simulations were in reasonable agreement with each other (37). These two simulations are complementary: 'real' simulations of the atomic events on simple Kossel like crystals can be simulated with a Monte Carlo simulation; the simulation of step-trains, occupying much larger surfaces can be simulated by solving coupled differential equations (these latter calculations take much less computer time).

One of the most important results of the step simulation was that perturbations always give a reduction of the flux of steps and hence a reduction in the rate of growth. Taking arbitrary

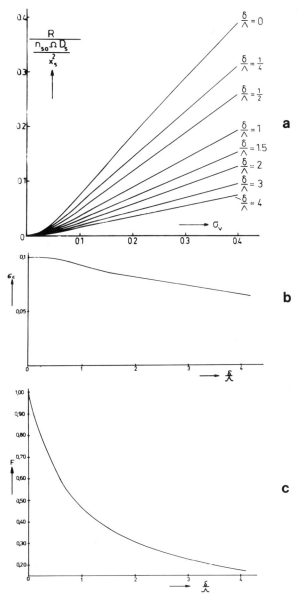

Fig. 6 (a) Dimensionless rate of growth versus bulk supersaturation σ_V. Volume diffusion resistance (proportional to δ/Λ) increases from upper (zero) to lower curves, (b) Value of σ_c (identical to σ_1 in BCF equation) as a function of δ/Λ (resistance for volume diffusion) obtained from Fig. 6(a) by a least square fitting program, (c) C in BCF equation as a function of volume diffusion resistance obtained from Fig. 6(a) by a least square program.

perturbations in a set of 100 steps it was found, however, that the flux of steps was only reduced by about 6% for $\sigma \simeq \sigma_1$. Below and above this value of σ the reduction is less. So considerable deviations from an ideal law in most cases cannot be attributed to perturbations in trains of steps (38).

So far we have discussed deviations from equidistant step-trains of straight steps, which are valid far enough from the centre of the spiral (see also Surek (39)). Approximate equations for the shape of the spiral were derived by BCF (8) and Cabrera and Levine (40). Recently Müller-Krumbhaar and Kroll (41) and Ohara and Reid (26) have solved the equation of the spiral numerically. The numerical calculation was in perfect agreement with the Cabrera-Levine result. This result was compared with a Monte Carlo simulation. More work concerning the shapes of spirals in dependence on surface diffusion, the shapes of the critical nucleus, etc. is now in progress.

In order to describe perturbations in very large steptrains a continuous kinematic wave theory originally designed for road traffic flows, was applied to 'traffic flows of steps' by Frank (42), Cabrera and Coleman (43) and Chernov (32).

Experiments on crystal growth rates as a function of supersaturation (see lowest part of Fig. 1)

We will now check what progress has been made recently concerning the experimental confirmation of the theories discussed above. It was shown that for the experimentalist it is a real challenge to decide whether a measured (R,σ) curve must be represented by a two-dimensional nuclation mechanism curve or by a BCF curve (44). (Compare Figs 3 and 5.) Birth and spread models give a formula:

$$R = A \sigma^{5/6} \exp(-B/\sigma)$$

As with the BCF curves B determines the curvature and A the 'slope'.

Experimental data of Mullin and Garside (34) for potash alum are plotted in Fig. 8. Using a least square fitting program both a BCF curve and a nucleation curve were determined from these points. It can be seen that except for the very low σ values both curves give a satisfactory description of the measured points. Also a power law $R = A'\sigma^P$ was determined. This gives a satisfactory description of the measured (R,σ) points as can be seen from Fig. 8.

This suggests that the use of purely empirical power laws in the world of industrial crystallization is justifiable. In order to distinguish between the two alternative models measurements must be carried out at very low supersaturations and over a

CRYSTAL GROWTH FROM SOLUTION

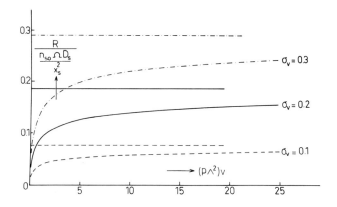

Fig. 7 Dimensionless growth rate versus constant times rate of flow v at constant σ. It can be seen that the asymptotic value for the pure rate of growth (without volume diffusion) is not reached even at high flow velocities.

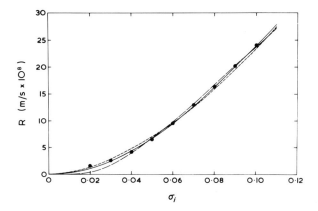

Fig. 8 Surface integration kinetics for single crystals of potash-alum at 32°C. Best BCF (———), polynucleation (– – –), and power law (-----) curves. The power p = 1.6. Here σ_i is σ at the interface.

sufficiently large range of supersaturations. On the basis of
theoretical considerations the most appropriate way to measure
rates of growth at different supersaturations can be planned.
But even under the most favourable conditions it will not be easy
to determine C, or A and σ_1, or B following from BCF or nucleation
curves respectively with an accuracy of better than 30% (34).

From rate versus supersaturation curves only two constants
can be determined, namely a slope (C or A), and a constant
determining the curvature (σ_1 or B). These constants consist
of subconstants, which often cannot be determined independently,
like for example the surface diffusion coefficient, the relaxation
time for entering the step, the edge energy of a critical nucleus,
etc. This makes a severe test of the theory and an unambiguous
determination of the growth mechanism for growth from solution
difficult. Apart from this difficulty we refer to Fig. 6a-c, where
it was shown that the influence of volume diffusion, especially in
the slope of an $R(\sigma)$ curve, cannot be neglected. In order to get
more experimental information to check models or theories,
additional information concerning the hydrodynamics around the
growing crystal, step patterns on crystal surfaces, the role of
impurities and especially the influence of dislocations on the
growth mechanism is required. Also activation energies obtained
from an Arrhenius plot need to be available. Then orders of
magnitude of C or A can be checked.

Notwithstanding all these difficulties, during the last few
years considerable progress has been made in the field of crystal
growth from solution. We will not repeat all the literature data
summarized in a previous paper (44), where it was shown that most
$R(\sigma)$ curves could be described by a BCF curve and some curves with
a nucleation curve, but we want to focus our attention on some
very important recent experimental work:

(a) Very precise $R(\sigma)$ curves were determined for the (110) and
(001) faces of TGS (triglycine sulphate) by Novotny, Moravec and
Solc (45) and Moravec and Novotny (46,47). These authors determined
both optically (by projecting an image of the crystal on a screen)
and by weighing the rate versus supersaturation in a three vessel
system of the Walker-Kohman type. The range of supersaturation
was $0 \leq \sigma \leq 10^{-2}$.

The hydrodynamic conditions, concentration of impurities and
distribution of dislocations on the crystal were all studied very
carefully. The results were interpreted on the basis of the BCF
theory adapted to growth from solution (48). σ_1 values were
determined as 1.0×10^{-3} and 5.0×10^{-3} for the (110) and (001)
faces respectively, leading to values of high mean displacements
x_s of 2.4×10^3 and 4.7×10^3 Å for the two respective faces.

(b) Other important experimental work is the measurement of R(σ) curves for paraffin side faces by Simon, Grassi and Boistelle (49, 50). These curves could only be fitted satisfactorily by a two-dimensional nucleation curve adapted for the high anisotropy of the side faces. The theoretical interpretation is of special value, among others, to the fact that the thermodynamics of the solution, which can be described as a regular solution, is included in the analysis. The role of one well defined impurity on the growth kinetics is also carefully investigated and explained. Recently a parabolic R(σ) law was found for the side faces of very thin paraffin crystals (51). This was interpreted as a second parabolic BCF law, where the rate determining step is the integration into the step.

(c) Finally we want to mention the work of Davey (52) and Davey and Mullin (53,54) on ADP (ammonium dihydrogen phosphate). Kinematic waves were observed on the prism faces. New rate versus supersaturation curves were measured for the pyramidal faces. Contrary to what was suggested before (1,44), the R(σ) curves can be better described by a BCF curve than by a two-dimensional nucleation curve.

THEORETICAL AND EXPERIMENTAL EVIDENCE FOR THE INFLUENCE OF DISLOCATIONS ON THE RATE OF GROWTH: GROWTH DISPERSION

It follows from the BCF theory that if the distance between dislocations is smaller than 9.5 r^* (r^* is the radius of the critical nucleus which is inversely proportional to σ) that a group of co-operating spirals will occur. The more spirals cooperate, the stronger the source of steps. In Figs 9a and b two spiral systems

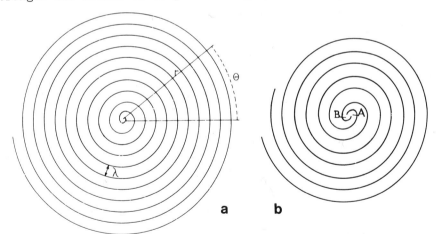

Fig. 9(a) Single and (b) Double spirals. The shape of the spiral is given by 2 r^* θ, where r^* is the radius of the critical nucleus.

are presented. Here the groups have one and two collaborating spirals arising from one and two screw dislocations emerging on the surface respectively. If we compare two crystallographic faces, which are dominated by the groups a and b respectively, it follows that

$$\sigma_1 = 9.5\, \gamma a / S k T x_s$$

where γ is the edge energy per growth unit in the edge, x_s the mean displacement and S the number of collaboration spirals: S is 1 and 2 for cases a and b respectively. This makes

$$\sigma_{1a} : \sigma_{1b} = 2 : 1$$

and the rates of growth for the two cases in the parabolic part ($\sigma < \sigma_1$) where the BCF equation becomes $R = C\, \sigma^2/\sigma_1$ are

$$R_a : R_b = 1 : 2$$

The reason for the two-times higher growth rate in case 2 compared with case 1 is the two-times higher density of steps emitted from the step source. In Fig. 5 curves 3-7 have different values of σ_1 so that for a given value of $\sigma < \sigma_1$ according to $R = C\, \sigma^2/\sigma_1$ the following relation holds (see caption of Fig. 5).

$$R_3:R_4:R_5:R_6:R_7 = \sigma_{1.7}:\sigma_{1.6}:\sigma_{1.5}:\sigma_{1.4}:\sigma_{1.3} = 5:4:3:2:1$$

These values 5, 4, 3.... can be considered to be the result of the fact that in case 3 the dominating group of spirals consists of 5 spirals, in case 4 of 4 spirals etc. In the case of a set of dislocations of equal sign on a grain boundary a second linear law occurs having a slope C^1 where $0 < C^1 < C$, where C is the slope of the BCF curve (see refs 30 and 44). Finally a special case concerns two closely adjacent dislocations of opposite sign: below a certain value of the supersaturation the growth rate is reduced, whereas above this value a normal BCF law holds.

So far we have only discussed the implications of the BCF theory concerning the dislocation structure of crystals and their influence on the variation of the rate of growth. We will now discuss empirical evidence for this effect. We first note that combinations such as presented above in Fig. 9 and mentioned above are all observed in reality (55-58).

Concerning rates of growth we note that it follows from the work of Berg (59) and Bunn (60) that there is a large variation in growth rate from crystal to crystal. The most striking example is the observation that when a crystal was broken into two parts one part continued to grow normally while the other did not grow at all. Also Booth and Buckley (61) observed occasional jumps in the rate of growth at constant supersaturation.

Authier (62) showed that bundles of distorted areas or dislocations originate from a 'bad' seed crystal and remain roughly perpendicular to the growing faces. Sudden changes in supersaturation, stirring conditions leading to solvent inclusions, impurities, etc. may give rise to distorted areas. Moravec and Novotny (46) also found that bundles of dislocations remain perpendicular to growing faces. These authors also discovered that the dislocation densities of dislocations ending in the (110) faces increase if the supersaturation increases.

Mussard and Goldstaub (63) observed that a crystal face which first grew very slowly, probably due to the absence of screw dislocations giving rise to a two-dimensional nucleation mechanism, started to grow faster as soon as a bundle of dislocations emerged on the surface.

Different growth rates were observed for different types of NaCl and KCl whiskers by Sichiri, Kinoshita and Kato (64). This observation must be partly attributed to different dislocation structures in different whiskers. Recently, it was shown by Sichiri (65) that on faces of whiskers on which no dislocations emerged, a two-dimensional nucleation mechanism operates, giving a zero growth rate for low supersaturations.

Important work has been carried out on dislocation structures in crystals grown from solution by Klapper et al. (66,67,68,69,70). Of special interest is the growth history of a particular crystal of lithium formate monohydrate (70) where it is shown that the value R_1/R_2 corresponding to the rates of growth of two different adjacent faces varied during the growth process owing to changes in the dislocation structure.

GROWTH AND SIZE DISPERSION, POPULATION DENSITY DISTRIBUTIONS

It is to be expected that for small crystals, e.g. crystals varying from 1 to 100μ in industrial crystallizers, the dislocation structure of dominant groups of spirals may vary strongly from one crystal to another and from one crystallographic face to another (where we compare equivalent crystallographic faces). Hence the $R(\sigma)$ curves may vary (see Fig. 5).

We can expect that there will be a variation of a factor of say five in the growth rates of equivalent crystallographic faces. (S may vary from 1 to 5 for 1 to 5 collaborating spirals; occasionally the growth rate of a certain face may become zero below a certain supersaturation for the case of two spirals of opposite sign or in case of absence of dislocations).

The phenomenon of variation in growth rate at the same σ will be called <u>growth dispersion</u>. If a fraction of crystals showing growth dispersion grows at a constant supersaturation a variation in size will occur leading to a <u>size dispersion</u>. Such size dispersions have been found by White and Wright (71,72) for sucrose, Natal'ina and Treivus (73) for KDP and Janse and De Jong (74) for $K_2Cr_2O_7$.

The occurrence of growth dispersion has very important implications for studies concerning population density distributions. Janse and De Jong show in their paper presented at this Symposium (74) that growth dispersion leads to an upward curvature in the population plot. In their calculation they assume that each crystal grows with a fixed growth rate (under constant supersaturation). Due to collisions with the stirrer and the crystallizer wall the dislocation structure of crystals may also change in time. This may give an increase of the rate of growth of certain crystallographic faces during the lifetime of a crystal in the crystallizer. Then, on the average, the growth rate of crystals increases if the size increases. This would give a higher upward curvature in the population plot than that calculated by Janse and De Jong (74). Implications of the fact that the rate of growth for each crystal is not constant are now under investigation.

INFLUENCE OF IMPURITIES, SOLVENT INCLUSIONS, DENDRITIC GROWTH

So far we have discussed mainly growth under mild and relatively pure conditions. It is generally known that small traces of impurities may considerably change the growth kinetics and the habits of crystals. Relatively little research has been carried out with the aim of explaining, on the basis of models, the role of impurities on growth kinetics. Some valuable data have, however, been summarized in the book edited by Kern (75) and recent publications have been made by Boistelle et al. (76,77,78). The influence of the impurity dioctadecylamine on paraffin was explained by generalizing the growth model applied to growth under pure conditions to growth for this single well-defined impurity (50). Small traces of lead may drastically influence the growth kinetics of KCl as shown by Botsaris et al. (79). For the influence of Cr^{3+} ions on ADP reference should be made to the work of Davey and Mullin (53,54).

It was shown by Carlson (80) that for large crystals, solvent inclusion may occur as 'starvation veils' due to an insufficient supply of fresh supersaturated solution to a part of the crystal surface having a certain distance from the leading edge. The hydrodynamic regime around the crystal is now recognized as extremely important for the prevention of solvent inclusions. Knowledge of the hydrodynamics has been applied by Scheel (81) to grow large and high quality crystals from the flux. Solvent inclusions may occur if crystal growth starts again on rounded corners. Then faceted pyramids limited by the natural F faces develop. During the competition between pyramids for survival solvent droplets are encaptured. An example was observed by Mischgofsky (82). NaCl crystals obtained from industrial crystallizers show quite often rows of solvent inclusions along the body diagonals of the cube-shaped crystals. This is probably due to dissolution and subsequent growth of the corners.

In the case of the growth of ADP and KDP single crystals on seed crystals with artificially introduced (001) faces, pyramids limited by (011) develop. Finally only one of these pyramids survives and from that moment on normal growth occurs. Before this closing process, however, numerous droplets are included in the crystal giving a milky appearance to the centre of the crystal. A systematic experimental study on liquid inclusions in hexamethylene-tetramine crystals was carried out by Denbigh and White (83). The process of solvent inclusions were explained in principle on the basis of the theories of Chernov.

From the few observations given above it can be concluded that dissolution and breaking of crystals should be prevented in crystallizers to prevent solvent inclusions.

Dendritic growth may occur under the influence of volume diffusion since corners and edges are favoured sites. For the transport of growth units from the solution to the crystal, step growth mechanisms have a high resistance to growth because gradients in the supersaturation over the surface are leveled out. For a treatment of conditions under which dendritic growth may occur, we refer again to the paper of Chernov (84). Since dendritic growth normally only occurs at high supersaturations it probably seldom occurs in industrial crystallizers, since these normally operate under low supersaturations, although an exception may be the first stages in batch crystallization.

CONCLUSION

Asselbergs and De Jong (85) present a formalism in which they show that the behaviour of a crystallizer is determined by the heat transfer rate, the suspension hydrodynamics and the crystallization kinetics. Apart from such problems as heat transfer, incrustation, hydrodynamic conditions and classification during the removal of crystals, the behaviour of a crystallizer is governed by two factors, viz. secondary nucleation and growth kinetics. In this paper we have only discussed the latter; the former is much less developed and probably much more complicated. However, a few very interesting empirical papers have been published (86,87,88), but a theoretical model explaining the phenomenon of secondary nucleation is lacking.

The rate of secondary nucleation B^o (m^{-3} s^{-1}) may be expressed by $B^o = k_B \varepsilon^h \sigma^r M^j$ and the rate of growth by $R = k_G \sigma^p$, where p, h, r and j are empirical constants, k_G an empirical constant for growth similar to C and A in the BCF equation and the nucleation equation, respectively, discussed above, k_B, the empirical constant for nucleation, ε the dissipated energy per unit of mass of the suspension (W/kg), M the slurry concentration (kg/m^3) and σ the relative supersaturation. Eliminating σ:

$$B^o = k_N \varepsilon^h R^i M^j$$

Also it can be shown that under certain conditions (85) the dominant size on a mass basis, L_D, is given by:

$$L_D = k_3 \tau^{(i-1)/(i+3)}$$

L_D is determined by the average residence time of crystals in the crystallizer τ, and $i = r/p$ is the ratio of the 'orders' of growth and nucleation; k_3 is a constant.

The use of power laws for the growth kinetics is justified. For the growth kinetics we refer to the section above (Fig. 7) where it was shown that both a BCF curve and a two-dimensional nucleation curve can be adequately described by such a law. In case of a BCF law p is 1 < p < 2 and in the case of a nucleation law p may vary from say 2 to 5 (39). (See also Nielsen (89).) It can be seen that the behaviour of a crystallizer depends very much on this power i. All the events taking place during the crystal growth processes are to a very large extent as it were summarized in the power p, and to a lesser extent in k_G. In p and k_G the worlds of scientists working in the field of crystal growth kinetics and chemical engineers working in the field of industrial crystallization come together.

Summarizing, the science of crystal growth can give the following contributions to the applied science of industrial crystallization:

(i) Theoretical and fundamental studies on crystal morphology, which play an important role in industrial crystallization, can clarify habits, crystal growth kinetics, the role of impurities, etc. - see also the work of Troost (90).

(ii) Rate versus supersaturation curves measured in the flow systems or by a weighing method, compared with rates of growth measured on crystals in a fluidized bed give a basis for k_G and p in power laws used in calculations for crystallizers - see the account by Mullin (91).

(iii) The role of dislocations in giving growth dispersion rates may clarify measured population density curves.

(iv) Experimental and theoretical evidence for the occurrence of solvent inclusions and the formation of dendrites may lead to the finding of ways for their elimination.

ACKNOWLEDGEMENTS

The author wishes to thank ir. A.H. Janse of the group of Prof. ir. E.J. de Jong for stimulating discussions and for critically reading the manuscript. Also we thank ir. C. van Leeuwen for checking the manuscript and Miss M. Wijnen and Mr. W. Verhoeff who typed and assembled it.

REFERENCES

1. P. Bennema, J. Cryst. Growth 24/25 (1974), 76.
2. P. Hartman in: Crystal Growth: An introduction edited by P. Hartman, North-Holland Publ. Comp. 1973, p. 367.
3. R. Cadoret, Thesis University of Caen, 1965.
4. R. Cadoret and J.C. Monier in: Adsorption et Croissance Cristalline (Ed. CNRS Paris, 1965), p. 559.
5. P. Bennema, J. Cryst. Growth 5 (1969), 29.
6. H. Schrijver (Private Communication).
7. E.P.K. Ottens, Thesis Technical University Delft, 1973.
8. W.K. Burton, N. Cabrera and F.C. Frank, Phil. Trans. Roy. Soc. Lond. A 243 (1951), 299.
9. K.A. Jackson in: Liquid Metals and Solidification (American Society for Metals, Metals Park, Ohio 1958), p. 174.
10. B. Mutaftschiev in: Adsorption et Croissance Cristalline (Ed. CNRS, Paris 1965), p. 231.
11. D.E. Temkin in: Crystallization Processes (Consultant Bureau New York 1966).
12. H.J. Leamy and K.A. Jackson, J. Appl. Phys. 42 (1971), 2121.
13. G.H. Gilmer and P. Bennema, J. Cryst. Growth 13/14 (1972), 148.
14. G.H. Gilmer and P. Bennema, J. Appl. Phys. 43 (1972), 1347.
15. F.L. Binsbergen, Kolloid Z Polymere 237 (1970), 289.
16. H. Müller-Krumbhaar, Phys. Rev. B 10 (1974), 1308.
17. H.J. Leamy, G.H. Gilmer and K.A. Jackson in Surface Physics of Crystalline Materials, edited by J.M. Blakeley. (Academic, New York, to be published).
18. S.W.H. de Haan, V.J.A. Meeussen, B.P. Veltman, P. Bennema, C. van Leeuwen and G.H. Gilmer, J. Cryst. Growth 24/25 (1974), 491.
19. C. van Leeuwen and J. van der Eerden, to be published.
20. G.H. Gilmer, H.J. Leamy, K.A. Jackson and H. Reiss, J. Cryst. Growth 24/25 (1974), 495.

21. D. van Dijk, C. van Leeuwen and P. Bennema, J. Cryst. Growth 23 (1974), 81.
22. G.H. Gilmer (private communication).
23. A.A. Chernov in: Crystal Growth, edited by H. Steffen, Peiser Pergamon Press (1967), p. 25.
24. H.J. Leamy and G.H. Gilmer, J. Cryst. Growth 24/25 (1974), 499.
25. C. van Leeuwen and F.H. Mischgofsky, J. Appl. Phys. 46 (1975), 1056.
26. M. Ohara and R.C. Reid, Modelling Crystal Growth Rates from Solution, Prentice Hall Inc. (1973), p. 190. (In this book also birth and spread models are described.)
27. C. van Leeuwen and P. Bennema, to be published in Surface Science.
28. P. Bennema, J. Cryst. Growth 1, (1967), 278.
29. P. Bennema in: Crystal Growth (see ref. 23), 1967, p. 413.
30. P. Bennema and G.H. Gilmer in: Crystal Growth: An introduction edited by P. Hartman, North-Holland series in Crystal Growth 1 (1973), Chapter 10, p. 263.
31. G.H. Gilmer, R. Ghez and N. Cabrera, J. Cryst. Growth 8 (1971), 79.
32. A.A. Chernov, Sov. Phys. Uspekki (Eng. Transl.) 4 (1961), 116.
33. R. Janssen-van Rosmalen, P. Bennema and J. Garside, to be published in J. Cryst. Growth.
34. J. Garside, R. Janssen-van Rosmalen and P. Bennema, to be published in J. Cryst. Growth.
35. O. Söhnel and M. Krpata, Coll. Czechoslov. Chem. Comm. 39 (1974), 2520.
36. R.L. Schwoebel, J. Appl. Phys. 40 (1969), 614.
37. C. van Leeuwen, R. Janssen-van Rosmalen and P. Bennema, Surf. Science 44 (1974), 213.
38. R. Janssen-van Rosmalen and P. Bennema, to be published.
39. T. Surek, J. Cryst. Growth 13/14 (1972), 19.
40. N. Cabrera and M.M. Levine, Phil. Mag. 1 (1956), 450.
41. H. Müller-Krumbhaar and D. Kroll, to be published.
42. F.C. Frank in: Growth and Perfection of Crystals (John Wiley, New York, 1958), p. 411.
43. N. Cabrera and R.V. Coleman in: The Art and Science of Growing Crystals, Ed. J.J. Gilman (Wiley, New York, 1963), p. 3.
44. P. Bennema, J. Boon, C. van Leeuwen and G.H. Gilmer, Kristall und Technik 8 (1973), 659.
45. J. Novotny, F. Moravec and Z. Šolc, Czechoslowak Academy of Sciences, Vol. B. 23 (1973), 261.
46. F. Moravec and J. Novotny, Kristall und Technik 6,(1971) 335.
47. F. Moravec and J. Novotny, Kristall und Technik 8,(1972), 891.
48. P. Bennema, Thesis Technical University Delft, 1965.
49. B. Simon, A. Grassi and R. Boistelle, J. Cryst. Growth 26 (1974) 77.
50. B. Simon, A. Grassi and R. Boistelle, J. Cryst. Growth 26 (1974) 90.
51. A.C. Doussoulin, Thesis University D'Aix, Marseille (1975).

52. R. Davey, Thesis University College London (1973).
53. R. Davey and J.W. Mullin, J. Cryst. Growth 23 (1974), 89.
54. R. Davey and J.W. Mullin (private communications, to be published).
55. W. de Keyser and S. Amelinckx, Les Dislocations et la Croissance des Cristaux (Masson, Paris, 1955).
56. A.R. Verma, Crystal Growth and Dislocations (Butterworth, London 1953).
57. H. Bethge, Crystal Growth, edited by H.S. Peiser, Pergamon Press (1966), 623.
58. H. Bethge in Nova Acta Leopoldina, Physik und Chemie der Kristalloberfläche, ed. J.A. Barth, Leipzig 1968, p. 50.
59. W.F. Berg, Proc. Roy. Soc. A 164 (1938), 79.
60. C.W. Bunn, Disc. Far. Soc. 5 (1949), 132.
61. A.H. Booth and H.E. Buckley, Nature 169 (1952), 367.
62. A. Authier, J. Cryst. Growth 13/14 (1972), 34.
63. F. Mussard and S. Goldstaub, J. Cryst. Growth 13/14 (1972), 445.
64. T. Shichiri, H. Kinoshita and N. Kato in: Crystal Growth (see ref. 23.) (1967), 385.
65. T. Shichiri, J. Cryst. Growth 24/25 (1974), 350.
66. H. Klapper, Phys. Stat. Sol. 14 (1972), 99.
67. H. Klapper, Phys. Stat. Sol. 14 (1972),443.
68. H. Klapper, J. Cryst. Growth 15 (1972),281.
69. H. Klapper and H. Küppers, Acta Cryst. Vol. A 29 (1973), 495.
70. H. Klapper, Zeit f. Naturf. 28a (1973).
71. P.G. Wright and E.T. White, 61a Proc. Queensland Soc. Sugar Cane Technol. 36th Conf., p. 299 (1969).
72. E.T. White and P.G. Wright, Chem. Eng. Progress. Symp. Series no. 110, 67, p. 81.
73. L.N. Natal'ina and E.B. Treivus, Sov. Phys. Cryst. (Eng.Transl.) 19 (1974), 389.
74. A.H. Janse and E.J. de Jong, see contribution to this symposium.
75. Adsorption et Croissance Cristalline, Ed. CNRS, Paris, 1965 edited by R. Kern (see also the paper of P. Hartman, p. 477).
76. R. Boistelle, M. Mathieu and B. Simon, C.R. Acad. Sc. Paris 274 (1972), 473.
77. R. Boistelle, G. Pèpe, B. Simon and A. Leclaire, Acta Cryst. B 30 (1974), 2200.
78. R. Boistelle, M. Mathieu and B. Simon, Surf. Science 42 (1974), 373.
79. C.D. Botsaris, E.A. Mason and R.C. Reid, J. Chem. Phys. 45 (1966) 1893.
80. A. Carlson in: Growth and Perfection of Crystals, Wiley New York, 1958, p. 259.
81. H.J. Scheel, J. Cryst. Growth 13/14 (1972), 560.
82. F.H. Mischgofsky, private communication.
83. K.G. Denbich and E.T. White, Chem. Eng. Science 21 (1966), 739.
84. A.A. Chernov, J. Cryst. Growth 24/25 (1974), 11.
85. J.C. Asselbergs and E.J. de Jong, see contribution to this symposium.

86. E.P.K. Ottens, A.H. Janse and E.J. de Jong, J. Cryst. Growth 13/14 (1972), 500.
87. H. Garabedian and R.F. Strickland-Constable, J. Cryst. Growth (1972), 506.
88. E.G. Denk Jr. and G.D. Botsaris, J. Cryst. Growth 13/14 (1972), 493.
89. A.E. Nielsen, Kinetics of Precipitation (Pergamon Oxford, 1964).
90. S. Troost, J. Cryst. Growth 13/14 (1972), 449.
91. J.W. Mullin, Crystallization, London Butterworths, second edition 1972, chapter 6, pp. 174-232.

COMPUTER SIMULATION OF CRYSTAL GROWTH FROM SOLUTION

T.A. CHEREPANOVA, A.V. SHIRIN AND V.T. BORISOV

Computing Centre, Latvian State University

Boulevard Raina 29, Riga, USSR

Studying crystal growth kinetics by computer simulation gives the possibility of investigating directly the accepted physical model without any additional assumptions characteristic of the usual analytical theories. Previous work (1)-(5) has been mainly concerned with one-component systems. The aim of the present work is to construct a model of crystal growth from a binary melt.

A section of the system near the crystal-melt interface is shown in Fig.1. Atoms belonging to solid and liquid phases are indicated by Latin and Greek symbols respectively. The crystal is assumed to have a simple cubic lattice and to consist of atomic columns parallel to the growth direction. The position of every atom is indicated by coordinates, i, j of an atom column and the height k in a column, counted off some general level (k = 0). it is assumed that any straight line parallel to the growth direction crosses the phases boundary once only. A solid phase atom z (z = a, b) is considered to be a surface atom if its neighbour above (along the growth direction) belongs to the fluid.

The following elementary transitions are allowed: an atom of the fluid ξ ($\xi = \alpha, \beta$) can add to a neighbouring surface atom z with the probability p_ξ; a surface atom z can transfer into the melt with the probability q_z. As at moderate growth rates diffusion transitions of atoms in the fluid occur more frequently (with characteristic time $\sim 10^{-10}$ sec) than resulting additions to the crystal ($\sim 10^{-8}$ sec for crystal growth velocity ~ 1 mm/sec, the fluid washing the crystal may be considered to be completely mixed. Besides, as the binding energy between atoms of the fluid $e_{\alpha\alpha}$ $e_{\alpha\beta}$ $e_{\beta\beta}$ generally speaking, does not differ much from corresponding binding energies $e_{a\alpha}$ $e_{a\beta}$ $e_{b\alpha}$ $e_{b\beta}$ between atoms of different phases, it may

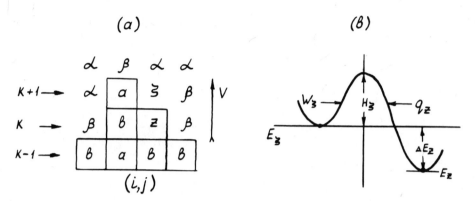

Fig. 1 (a) Model of crystal, (b) elementary transitions of particles

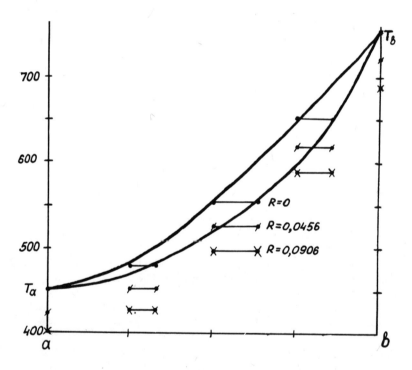

Fig. 2 Simulation of a binary system: ● = equilibrium diagram points, ◢ and ✘ = kinetic diagram points at supercoolings σ = 0.05 and 0.1 respectively

be assumed that the energy E_ξ of a particle ξ in the melt and its activation energy H_ξ of addition to the crystal are determined only by concentration c_ξ in the melt and do not depend on the number of neighbours in the solid phase. Thus, for the probabilities of addition to the crystal the expressions

$$p_\alpha = c_\alpha w_\alpha, \quad p_\beta = c_\beta w_\beta$$

are assumed, where w_α, w_β, are the transition probabilities provided that the type of the adherent atom is determined. According to the scheme in Fig.1b:

$$w_\alpha = \nu_\alpha \exp[-H_\alpha/kT]; \quad w_\beta = \nu_\beta \exp[-H_\beta/kT]$$

where ν_α, ν_β are corresponding frequency (entropy) factors,

The energy of a surface atom belonging to the solid phase, depends on the relief of the phase boundary and the component composition of the adjacent surroundings. For the considered quasi-chemical model in approximation of the nearest neighbours we can write

$$E_z = \ell_{za} e_{za} + \ell_{zb} e_{zb} + \ell_{z\alpha} e_{z\alpha} + \ell_{z\beta} e_{z\beta}$$

where e_{aa}, e_{ab}, e_{bb} are the binding energies of particles in the crystal, and coefficients ℓ_{zj} are the numbers of neighbours of a corresponding sort. With the atom transition into the melt these coordination numbers do not change (as a simultaneous proceeding of several elementary processes is not allowed in simulation) and its energy in the fluid is

$$E_\xi = \ell_{za} e_{\xi a} + \ell_{zb} e_{\xi b} + \ell_{z\alpha} e_{\xi\alpha} + \ell_{z\beta} e_{\xi\beta}$$

Here $\xi = \alpha$ if $z = a$; $\xi = \beta$ if $z = b$; all values e_{ij} are assumed to be symmetrical with respect to transposition of indices.

Introducing the difference of energy levels $\Delta E_z = E_\xi - E_z$ (Fig.1b) and considering that in the described model the activated states for processes of addition and detachment of particles are the same, we get (according to the theory of absolute velocities of reactions) expressions for frequencies of detachment from the crystal of particles of any sort:

$$q_a = \nu_a \exp[-(H_\alpha + \Delta E_a)/kT]$$

$$q_b = \nu_b \exp[-(H_\beta + \Delta E_b)/kT]$$

Here ν_a, ν_b are frequency (entropy) factors of the crystalline phase. Taking account both of complete intermixing in the fluid and the crystallographic structure of the model, we obtain the correlations

$$\ell_{z\alpha}/\ell_{z\beta} = c_\alpha/c_\beta; \quad \ell_{za} + \ell_{zb} + \ell_{z\alpha} + \ell_{z\beta} = 6;$$

$$c_\alpha + c_\beta = c_a + c_b = 1,$$

and the expression for ΔE_z transforms to:

$$\Delta E_z = \ell_{za} \phi_{z1} + \ell_{zb} \phi_{z2} + 6 \varepsilon_z$$

$$\phi_{z1} = e_{\xi a} - e_{za} - \varepsilon_z; \quad \phi_{z2} = e_{\xi b} - e_{zb} - \varepsilon_z$$

$$\varepsilon_z = c_\alpha (e_{\xi\alpha} - e_{z\alpha}) + c_\beta (e_{\xi\beta} - e_{z\beta})$$

Here the terms, dependent on the phase boundary configuration are written out. ε_z, which clearly depends on the melt composition, affects the process only slightly since, as mentioned above, differences of the kind $e_{\xi\alpha} - e_{z\alpha}$ are usually small in comparison with $e_{\xi a} - e_{za}$. Given the constants c_α, c_β, T the values of w_α, w_β are also constant. Thus, the result of simulation depends only on the correlations:

$$p'_\alpha = p_\alpha/w_\alpha, \quad p'_\beta = p_\beta/w_\alpha, \quad q'_a = q_a/w_\alpha, \quad q'_b = q_\beta/w_\alpha$$

Then

$$p'_\alpha = c_\alpha, \quad p'_\beta = w c_\beta, \quad q'_a = \exp[\theta_a - \Delta E_a/kT],$$

$$p'_b = \exp[\theta_b - \Delta E_b/kT]$$

Generally speaking, the values $\theta_a = \ln(\nu_a/\nu_\alpha)$, $\theta_b = \ln(\nu_b/\nu_\beta)$, $w = w_\beta/w_\alpha$ rather feebly depend upon the melt composition and temperature. For the accepted model the exponent indices in the expressions for p'_ξ, q'_z actually represent the differences of local values (considering the adjacent surroundings) of chemical potentials of components without entropy configurational parts.

Thus, binary crystal growth kinetics are uniquely determined by giving 11 constants (or feeble functions c_α, c_β, T) in the general case; two entropy factors θ_a, θ_b (determining, judging by one-component simulation experience, the type of crystal growth mechanism); a kinetic factor w that represents the correlation of the addition probabilities of particles of various sorts; energy parameters of interaction $e_{aa} - e_{a\alpha}$; $e_{ab} - e_{\alpha b}$, $e_{ba} - e_{\beta a}$,

$e_{bb} - e_{b\beta}$, $e_{aa} - e_{\alpha a}$, $e_{a\beta} - e_{\alpha\beta}$, $e_{b\alpha} - e_{\beta\alpha}$, $e_{b\beta} - e_{\beta\beta}$. In the most significant case the four latter differences are negligibly small. In simulation we shall consider them equal to zero. The crystal state at every moment of time is given by the structure function U(i, j, k), assuming in every site of crystal lattice (i, j, k) the value 1 or 2, depending on sort of the atom occupying this site. The cyclic boundary conditions are given. The initial state of the crystal is arbitrary.

Under simulation physical time is divided into equal intervals of duration τ. During one interval only one of possible elementary events may be realised. The working cycle of a simulative programme begins with an equally-probable choice of a column (i, j), in which the event will be realised. After the choice of (i, j) with the help of the function U(i, j, k) the type of a surface atom and its adjacent surroundings are determined. This permits calculation of the energies $\Delta E_z = \Delta E_a$, ΔE_b and the probabilities of detachments. If z turns out to be equal to a, the possible events are p'_α, p'_β, q'_a, 0. For z = b allowable are p'_α, p'_β, q'_b, 0 (0 being a blank event). Interval τ is chosen as small as possible:

$$\tau(p'_\alpha + p'_\beta + \max(q'_a, q'_b)) = 1$$

Here the maximum value of the corresponding values for all possible configurations of the phase boundary during crystal growth is implied. The above-introduced quantities, q'_ξ, p'_z are the probability densities. The probabilities of realising elementary transitions during the interval τ have the form

$$P_\alpha = \tau c_\alpha, \quad P_\beta = \tau w c_\beta, \quad Q_a = \tau \exp[\theta_a - \Delta E_a/kT]$$

$$Q_b = \tau \exp[\theta_b - \Delta E_b/kT]$$

where $P_\alpha + P_\beta + Q_z \leq 1$.

The described model was studied on the following example. The crystal section is a square of a size 50 x 50 atoms ($0 \leq i, j \leq 49$, the number of atoms on the plane n_0 = 2500), $\theta_a = \theta_b = 1$, $w = 1$, $e_{\alpha\alpha} - e_{aa} = e_{\alpha\beta} - e_{ab} = 300$, $e_{\beta\beta} - e_{bb} = 500$ cal/mol.) Other differences of interaction parameters are equal to zero. The accepted physical conditions correspond approximately to an easily fusible metal alloy.

The equilibrium diagram of the alloy conditions obtained by means of simulation is given in Fig.2 (points in Fig.2). The equilibrium was studied with the melt concentrations c = 0; 25; 50; 75; 100%. For every composition the temperature was chosen so that non-dimensional crystal growth rate R = $\Delta M/N\tau$ (M = increase of the number of particles in the crystal in N tests) turned into zero. For the points indicated on the diagram the error $|\Delta R| \leq$ 0.006. Corresponding temperatures are 450, 480, 555, 675, 750 K.

Let us consider in detail the equilibrium at 555 K. We indicate by η_a^k, η_b^k, η_e^k the numbers of a, b, and fluid atoms occurring at the atom layer k; $(\eta_a^k + \eta_b^k + \eta_e^k = \eta_o)$. The beginning of reading k corresponds to some fixed layer, belonging to the solid phase. Fig.3 illustrates the aggregate and component structure of a transition zone between phases. It shows the solid atom concentration $c(k)$ and the concentration $c_a(k)$ of atoms a in the solid phase, defined by

$$c(k) = (\eta_a^k + \eta_b^k)/\eta_o; \quad c_a(k) = \eta_a^k/(\eta_a^k + \eta_b^k)$$

and corresponding dispersion (the dotted line is produced in the region adjacent to the crystal, where the influence of the chosen initial model condition is still felt). The $c_a(k)$ concentration, on the whole, is decreasing into the depth of the zone; the great dispersion at the beginning is caused by dynamic instability of its external edge. The dimension of the interface zone will be characterised by the number of atom planes whose mean concentration values are within the limits of $0.05 < c(k) < 0.95$. In the considered case, corresponding to the metal, the transition region is diffused over 14 atom layers. The degree of its roughness is expressed by the value $S = 2.87$ (Fig.4). This characteristic expresses the mean (over the surface and time), difference of heights $k(i, j)$ of the corresponding neighbouring surface atoms:

$$S = \frac{1}{4} < \sum_{i'j'}^{4} |k(i, j) - k(i', j')| >$$

Here averaging takes place over all atom columns i, j. For every i, j summing over i', j' extends over four adjacent columns.

The solid phase, being in equilibrium with the melt, differs from it in composition ($c_a = 0.37$; $c_\alpha = 0.5$) and is characterised by the presence of the near-order caused by differences of partial heats of melting : $e_{\alpha\alpha} - e_{aa}$ and $e_{\beta\beta} - e_{bb}$. To describe the near order we use the average numbers of every sort of neighbours existing round the atom a : $<\ell_{az}>$, round the atom b : $<\ell_{bz}>$; the deviations of these values $<\ell_{xz}> - \ell_{xz}^o$ from their ideal values ℓ_{xz}^o (x, z = a, b) which correspond to the complete intermixing of atoms a and b. For the considered example $<\ell_{aa}> = 2.34$; $<\ell_{ab}> = 3.65$; $<\ell_{ba}> = 2.16$; $<\ell_{bb}> = 3.84$; $<\ell_{aa}> - \ell_{aa}^o = 0.14 = \ell_{ab}^o - <\ell_{ab}>$; $<\ell_{bb}> - \ell_{bb}^o = 0.05 = \ell_{ba}^o - <\ell_{ba}>$. This means that an increase in the number of one-type inter-atomic bindings is observed in the crystal at the expense of different-type bindings.

In addition to these characteristics we must note that, as the study of the solid phase structure has shown, the atoms of every sort form bound groupings of very different configurations - from linear chains to rounded formations with the average number of atoms in a group equal to 2.7 and 5.0 for atoms a and b respectively.

The dimensions of interface zones and the corresponding roughness lie in the intervals δ = 14-11, S = 2.87-2.75. Since the near-order in the solid phase for all values of concentrations manifests itself rather feebly, it can be expected that experimental results may be described by means of the thermodynamic diagram theory for regular solutions. The characteristics of the model, T_a = 450, T_b = 750 K, q_a = 3 $(e_{\alpha\alpha} - e_{aa})$ = 900, q_b = 3 $(e_{\beta\beta} - e_{bb})$ = 1500 cal/mol, were used for finding the equilibrium diagram (T_z, q_z are the melting temperatures and heats of pure components). The mixing energies of the solid and liquid phases ω_s = e_{ab} - $(e_{aa} + e_{bb})/2$, ω_e = $e_{\alpha\beta}$ - $(e_{\alpha\alpha} + e_{\beta\beta})/2$, which are also necessary for calculation are not by themselves defined by the entrance parameters of the model. Therefore they were varied till the best result was obtained. The continuous lines of liquidus and solidus in Fig.2 correspond to values ω_s = -41, ω_e = -570 cal/mol. The calculation according to the regular solution theory is in good agreement with the direct simulation results. Some discrepancy, consisting the obtained values being $\omega_s - \omega_e$ = 529, whereas, according to simulation $\omega_s - \omega_e$ = 100 cal/mol is required, may be attributed to the presence of the near-order in the crystal which is not taken into account.

The growth kinetics was considered for supercoolings σ = 0.05; σ = 0.1. The points corresponding to liquidus and solidus of kinetic diagrams are shown in Fig.2. It is evident that for the given model parameters no noticeable change of the growing solid phase composition is observed in the region of supercoolings $\Delta T = T_e - T < 50°$. The same applies to the near-order of the system. Under sufficiently great supercooling, however, the lines of liquidus and solidus must draw together, according to the expressions for the probabilities of addition and detachment. Hence:

$$Q_a \to 0, \quad Q_b \to 0 \quad p_\beta/p_\alpha \to c_b/c_a$$

at $T \to 0$. The latter condition provides the binding between the composition of the growing solid phase and the composition of the initial fluid, taking place at low temperature:

$$c_b = w \, b_\beta/[1 + (w - 1)c_\beta]$$

for the simulation case w = 1, and $c_b \to c_\beta$ with $T \to 0$.

The behaviour of other characteristics of the model is shown in Fig.4. Roughness and width of the transition region δ have a

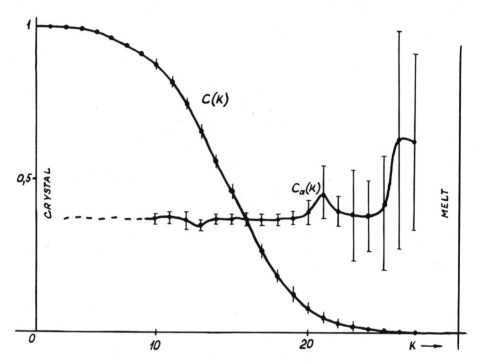

Fig. 3 Structure and composition of the transition zone between phases at $c_\alpha = 0.5$ and $\sigma = 0$

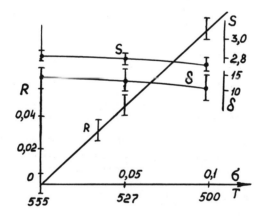

Fig. 4 Dependence of crystal growth rate and characteristics of transition zone on supercooling

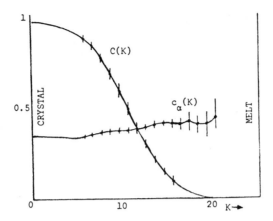

Fig. 5 Structure and composition of the transition zone between phases at $c_\alpha = 0.5$ and $\sigma = 0.1$.

feebly marked tendency to decreasing with increasing the supercooling. The growth rate with rather a high accuracy depends linearly upon the relative supercooling $\sigma = \Delta T/T_o$. That is, with the values θ_a, $\theta_b = 1$, corresponding to metal alloys, the normal crystal growth mechanism prevails. The simulation affords the possibility to estimate for the alloy the absolute value kinetic coefficient of the crystal growth found by the expression $v = k\,T$ (v is dimensional crystal growth rate). From the above:

$$K = dv/d\Delta T = (dR/d\sigma) \cdot \lambda \nu_\alpha \exp(-H_\alpha/k\,T_o)/T_o$$

$$= (dR/d\sigma) \cdot D_\alpha/\lambda T_o$$

where λ is the lattice parameter, D_α the diffusion coefficient, α atoms in the melt. From Fig.3 and 4, $dR/D\sigma = 0.91$, $T_o = 555$ K. Supposing $\lambda = 3.10^{-8}$ cm, $D_\alpha = 5.10^{-s}$ cm^2/sec, we obtain $K = 2.7$ cm/sec.grad, which by the value order corresponds to values of coefficients according to the theory of normal crystal growth for pure metals.

REFERENCES

1. Leamy, H.Y. and Jackson, K.A., J.Appl.Phys., 42 (1971) 2121.
2. Solovyev, V.V. and Borisov, V.T., DAN USSR, 102 (1972) 329.
3. Yesin, V.O., Danilyuk, V.I., Plishkin, Y.M. and Podchinenova, G.L., Crystallography (USSR), 18 (5) (1973) 920.
4. Cherepanova, T.A. and Kiseljov, V.F., Compil. 'Problems of Crystallisation Theory II" (USSR), Riga, 61 (1975).
5. Borisov, V.T., DAN USSR, 142 (1) (1962) 69.
6. Petrovsky, V.A. and Borisov, V.T., DAN USSR 204, (6) (1972) 1343.

SURFACE PERFECTION AS A FACTOR INFLUENCING THE BEHAVIOUR OF GROWTH LAYERS ON THE {100} FACES ON AMMONIUM DIHYDROGEN PHOSPHATE (ADP) CRYSTALS IN PURE SOLUTION

V.R. PHILLIPS AND J.W. MULLIN

Department of Chemical Engineering

University College London, Torrington Place
London WC1E 7JE, England

INTRODUCTION

Davey and Mullin (1,2) used reflection microscopy to study the layers which move across the {100} faces of ammonium dihydrogen phosphate (ADP) crystals (Fig. 1) growing from pure aqueous solution. After further experimental work, the present authors concluded that the degree of surface perfection governed the type of layers on the {100} faces. Specifically, the layer behaviour reported by Davey and Mullin (2) applies only to faces which exhibit some macro-imperfection (interpreted as any surface feature such as a crack, scratch or fracture), while surfaces with no macro-imperfection show a very different type of layer behaviour. We will now consider the relationship between the two types of layer.

Although it has long been realized that surface imperfections affect the growth of crystals (3,4,5), less perfect crystals in general growing faster than more perfect crystals, there is still little systematic knowledge of the effect of the degree of surface perfection on growth rates or on layer behaviour.

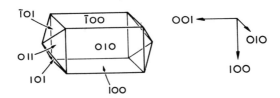

Fig. 1 The morphology of ADP

Surface imperfections cause lattice strain, which in turn causes dislocations to form, as the lattice attempts to relieve the strain (6,7,8,9). Thus surface imperfections facilitate Burton, Cabrera and Frank type growth, in which spiral steps are generated from screw dislocation sources (10).

The original Burton, Cabrera and Frank analysis envisaged these steps only as monolayers, but, experimentally, steps can be seen on some crystals even at low magnifications. Such steps are probably bunches of many monolayers. Suggestions as to why such bunches should form include orbiting of the source (11, 12) and kinematic wave formation (13).

If N elementary screw dislocations of the same hand emerge on a surface within some critical distance of each other, they can co-operate to form a single source which generates, rather than simple spirals, one multiple spiral with N branches (10). Again, two neighbouring emergent screw dislocations of equal magnitude (equal Burgers vectors) but of opposite hand may co-operate to generate closed loops rather than open spirals (10).

Morgan and Dunning (14) observed layers on the surfaces of carbon tetrabromide crystals growing from vapour. The layers were generated at such imperfections as grain boundaries or regions where two adjacent crystals had intergrown, but also sometimes from regions of surface seemingly without macro-imperfections. Pilkington and Dunning (15) observed layers on freshly cleaved {100} faces of potassium chloride growing from a mother liquor containing small amounts of lead chloride. Layers were generated at such imperfections as cleavage steps and, as above, from regions of surface seemingly without macro-imperfections. Neither of these pairs of workers, however, reported any difference between the layers generated at macro-imperfections and those generated from regions of surface without macro-imperfections.

To return to the work of Davey and Mullin (2), these workers found three layer regimes for the {100} faces of ADP crystals. When supersaturation $\sigma = (c-c^*)/c^*$ ($[kg/m^3]/[kg/m^3]$) was below about 0.03, no layers were seen. When σ was between about 0.03 and 0.07, regular-shaped elliptical layers were seen. The layers were concentric and spread from point sources. The major axes of the ellipses were in the [010] direction and their axial ratio was about 1.5. The spreading velocity of the ellipses was a linear function of σ and was of the order $(1 - 10) \times 10^{-7}$ m/s. In the present paper, these layers will be called Type S layers, because they are relatively slow-moving. When σ was above about 0.07, parallel layers were seen. These layers spread along the {100} faces in the [001] direction, but had no definite sources. Their spreading velocities were of the order $(1 - 10) \times 10^{-5}$ m/s. The range of

estimated velocity of the solution across the crystal surface was
1 - 18 mm/s and in this range there was no noticeable effect of
solution velocity on the behaviour of either type of layer.

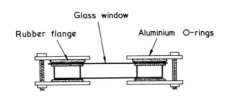

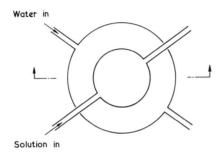

Fig. 2.   The flow cell

EXPERIMENTAL

The apparatus and method were basically those of Davey and
Mullin (1), which have been described in detail by Davey (16).
Briefly, crystals were observed while growing in a small glass-
covered flow cell (Fig.2) under conditions of controlled supersaturation, temperature and solution flow-rate. The crystals
received oblique illumination by a thin pencil of light and were
viewed through a microscope by the reflected rays, so that layer
fronts appeared as dark lines, while the rest of the crystal
surface appeared bright. Considerable manoeuvring of the cell was
often necessary before the optimum reflecting position was located.
A travelling microscope barrel was used, with a X10 filar
micrometer eyepiece and a X2.4 or a X8.5 objective. Photographs
were taken through the eyepiece, using a complete Exakta Varex IIb
single lens reflex camera with the lens focussed at infinity.

A stock of feed solution was stirred with excess crystals in a
1000 cm^3 flask at a closely controlled and accurately known
temperature. The dominant error in the measurement of this
temperature was the calibration error of the thermometer used.
This error was not greater than 0.05°C. Saturated solution was
withdrawn from the flask through a sintered glass filter stick
(porosity 90 - 150 μm) and pumped to the cell by a variable speed
peristaltic pump at a rate of 70 mm^3/s. Immediately before
reaching the cell, the solution passed through a glass coil heat

exchanger, which cooled the solution to the growth temperature. A growth temperature of 25.0°C was used for almost all the present work. After passing through the cell, the solution was returned to the stock flask. Immediately before reaching the flask, it passed through another glass coil heat exchanger which reheated it to the temperature of the stock solution. The sections of tubing between the stock flask and the first heat exchanger and between the cell and the second heat exchanger were heated gently (not higher than 45°C) by heating tapes, to prevent unwanted crystallization in these sections. The temperature of the solution inside the growth cell was monitored by a copper-constantan thermocouple which passed through a small hole in the floor of the cell and was both sealed in place by and insulated by epoxy resin. Variations in the steady state temperature recorded by this thermocouple were never more than ± 0.1°C. No temperature rise due to radiation from the illuminating lamp could be detected.

Analytical grade ADP was used, together with water once distilled from glass (specific resistance not less than 0.3 MΩ. cm). All tubing in contact with the solution was of silicone rubber and no grease contacted the solution. After an experiment, the circuit was first flushed with distilled water and then by air until dry. Supersaturations were calculated using the solubility data of Mullin and Amatavivadhana (17). The error $\Delta\sigma$ on these supersaturations was estimated to be not more than 0.003.

Crystals for observation were nucleated inside the cell, with the solution flow temporarily stopped. Various nucleation techniques were used. When nucleation was induced by injecting air into the cell, crystals <u>with</u> surface macro-imperfections were usually produced. When nucleation was induced by surrounding the cell temporarily with ice, crystals <u>without</u> surface macro-imperfections were produced. (Various examples of macro-imperfections are given later.) The reason <u>why</u> the different nucleation techniques should produce crystals with these differences remains unclear.

A convenient size of crystal for observation was about 0.5 - 1.0 mm and a convenient total number of crystals in the cell was about 20. Under these conditions the fall-off of supersaturation across the cell was estimated to be not more than 1%. The crystals grew resting on the floor of the cell, almost invariably with a {100} face uppermost and parallel to the cell floor (the position shown in Fig.1). This face would then be the one which was observed. Layer spreading velocities were measured by tracking selected layer fronts using the filar micrometer eyepiece, or by timing layer fronts between divisions of the eyepiece micrometer scale.

In some tests, crystal faces which previously showed no

macro-imperfections were deliberately damaged by scratching or fracturing while still immersed in solution in the cell. This was done by using a cell with an acrylic cover, in which a 5 mm diameter central hole was drilled, and by passing through the hole an ordinary steel needle with a jagged broken end. When not in use, the hole was closed by a small rubber plug.

RESULTS

When nucleation of crystals in the cell was induced by injecting air into the cell, the crystals produced usually possessed some macro-imperfections. Examples of what is meant by the term "macro-imperfection" follow below. Spontaneously occurring macro-imperfections are such features as cracks and terraces. Other possible macro-imperfections include fractures, scratches, crystallites which have landed on the main crystal, intergrowths of two adjacent crystals, and, on crystals which were removed from solution on a previous occasion, surface residues from the drying-on of incompletely removed mother liquor. For crystals with macro-imperfections, the layer behaviour reported by Davey and Mullin (2) was verified (Fig.3), except that at σ greater than 0.07, the "parallel" layers were sometimes observed to be generated as ellipses at point sources, although they rapidly became effectively parallel as they began to spread from the sources.

On the other hand, when nucleation of crystals in the cell was induced by surrounding the cell temporarily with ice, the crystals produced possessed, after a brief healing period, upper {100} faces which were completely without macro-imperfections: no flaws at all could be seen on these faces when viewing with the X85 magnification which was the maximum used in this work. For such faces, very different layer behaviour was observed (Fig.3). At supersaturations of up to about 0.06, the faces remained completely featureless and mirrorlike, no growth layers of any type being observed on them. When σ was above about 0.06, irregular-shaped elliptical layers were seen on these faces. These layers were concentric and spread from definite point sources, but there were no visible flaws at the sources. An example of these layers, which we shall call Type F layers, because they are relatively fast moving, is shown in Fig. 4a and, for comparison, an example of Type S layers is shown in Fig. 4b. The pale rectilinear outline within the face in Fig. 4a. results from an undesirable internal reflexion and should be ignored. The thickness of the Type F layers, at least as judged by their visibility was much less than that of Type S layers. (In future work we hope to use an interference objective to measure layer thicknesses accurately; if possible, while the crystal is still growing in the cell.)

The spreading velocity of the Type F layers in the $[001]$ direction was in the range $(2 - 5) \times 10^{-5}$ m/s (Fig.5). These

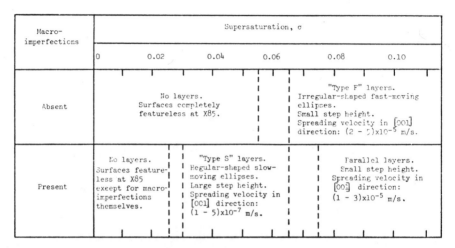

Fig. 3 Table summarizing layer behaviour on {100} faces

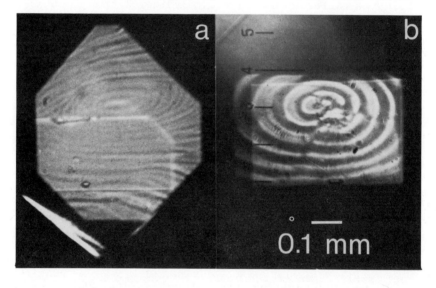

Fig. 4.   (a) Type F layers. T = 23.1°C, σ = 0.092
         (b) Type S layers. T = 24.3°C, σ = 0.048
         Both photographs to same scale

spreading velocities are about a hundred times those of the Type S layers seen on surfaces with macro-imperfections. This is a far greater difference of spreading velocity than can be accounted for by differences in supersaturation. Spreading velocity did not differ significantly from crystal to crystal, nor did it vary significantly with position on a given crystal surface, nor with crystal size. Again, spreading velocity did not vary significantly

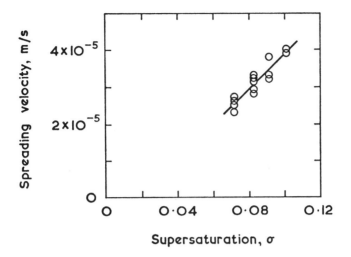

Fig. 5. Spreading velocity of Type F layers in the $[00\bar{1}]$ direction as a function of supersaturation, at $25.0°C$. The line is a least squares fit.

with position of a crystal within the cell. Since the estimated solution velocity varied with position in the cell, from 1 to 18 mm/s, this latter observation indicates that there was no significant mass transfer resistance to growth.

There were usually one or two or three sources of Type F layers on any given {100} surface at a given time, but new sources would rapidly spring up in other positions on the face, while old sources would cease to generate and disappear. No tendency was noticed for sources to occur more on certain parts of a face and less on other parts. Provided supersaturation was maintained, generation of Type F layers on a given surface continued for as long as observations were made (up to five hours in some cases). It is emphasised that no flaws were visible at sources of Type F layers. A few brief tests were made at very high supersaturations ($T \simeq 10°C$, $\sigma \simeq 0.50$) and Type F layers were observed on faces without macro-imperfections even at these very high supersaturations.

In another series of tests, {100} faces which had previously showed no macro-imperfections were deliberately damaged by scratching or fracturing, using a needle. At supersaturations up to about 0.03, the effect was simply local mechanical damage : no layers developed as a result of the damage and the rest of the surface remained featureless. At supersaturations between about 0.03 and about 0.06 thick slow-moving regular-shaped elliptical layers were generated from the damaged region. Examples of this behaviour are given in Figs. 6 and 7. The spreading velocities of these regular-shaped elliptical layers generated at deliberately damaged regions were measured as a function of supersaturation. These spreading velocities were similar to those reported by Davey and Mullin (1) for Type S layers on {100} faces which spontaneously had macro-imperfections (Fig.8), thus helping to confirm that the layers generated at deliberately damaged regions were indeed Type S layers and also that Type S layers are associated with surfaces having macro-imperfections. It also confirms that the lack of layers on the surfaces prior to damaging was not due to erroneously low supersaturations.

When a {100} face which had previously showed no macro-imperfections was damaged at $\sigma > 0.06$, that is, when it was showing Type F layers, the damaged region rapidly healed without generating Type S layers, although it sometimes generated Type F layers. Meanwhile, generation of Type F layers from their original source or sources would continue unchanged.

INTERPRETATION OF RESULTS

In this section, we attempt to explain why the layer behaviour of surfaces without macro-imperfections differs from that of surfaces with macro-imperfections. It appears that the different types of layers seen on surfaces with and without macro-imperfections are generated from sources of different strengths. When a macro-imperfection is present, it provides a strong source, one which may generate visible layers at $\sigma > 0.03$. However, when no macro-imperfection is present, the surface possesses only weak sources, which can only generate visible layers when σ rises above 0.06. The strength of a source can be defined by the number and magnitude (Burgers vector) of the emergent screw dislocations within it.

A dislocation source with low Burgers vector might be expected to lead to thin layers and a source with high Burgers vector to thick layers. This agrees with the observation that thickness, at least as judged by visibility, is much less for Type F layers than for Type S layers.

The character of the layer fronts observed during these experiments is open to debate. Since neither Type S nor Type F layer fronts were polygonized in shape, neither are likely to be true macrosteps. They may on the other hand be bunches of monosteps, the bunching occurring as a result of some periodic disturbance at the source (11, 12). Frank (13), using kinematic wave theory, has predicted that if monostep spreading velocity depends only on monostep spacing, then "shock waves" (discontinuities in monostep

SURFACE PERFECTION

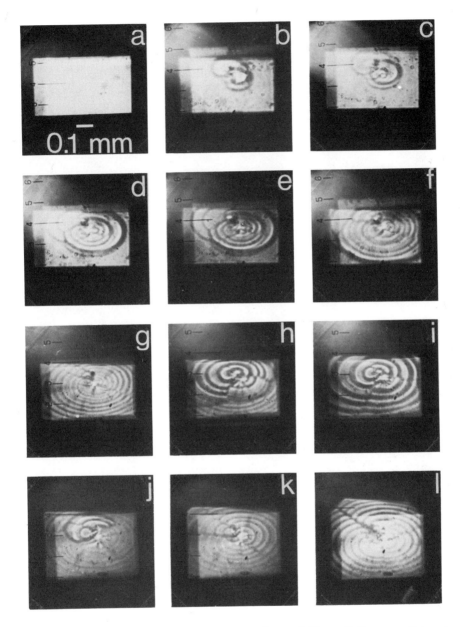

Fig. 6 Photo sequence showing spreading of Type S layers from a deliberate scratch. T = 24.3°C, σ = 0.048. (a) before scratching, (b) 3 min. after scratching (c) 4 min. (d) 6 min. (e) 7 min. (f) 8 min. (g) 21 min. (h) 37 min. (i) 45 min. (j) 65 min. (k) 76 min. (l) 117 min.

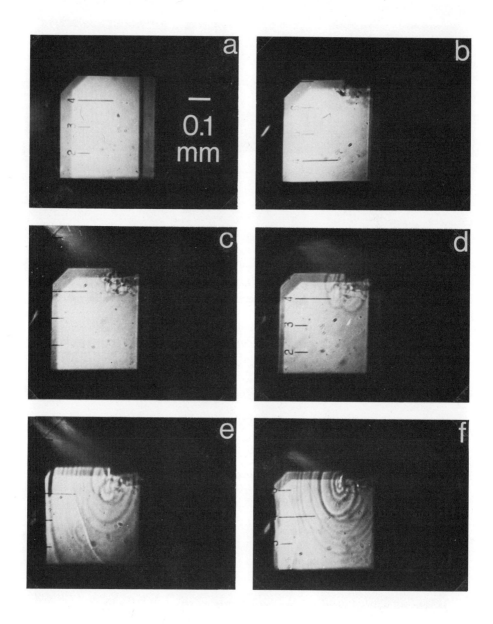

Fig. 7 Photo sequence showing spreading of Type S layers from a deliberate fracture. T = 25.0°C, σ = 0.047. (a) before fracturing, (b) 6 min. after fracturing, (c) 11 min. (d) 24 min. (e) 60 min. (f) 70 min.

spacing) should occur in these bunches as they spread. These shock waves should build up and then degenerate so that the appearance of a given layer front is predicted to change as it spreads. This was not observed during the present experiments: the appearance of a given layer front was found to be unchanged as it spread. Thus the simple kinematic wave theory seems not to apply to the present observations, perhaps because the condition that monostep spreading velocity should depend only on monostep spacing is not fulfilled. In summary, the layer fronts observed in the present work are probably bunches of monosteps, but their spreading is a steady state process, so they do not conform with kinematic wave theory.

The number of elementary dislocations of the same hand emerging on a face, within the critical distance of one another for co-operation (10), should be greater for a macro-imperfection source than for a non-macro-imperfection source and hence the multiple spiral generated by the former will have more branches than that generated by the latter. If bunching of all the branches of these spirals then occurs, the former can produce thicker layers than the latter, which is again in agreement with experimental observations. Since closed loop rather than open spiral layers were always observed during these experiments, the existence of sources in pairs must also be postulated (10).

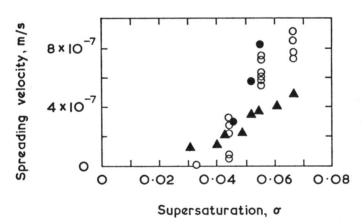

Fig. 8 Comparison of spreading velocities, in [001] direction, of Type S layers on (a) {100} faces which spontaneously have macro-imperfections, and (b) {100} faces which were previously without macro-imperfections but which have been deliberately damaged.

 = Category (a), 24.0°C, "low growth régime"
 = Category (a), 24.0°C, "high growth régime"
 = Category (b), 25.0°C

In the present work, the thin, Type F, layers were found to spread about one hundred times as fast as the thick, Type S, layers. It is physically not unreasonable for layer spreading velocity to vary inversely with layer thickness if we consider layer fronts as bunches of line sinks which must be fed by bulk and/or surface diffusion (18). Some previous workers (14, 19, 20) report experimental evidence that layer spreading velocity does vary inversely with layer thickness, although others (1,21) report spreading velocity to be independent of layer thickness.

REFERENCES

1. Davey, R.J. and Mullin, J.W., J. Cryst. Growth, 26 (1974) 45.
2. Davey, R.J. and Mullin, J.W., J. Cryst. Growth, 29 (1975) 45.
3. Strickland-Constable, R.F., "Kinetics and Mechanism of Crystallization", Academic Press, London, 1968, p.213.
4. Williams, A.P., Phil. Mag. [8], 2 (1957) 635.
5. Krueger, G.C. and Miller, C.W., J. Chem. Phys., 21 (1953) 2018.
6. Fisher, J.C., Fullman, R.L. and Sears, G.W., Acta Met., 2 (1954) 344.
7. Mitchell, J.W. in "Growth and Perfection of Crystals" (Eds. R.H. Doremus et al.), Wiley, New York, 1958, p.386.
8. Thomas, J.M., Endeavour, 29 (1970) 151.
9. Powers, H.E.C., Rev. Sugar Technol., 1 (1969/70) 95.
10. Burton, W.K., Cabrera, N. and Frank, F.C., Phil. Trans. Roy. Soc., A243 (1951) 299.
11. Amelinckx, S., Bontinck, W. and Dekeyser, W., Phil. Mag., [8], 2 (1957) 1264.
12. Dunning, W.J., Jackson, R.W. and Mead, D.G., "Adsorption et Croissance Cristalline", C.N.R.S., Paris, 1965, p.314.
13. Frank, F.C. in "Growth and Perfection of Crystals" (Eds. R.H. Doremus et al.), Wiley, New York, 1958, p.411.
14. Morgan, A.E. and Dunning, W.J., J. Cryst. Growth, 7 (1970) 179.
15. Pilkington, G. and Dunning, W.J., J. Cryst. Growth, 13/14 (1972) 454.
16. Davey, R.J., Ph.D. Thesis, University of London, 1973.
17. Mullin, J.W. and Amatavivadhana, A., J. Appl. Chem., 17 (1967) 151.
18. Chernov, A.A., Soviet Physics Uspekhi, 4 (1961) 116.
19. Dukova, E.D. and Gavrilenko, E.V., Soviet Physics Crystallography, 14 (1970) 736.
20. Solc, Z. and Söhnel, O., Krist. Tech., 8 (1973) 811.
21. Albon, N. and Dunning, W.J. in "Growth and Perfection of Crystals" (Eds. R.H. Doremus et al.), Wiley, New York, 1958, p.446.

GROWTH RATE OF CITRIC ACID MONOHYDRATE CRYSTALS IN A LIQUID
FLUIDIZED BED

C. LAGUERIE AND H. ANGELINO

Institut du Genie Chimique, Laboratoire Associé CNRS 192

Chemin de la Loge, 31078 Toulouse Cedex, France

INTRODUCTION

Continuous classifying crystallizers are being increasingly used in industry. They exhibit many advantages over the MSMPR-type crystallizers: crystal classification especially provides a narrow product size distribution (1,2). Moreover, collision breeding (secondary nucleation) may be reduced because crystal-crystal or crystal-wall collisions in a fluidized bed are much less violent than crystal-agitator collisions. Nevertheless, crystallization studies carried out in fluidized beds are not very numerous (3-9). For citric acid monohydrate, crystal growth studies have been carried out with single crystals (10) and in agitated vessels (11-13). In the present paper, citric acid monohydrate crystal growth rates in a fluidized bed at 25°C are presented.

THEORETICAL ANALYSIS

Crystal growth results from two steps (1-15): diffusional mass transfer of the solute from the bulk solution to the crystal surface followed by integration of the solute into the crystal lattice. The overall process can be described by

$$(dm_c/dt) = k_G \, a_c (C-C^*)^g \tag{1}$$

The growth rate depends on the supersaturation of the solution and its temperature, and generally on the crystal-solution relative velocity and also on the crystal size. It is also influenced by crystal defects and impurities in the solution. In most cases

(1), the value of exponent g (the growth rate order) varies between 1 and 2. When g = 1, equation 1 can be written as

$$(dm_c/dt) = k\, U^m\, L^n\, a_c(C-C^*) \qquad (2)$$

where the liquid velocity and crystal size dependence are shown. For experiments carried out in a fluidized bed with a batch of uniform sized crystals, assumptions of perfect mixing of crystals and of piston flow of liquid lead to the equation

$$(dM/dt) = k\, U^m\, L^n\, A(C-C^*)_{lm} \qquad (3)$$

which is valuable for a short interval of time and over the whole bed.

By defining the volume and surface shape factors, α and β, with respect to the volume equivalent sphere as

$$M = N\, \alpha\, \rho_c\, L^3 \qquad (4a)$$

and

$$A = N\, \beta\, L^2 \qquad (4b)$$

then equation 3 can be written

$$(3\, \alpha\, \rho_c/\beta)(dL/dt) = k\, U^m\, L^n\, (C-C^*)_{lm} \qquad (5)$$

Integrating equation 5 over the duration of an experiment, it can be shown that

$$k\, U^m \left[\frac{1-n}{L_2^{1-n} - L_1^{1-n}} \right] = \frac{3\, \alpha\, \rho_c}{\beta\, t\, (\overline{C-C^*})_{lm}} \qquad (6)$$

in which

$$(\overline{C-C^*})_{lm} = \frac{1}{t} \int_0^t (C-C^*)_{lm}\, dt \qquad (7)$$

The k, m and n parameters can be determined by using a numerical method of identification based on minimisation of the following criterion

$$CR = \sum_{j=1}^{n_{exp}} \left[k_j\, U_j^m\, \frac{1-n}{L_{2j}^{1-n} - L_{1j}^{1-n}} - \frac{3\, \alpha\, \rho_c}{\beta\, t_j\, (\overline{C-C^*})_{lmj}} \right]^2 \qquad (8)$$

This analysis can also be applied to the process of dissolution and it could be generalized for a growth rate order different from 1.

APPARATUS AND PROCEDURE

The apparatus is shown schematically in Fig. 1. The main part is the crystallizer which consists of a Perspex column 9.4 cm diam. with a conic section at its lower part and a symmetrical overflow at its upper part. The 1.5 mm thick perforated plate distributor can move around a diametral horizontal axis so that the magma can be removed from the column base. The solution is circulated by a small stainless steel centrifugal pump. The flow rate can be adjusted by means of a bypass and it is measured by a flow meter. A 35 ℓ stirred tank is maintained at 30°C by a coil in which hot water is circulating and provides a constant concentration of the column inlet solution during an experiment. Supersaturation is achieved by cooling the solution in a coil exchanger. The bed temperature is indicated on a thermometer graduated to 0.1°C and the cooling water flow rate is regulated to maintain this temperature at about 25°C. More details can be found elsewhere (16,17).

A solution of desired concentration was prepared by adding either citric acid or water to the stirred tank. Crystals to be grown were sieved, weighed and counted by taking a sample of three hundred of them. They were introduced at the top of the column. The inlet concentration was continuously measured by means of a conductimeter while the outlet concentration was determined by caustic soda titration, every eight minutes. At the end of the run, the crystals were filtered, washed in ethyl ether, air-dried, weighed and counted.

EXPERIMENTAL RESULTS

A set of 13 experiments was carried out using the same solution velocity ($U = 0.246$ cm/s), for approximate characteristic sizes, by varying the solution supersaturation between 0.02 and 0.14 kg of hydrate per kg of water.

It has been assumed that the size variation during an experiment does not affect the growth rate and our results have been presented as the mass of crystal deposited per hour and per unit surface area:

$$R_g = \Delta M/At \tag{9}$$

The results and experimental conditions are shown in Table 1. R_g

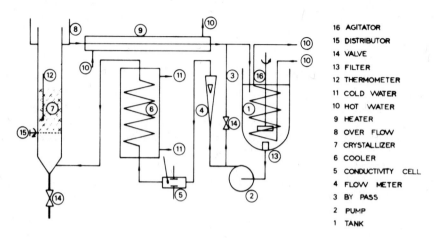

Fig. 1

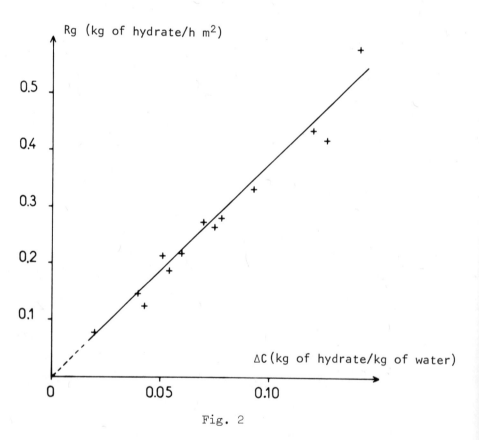

Fig. 2

Table 1. Effect of supersaturation on growth rate (U = 0.246 cm/s)

Initial mass of crystals (kg)	Temp. (°C)	Average size (cm)	Supersat. kg hydrate kg water	R_g kg hydrate/ h m²
0.250	25.00	0.186	0.020	0.077
0.220	24.95	0.146	0.030	0.146
0.250	24.85	0.172	0.033	0.123
0.220	24.65	0.161	0.051	0.213
0.220	25.00	0.148	0.054	0.187
0.250	25.25	0.164	0.060	0.216
0.220	24.80	0.150	0.070	0.273
0.220	24.95	0.150	0.075	0.264
0.220	25.00	0.153	0.078	0.280
0.250	25.05	0.164	0.093	0.332
0.220	24.90	0.160	0.120	0.435
0.220	24.55	0.161	0.126	0.418
0.250	24.80	0.158	0.141	0.579

Table 2. Effects of solution velocity and characteristic size (Initial mass = 0.250 kg)

U (cm/s)	L_1 (cm)	L_2 (cm)	T (°C)	$(C-C^*)_{lm}$	t (h)	kg
0.207	0.138	0.158	25.1	0.048	0.66	3.45
0.402	0.126	0.138	25.4	0.029	0.66	3.69
0.439	0.142	0.164	25.3	0.049	0.66	3.76
0.301	0.162	0.180	25.3	0.042	0.66	3.61
0.326	0.176	0.201	25.0	0.057	0.66	3.69
0.412	0.168	0.182	25.4	0.031	0.66	3.80
0.248	0.198	0.221	25.3	0.054	0.66	3.58
0.152	0.236	0.260	25.1	0.061	0.66	3.32
0.228	0.220	0.244	25.3	0.059	0.66	3.49
0.326	0.261	0.287	25.2	0.061	0.66	3.78

is plotted versus ΔC in Fig. 2 and all the data can be correlated by an average straight line, the equation of which, under these experimental conditions, is

$$R_g = 3.75 \, \Delta C \tag{10}$$

Therefore, the growth rate is first-order with respect to supersaturation.

Altogether 85 experiments were carried out for solution velocities in the range 0.14 - 0.46 cm/s, for characteristic sizes 0.112 - 0.462 cm and for supersaturations 0.02 - 0.14 kg of hydrate per kg of water. Temperatures were almost constant between 24.55°C and 25.55°C.

The identification of k, m and n parameters was achieved by using a numerical computer, and the following equation was deduced:

$$k_g = 5.37 \, U^{0.18} \, L^{0.10} \tag{11}$$

The standard deviation between $k \, U^m (1-n)/(L_2^m - L_1^n)$ and $3 \, \alpha \, \rho_c/\beta \, t(\overline{C-C^*})_{lm}$ is 6.5% and the maximum deviation is smaller than 17% which is the upper limit of experimental errors (16).

Some experimental conditions and average values of k_g are given in Table 2. It can be noted that the difference between the numbers of crystals at the beginning and end of a run is less than 6%. So we have considered that no nucleation occurred.

DISCUSSION

In order to determine whether the diffusional step is the limiting step or not, the results for crystallization were compared with those obtained for the dissolution of citric acid monohydrate crystals in a fluidized bed at 25°C (17). Dissolution led to the following relation:

$$k_D = 4.79 \, U^{0.21} \, L^{0.001} \tag{12}$$

with a standard deviation of 5.9%.

Relations 11 and 12 appear fairly different, but for the experimental conditions used, the ratio k_D/k_g remains in the range 0.95 - 1.08. Thus, it can be considered that the growth rate of citric acid monohydrate crystals in a fluidized bed at 25°C is limited by the diffusional mass transfer step. However, it can be recalled that Nývlt and Vaclavu (12) found that in an agitated

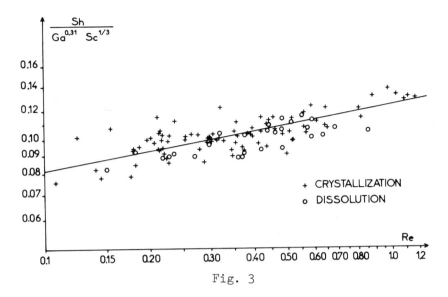

Fig. 3

vessel at 32°C the resistance to the integration of citric acid monohydrate molecules into the crystal lattice cannot be neglected with respect to diffusional transfer. However, the experimental conditions were quite different from ours and it is difficult to conclude.

Our results have been presented according to Damronglerd's correlation (18):

$$Sh = 0.124 \, Re^{0.18} \, Ga^{0.31} \, Sc^{1/3} \qquad (13)$$

Variations of $Sh/Ga^{0.13} \, Sc^{1/3}$ are plotted versus Re in Fig. 3, for all the results of crystallization and dissolution. A good agreement can be noted between the results of both operations.

CONCLUSION

A new method of analysing growth results has been developed. The growth rate of citric acid monohydrate crystals in a fluidized bed at 25°C obeys the relation

$$R_g = 5.37 \, U^{0.18} \, L^{0.10} \, \Delta C$$

The growth process is limited by diffusional mass transfer.

NOMENCLATURE

A	total surface area of crystals in the bed (m^2)
a_c	surface area of a crystal (m^2)
C	solution concentration (kg of hydrate per kg of water)
C^*	solubility of citric acid in water (kg of hydrate per kg of water)
$(C-C^*)_{lm}$	log mean of concentration difference in the bed (kg of hydrate per kg of water)
g	order of the growth rate
j	index (no. of an experiment)
k	dimensional constant (relation 2)
k_D	mass transfer coefficient for dissolution (kg of hydrate/h m^2 (kg of hydrate/kg of water)
k_g	overall coefficient for growth (kg of hydrate/h m^2 (kg of hydrate/kg of water))
L	characteristic size (cm)
L_1	characteristic size at the beginning of the run (cm)
L_2	characteristic size at the end of the run (cm)
M	total mass of crystals in the bed (kg)
m	exponent of U
m_c	mass of a crystal (kg)
N	number of crystals in the bed
n	exponent of L
n_{exp}	number of experiments
R_g	growth rate defined by relation 9 (kg of hydrate/h m^2)
T	temperature (°C)
t	time, duration of a run (h)
α	volume shape factor
β	surface shape factor
ΔC	supersaturation (kg of hydrate/kg of water)
ΔM	mass variation during a run (kg)
ρ_c	density of crystals (kg/cm^3)
Ga	Galileo number
Re	particle Reynolds number
Sc	Schmidt number
Sh	Sherwood number

REFERENCES

1. Mullin, J.W., Crystallization, 2nd edition, Butterworths, London, 1972.
2. Nývlt, J., Industrial Crystallization from Solutions, Butterworths, London, 1971.
3. Phillips, V.R. and Epstein, N., A.I.Ch.E.Jl, 20 (1974) 678.
4. Mullin, J.W. and Gaska, C., Can. Jl. Chem. Eng., 49 (1967) 483.

5. Rumford, F. and Bain, J., Trans. Instn Chem. Engrs, 38 (1960) T10.
6. Mullin, J.W. and Garside, J., Trans. Instn Chem. Engrs, 45 (1967) T291.
7. Rosen, H.N. and Hulburt, H.M., Chem. Eng. Prog. Symp. Series, 67 (110) (1971) 27.
8. Bransom, S.H. and Palmer, A.G.C., Brit. Chem. Engng, 9 (10) (1964) 673.
9. Aoyama, Y. and Toyokura, K., 4th CHISA, 5th Symp. Industrial Crystallization, 1972, I.4.1.
10. Cartier, R., Pindzola, D. and Bruins, P.F., Ind. Engng Chem., 51 (1959) 1409.
11. Mullin, J.W. and Leci, C.L., A.I.Ch.E. Symp. Series, 68 (121) (1972) 8.
12. Nývlt, J. and Vaclavu, V., Coll. Czech. Chem. Commun., 37 (1972) 3664.
13. Nývlt, J., 4th CHISA, 5th Symp. Industrial Crystallization, 1972, I.3.2.
14. Strickland-Constable, R.F., Kinetics and Mechanism of Crystallization, Academic Press, London, 1968.
15. Garside, J., Chem. Eng. Sci., 26 (1971) 1425.
16. Laguerie, C. and Angelino, H., to be published in Chem. Eng. Jl. 1975.
17. Laguerie, C. and Angelino, H., Conf. Engng Foundation, Asilomar, 1975.
18. Damronglerd, S., Doctor Ing. Thesis, Univ. Paul Sabatier, Toulouse, 1973.

THE OCCURRENCE OF GROWTH DISPERSION AND ITS CONSEQUENCES

A.H. JANSE AND E.J. DE JONG

Laboratory for Chemical Equipment, Delft University of Technology, Mekelweg 2, Delft, The Netherlands

INTRODUCTION

In our laboratory some experiments with $K_2Cr_2O_7$ crystals in a fluidized bed have been carried out in order to determine the influence of the supersaturation and the size of the crystals on the growth rate. It was found that the growth rate of the crystals increased with size (see Table 1). Furthermore it appeared that after an experiment a spread in sizes was present. This was observed when the increase in size was large (Fig. 1). As seed crystals with a narrow size range were used it can be concluded that the increase in size and as a consequence the growth rate differs from crystal to crystal. As the supersaturation was kept constant within 5% the only reasonable explanation is that crystals of one size exhibit a spread in growth rates. We have called this phenomenon 'growth dispersion'.

Some years ago White and Wright (1,2) found results for sucrose crystals similar to those shown in Fig. 1. They called the spread in sizes 'size dispersion'. Recently Natal'ina and Treivus (3) published results which show growth dispersion for KDP crystals.

The occurrence of different growth rates under identical conditions (among others to the volume diffusion step) can only be caused by differences in the surface integration step due to differences in dislocation structure. Theoretical and experimental evidence of the occurrence of growth dispersion is given by Bennema (4). In this paper we will consider the occurrence of growth dispersion from a practical point of view and discuss for an assumed growth dispersion model the consequences for:

- the population density distribution of a steady state continuous crystalliser,
- the interpretation and experimental set up of growth experiments.

INTRODUCTION OF THE MODIFIED POPULATION DENSITY

As we take into account the occurrence of growth dispersion, we define a distribution function over length and growth rate $f(L,G)$

$$f(L,G) = \lim_{\Delta L, \Delta G \to 0} \frac{\Delta N}{\Delta L \cdot \Delta G} \qquad (1)$$

where ΔN is the number of crystals per m^3 with size between L and $L = \Delta L$ and growth rate between G and $G + \Delta G$. The function $f(L,G)$ will be called the modified population density. From eq. 1 and the definition of the population density $n(L)$

$$n(L) = \lim_{\Delta L \to 0} \frac{\Delta N}{\Delta L} = \frac{dN}{dL} \qquad (2)$$

it can be concluded that the following relation between $f(L,G)$ and $n(L)$ exists

$$n(L) = \int_0^\infty f(L,G) \, dG \qquad (3)$$

Furthermore the average growth rate of crystals with size L is:

$$\bar{G}(L) = \frac{\int_0^\infty f(L,G) \cdot G \, dG}{\int_0^\infty f(L,G) \, dG} = \frac{\int_0^\infty f(L,G) \cdot G \, dG}{n(L)} \qquad (4)$$

If the modified population density $f(L,G)$ is known the population density $n(L)$ and the average growth rate $\bar{G}(L)$ can be calculated. In the next paragraph it will be shown that $f(L,G)$ can be determined from the number balance. To obtain simple expressions we limit ourselves to the following conditions:
- steady state continuous crystallizer, that means that the distribution and the growth rates do not change with time,
- ideal product removal,
- growth dispersion occurs, but all the crystals growth size independent. In other words, the growth rate of a certain crystal does not change during the life time of that crystal.

In connection with the last point it has to be noted that four cases can be distinguished. Considering a single crystal the growth rate can be size dependent or size independent. Considering crystals of a certain size they can exhibit growth dispersion or not. So we can distinguish the following cases:
1. no growth dispersion ; no size dependent growth rate,
2. no growth dispersion ; size dependent growth rate,

GROWTH DISPERSION

TABLE 1 Growth rates of $K_2Cr_2O_7$ crystals

Initial size	Growth rate	Remarks
165 μm	$G = 0.155\ \Delta C^{1.44}$	$G = \frac{dL}{dt}$ in μm/s
230 μm	$G = 0.18\ \Delta C^{1.45}$	ΔC in g/100g H_2O
387 μm	$G = 0.255\ \Delta C^{1.36}$	$0.4 < \Delta C < 1.7$
460 μm	$G = 0.251\ \Delta C^{1.57}$	Temperature 31°C

TABLE 2 Values of q and r

q μm/s	r	average growth rate at zero size	spread in growth rate at zero size	curve no.
0.1	4	0.05 μm/s	0.0025 $μm^2/s^2$	1
0.4	10	0.05 μm/s	0.00036 $μm^2/s^2$	2
0.01	4	0.005 μm/s	0.000025 $μm^2/s^2$	3
0.04	10	0.005 μm/s	0.0000035 $μm^2/s^2$	4
--	-	0.05 μm/s	0 $μm^2/s^2$	5
--	-	0.005 μm/s	0 $μm^2/s^2$	6

TABLE 3 The ratio of the growth rates $\bar{G}_{exp}/\bar{G}(L_o)$.

r \ y	0	$\frac{1}{4}$	$\frac{1}{2}$	$\frac{3}{4}$	1	1.5	2
10	1	1.03	1.05	1.07	1.08	1.09	1.10
8	1	1.04	1.07	1.09	1.11	1.13	1.15
6	1	1.07	1.13	1.17	1.20	1.24	1.26
5	1	1.12	1.22	1.30	1.35	1.42	1.47
4.5	1	1.42	1.85	2.13	2.30	2.51	2.64

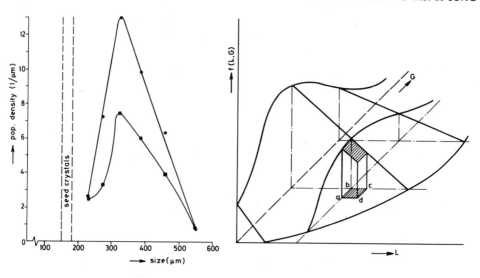

FIG. 1 The spread in sizes after a growth experiment.

FIG. 2 The modified population density.

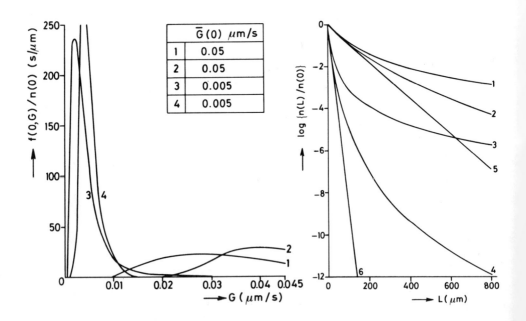

FIG. 3 The influence of r and q on the growth dispersion at zero size.

FIG. 4 The calculated population density curves.

3. growth dispersion   ; no size dependent growth rate,
4. growth dispersion   ; size dependent growth rate.

Cases 1 and 2 are known from the literature (5,6,7). In case 2 the size dependences of the growth rates of different crystals have to be the same. Otherwise growth dispersion would result. In case 4 the size dependences can be different. Consequently case 4 is difficult to treat mathematically. In order to show the importance of growth dispersion we limit ourselves to case 3.

THE NUMBER BALANCE

The number balance over a $\Delta L \cdot \Delta G$ interval (Fig. 2) can be treated as follows: As we assume size independent growth rate the only way to grow into the interval is to pass a b and the only way to grow out is to pass c d. Growing in and out via a d and b c does not occur. Therefore the number balance becomes:

$$f(L,G) \cdot \Delta G \cdot G \cdot V = f(L + \Delta L, G) \cdot \Delta G \cdot G \cdot V + \phi_v \cdot f(\bar{L},\bar{G}) \cdot \Delta G \cdot \Delta L \quad (5a)$$

The first term represents the number of crystals growing into the interval, the second the number of crystals growing out of the interval and the third the number of crystals removed by the product stream. For ΔL goes to zero eq. 5a can be written as:

$$\frac{d\, f(L,G)}{dL} + \frac{f(L,G)}{G\,\tau_1} = 0 \quad (5b)$$

Integration between 0 and L gives:

$$f(L,G) = f(0,G) \cdot \exp(-L/G\,\tau_1) \quad (6a)$$

with $f(0,G)$ the modified population density at zero size. This equation can be compared with the equation of the population density $n(L)$ in the case that growth dispersion does not occur (case 1):

$$n(L) = n(0) \exp(-L/G\,\tau_1) \quad (6b)$$

In order to obtain $f(L,G)$, $n(L)$, and $\bar{G}(L)$ we assume that the modified population density at zero size $f(0,G)$ is known and that it can be described by:

$$f(0,G) = n(0) \cdot \frac{q^{r-1}}{\Gamma(r-1)} \cdot G^{-r} \cdot \exp(-\frac{q}{G}) \quad (7)$$

where q and r are parameters which determine the shape of the $f(0,G)$ distribution (Fig. 3) and $n(0)$ is the nuclei population density. This model satisfies the following conditions:
- $f(0,0) = 0$
- $\lim_{G \to \infty} f(0,G) = 0$
- in combination with the number balance equation analytical solutions for eqs. 3 and 4 can be obtained.

With eq. 7 f(L,G), n(L) and $\bar{G}(L)$ can be calculated from eqs. 6a, 3, and 4:

$$f(L,G) = n(L) \cdot \frac{q}{\Gamma(r-1)} \cdot \frac{q\tau_1 + L}{q\tau_1} \cdot G^{-r} \cdot \exp\left(-\frac{q\tau_1 + L}{q\tau_1}\right) \quad (8)$$

$$n(L) = n(0) \cdot \left(1 + \frac{L}{q\tau_1}\right)^{1-r} \quad (9)$$

$$\bar{G}(L) = \frac{q\tau_1 + L}{(r-2)\cdot\tau_1} = \frac{q}{(r-2)} \cdot \left(1 + \frac{L}{q\tau_1}\right) \quad (10)$$

Eqs. 7,8,9, and 10 have been plotted in Figs. 3 to 5 for τ = 1000 s and for different values of the parameters q and r (see Table 2). On the basis of these figures the results will be discussed.

DISCUSSION OF THE RESULTS

The different values of the parameters r and q are summarized in Table 2. The first four cases representing
- high average growth rate ; high spread in growth rates
- high average growth rate ; small spread in growth rates
- small average growth rate ; high spread in growth rates
- small average growth rate ; small spread in growth rates

can be easily distinguished from Fig. 3. In this figure the ratio of the modified population density at zero size f(0,G) and the nuclei population density n(0) has been plotted against the growth rate.

If we look at the population density distributions shown in Fig. 4, we note the effect of r and q on the shape of the distribution. The higher the average growth rate and the higher the spread in growth rates at zero size, the smaller the decrease in population density with size. Curves 5 and 6 represent the distributions with growth rates of 0.05 and 0.005 µm/s respectively, but without growth dispersion. The values of the growth rates are euqal to the average growth rates of the cases 1 - 4 (see Table 2). From eqs. 8 to 10 it can be seen that f(L,G), n(L), and $\bar{G}(L)$ are also determined by the liquid residence time τ_1. In general a change in τ_1 results in new values for the supersaturation and the growth rates. This means that the value of the parameter q which determine the average growth rate (see eq. 10) changes. It is likely that the product $q\tau_1$ does not remain constant. Therefore it can be concluded from eq. 10 that τ_1 changes the size dependence of the average growth rates. This means that in growth experiments seed crystals from different distributions (different τ_1) can give different growth rates.

GROWTH DISPERSION

In this paper we have taken into account the occurrence of growth dispersion and have calculated the distributions 1 - 4 (Fig. 4). If these curved distributions were measured distributions, and we did not know that growth dispersion had occurred, we should have concluded that the growth rate increases with size. It can be proved that the same population density distribution n(L) (eq. 9) is obtained under the following condition: Growth dispersion does not occur, but the growth rate G(L) is size dependent (case 2), with the same size dependence as the average growth rate $\bar{G}(L)$ (see eq. 10):

$$\bar{G}(L) = \frac{q}{r-2} \cdot \left(1 + \frac{L}{q\tau_1}\right) \tag{11}$$

As the steady state distribution under the above mentioned condition is identical to eq. 9, it can be concluded that:
- from curved distribution the size dependent growth rate $\bar{G}(L)$ in case 2 (no growth dispersion, size dependent growth rate) or the average growth rate $\bar{G}(L)$ in case 3 (growth dispersion, no size dependent growth rate) can be calculated. It can be proved that this conclusion holds for each f(0,G) growth dispersion model.
- from such a curved distribution it cannot be concluded whether growth dispersion occurs or size dependent growth rate.

The occurrence of growth dispersion results in a size dependent average growth rate (eq. 10 and Fig. 5a). In Fig. 5b the same results of the calculations are plotted on double logarithmic paper. It appears that above 200 µm a power law dependence on size is found, with exponents ranging from 0.55 to 1.0. In the literature (8,9,10) a lot of data show size dependent (average?) growth rates, with size dependencies ranging grom $L^{0.63}$ to $L^{1.1}$. As the exponents reported in the literature are in the same range as our calculated exponents, the size dependent growth rates from the literature can be explained by the occurence of the growth dispersion. So size dependent average growth rates and growth dispersion are linked.

In this section it has been shown that growth dispersion has a tremendous influence on the crystal size distribution and that it causes a size dependent average growth rate. Therefore it is important to know whether growth dispersion occurs or not. If the sieve analysis of grown crystals shows size dispersion it can be concluded that growth dispersion (case 3 or 4) occurred. To determine whether the growth rate is size independent (case 3) or size dependent (case 4) single crystal growth experiments can be used.

CONSEQUENCES OF GROWTH DISPERSION FOR GROWTH EXPERIMENTS

The consequences of the occurrence of growth dispersion for growth experiments will now be discussed. We consider the growth of a fraction of crystals with size between L_o and $L_o + \Delta L$ with $\Delta L \ll L_o$.

As growth dispersion occurs, the mass of the crystals with a high growth rate increases strongly. The mass of the crystals with a small growth rate increases only a small amount. As crystals with a high growth rate contribute relatively more to the total mass, the total mass increases more than would be expected on basis of the average growth rate. If this is not taken into consideration, a too large value of the growth rate is found as will be demonstrated now.

The mass of the crystals which grow under constant supersaturation for a given time t can be calculated if the spread in growth rates $f(L_o,G)/n(L_o)$ is known:

$$\frac{M(t)}{M(0)} = \frac{\int_0^{\infty} f(L_o,G) \cdot (L_o + Gt)^3 \, dG}{\int_0^{\infty} f(L_o,G) \cdot L_o^3 \, dG} \tag{12a}$$

In combination with eqs. 8 and 10 we obtain:

$$\frac{M(t)}{M(0)} = 1 + 3y + 3y^2 \cdot \frac{r-2}{r-3} + y^3 \cdot \frac{(r-2)^2}{(r-3)(r-4)} \tag{12b}$$

with $y = \bar{G}(L_o) \cdot t/L_o$. Normally the experimentally determined growth rate (averaged over all the crystals) is calcucated from:

$$\bar{G}_{exp} = \frac{L_o}{t} \left\{ \sqrt[3]{\frac{M(t)}{M(0)}} - 1 \right\} \tag{13}$$

From eqs. 12 and 13 the ratio $\bar{G}_{exp}/\bar{G}(L_o)$ can be calculated. The results of such a calculation are given in Table 3 for different values of r (characterizing the spread in growth rates) and y (the ratio of the average increase in size and the initial size). It follows from the calculation that for a large value of y and a large spread in growth rates the ratio of the growth rates is much larger than unity. There is a strong increase in the ratio of the growth rates for r < 5. This is caused by the fact that according to the expression of the growth dispersion model (eq. 7) for r < 5 a large number of crystals with a very high growth rate is present. From a physical point of view this is impossible because the growth rate is limited by at least the volume diffusion step. Not withstanding the limitation of the growth dispersion model it can be concluded that in order to obtain accurate growth rates the increase in size has to be as low as possible. On the other hand, by letting the crystals grow for a long time the spread in sizes can be determined (see Fig. 1). This gives an indication to what extend growth dispersion occurs.

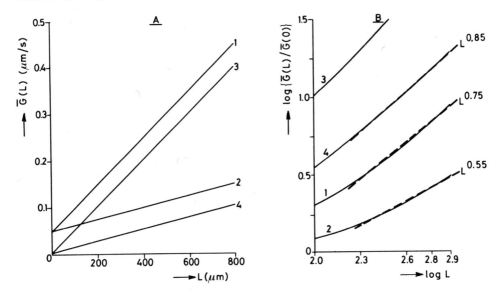

FIG. 5a, b The average growth rate calculated as a function of size.

To determine the total mass of the grown crystals we have used eq. 8 for the spread in growth rates. This is correct if we take a sieve fraction of crystals from a distribution. If we take a number of crystals from a distribution and let them grow in order to obtain larger seed crystals, eqs. 8 and 12 do not hold. If the grown crystals are sieved in several size ranges it means that the crystals with the highest growth rate are present at the larger size. Therefore the measured growth rate of such seed crystals increases with size as a result of the seed crystal preparation procedure. From the above mentioned it is clear that growth dispersion has to be taken into account in the interpretation and experimental set up of growth experiments.

CONCLUSION

It has been shown that growth dispersion can have a tremendous effect on the crystal size distribution. As a result of the occurrence of growth dispersion the average growth rate increases with crystal size. Therefore growth dispersion provides an explanation for the often reported size dependent growth rates. Because the important consequences for the crystal size distribution and also for the growth experiments in the future more attention has to be paid to the growth dispersion phenomenon.

ACKNOWLEDGEMENT

The authors wish to express their appreciation to Prof. Dr J.W. Sieben of the Department of Mathematics (Delft University of Technology) for his help in the mathematical description of growth dispersion. And to Dr P. Bennema of the Laboratory of Physical Chemistry (Delft University of Technology) for fruitful discussions during the preparation of this paper.

NOMENCLATURE

$f(L,G)$	modified population density	$m^{-1}.sm^{-1}.m^{-3}$
$G(L)$	size dependent growth rate	m/s
$\bar{G}(L)$	average growth rate	m/s
L	crystal size	m
$M(t)$	mass of the crystals	kg
$n(L)$	population density	$m^{-1}.m^{-3}$
q	parameter (see eq. 7)	m/s
r	parameter (see eq. 7)	-
V	volume of the crystallizer	m^3
y	$\bar{G}(L_o).t/L_o$ (eq. 12)	-
ϕ_v	discharge flow rate	m^3/s
τ_l	average residence time of the liquid	s
$\Gamma(r-1)$	gamma function	-

REFERENCES

1. Wright, P.G. and White, E.T.; Proc. Queensland Soc. Sugar Cane Technology, 36th Conf.; p. 299 (1969).
2. White, E.T. and Wright, P.G.; Chem. Eng. Progr. Sympos. Ser., (110) 67 (1970) 81.
3. Natal'ina, L.N. and Treivus, E.B.; Sov. Phys. Chrystallography, 19 (3), Nov.-Dec. (1974).
4. Bennema, P.; Paper B1; This Symposium.
5. Larson, M.A. and Randolph, A.D.; Chem. Eng. Progr. Sympos. Ser. (95) 65 (1969) 1.
6. Abegg, C.F.; Steevens, L.D., and Larson, M.A.; A.I.Ch.E.J., 14 (1968) 118.
7. Canning, T.F. and Randolph, A.D.; A.I.Ch.E.J., 13 (1967) 5.
8. White, E.T., Bendig, L.L., and Larson, M.A.; Paper presented at the 67th Annual A.I.Ch.E. Meeting, Washington, December 1974.
9. Mullin, J.W. and Gaska, C.; Canad. J. Chem. Eng. 47 (1969) 483.
10. McCabe, W.L. and Stevens, R.P.; Che. Eng. Progr. (4) 47 (1951) 168.

KINETICS OF TiO$_2$ PRECIPITATION

O. SÖHNEL AND J. MARAČEK

Research Institute of Inorganic Chemistry

400 60 Usti nad Labem, Czechoslovakia

INTRODUCTION

The quality of TiO$_2$ pigment prepared from ilmenite is to a great extent determined by the properties of hydrated TiO$_2$ gel obtained by the hydrolysis of titanylsulphate solution. It is not known which property of the gel plays the decisive role, but certainly chemical composition, structure and particle size are the factors on which the quality of titanium white is strongly dependent. The above mentioned properties are determined by the method of gel preparation, i.e. by the conditions during hydrolysis.

Despite the great attention which has been given in the literature to the study of titanium solution hydrolysis, both from the viewpoint of precipitation kinetics (1-4) and the structure of the solution (5,6) and the gel (1,2,7-9), no unambiguous result has been found so far. We have studied the kinetics of TiO$_2$ precipitation from titanylsulphate solutions during thermal seeded hydrolysis and tried to characterize the resulting TiO$_2$ gel.

EXPERIMENTAL

A titanylsulphate solution of approximate composition(g/ℓ): 220 TiO$_2$, 4 Ti^{3+}, 65 Fe^{2+} and 550 H$_2$SO$_4$, nucleated by TiO$_2$ hydrosol was heated to boiling (∼ 110°C). The TiO$_2$ hydrosol was prepared by neutralizing dilute titanium solution with NaOH to pH = 3.5. Samples taken at regular intervals after the solution had reached its boiling point were analysed for Ti content in the liquid phase. The degree of hydrolysis α was calculated from these data according to the equation

$$\alpha = (c_o - c)/c_o$$

The whole hydrolysis consists of 3 periods: (i) the boiling period, when the seeded solution is maintained at the boiling point; (ii) the diluting period, when hot water is continuously added to the boiling solution, and (iii) a period in which the diluted suspension is boiled until $\alpha \sim 0.98$.

A typical kinetic curve of hydrolysis is shown in Fig.1. Time is measured from the moment when the solution reached its boiling point. The number of particles do not change during the hydrolysis so that we can use 'chronomal analysis' suggested by Nielsen (10) for determination of the solid phase growth mechanism.

The experimental α values were converted into 'kinetic integral values' using the three expressions

$$t = K_p \int_0^\alpha x^{-2/3} (1-x)^{-1} dx = K_p I_p$$

$$t = K_D \int_0^\alpha x^{-1/3} (1-x)^{-1} dx = K_D I_D$$

$$t = K_m \int_{\alpha_o}^\alpha x^{-4/3} (1-x)^{-m} dx = K_m I_m^+$$

Integral values obtained in such a way are plotted versus time in Fig.2. The linear plot gives only the polynuclear integral with $p = 1$ in the region $0.1 < \alpha < 0.7$ (the boiling period). It is sometimes difficult to decide whether I_D or I_1 fits the linear dependence better. But in no case do I_{2-4} or I_{1-4}^+ give a linear plot.

It is clear from Fig.2 that the linear function $I_1 - t$ does not go through the origin in our time scale. It proves that the precipitation of TiO_2 had started before the boiling point was reached. The real beginning of precipitation can be determined from the condition $I_1 = 0$ for $\tau = 0$ provided that the mechanism of the reaction is the same.

Based on the given facts and also some others inferred from the literature (11), we conclude that the kinetics of TiO_2 precipitation from titanylsulphate solution during thermal seeded hydrolysis is controlled by the polynuclear growth of seeds with the kinetic order of unity. Because the expression

$$p = (m + 2)/3$$

KINETICS OF TiO₂ PRECIPITATION

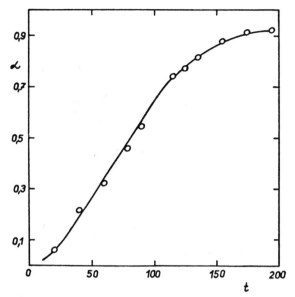

Fig. 1 The kinetic curve of hydrolysis

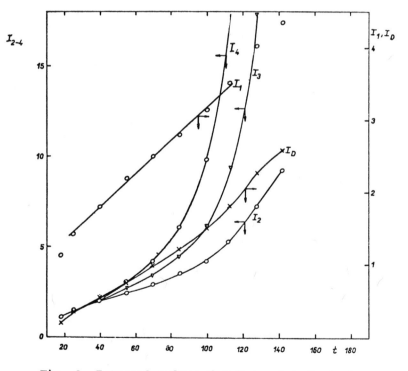

Fig. 2 Integral values for thermal hydrolysis

holds according to the nucleation theory, the critical surface nucleus is formed by one molecule in the present case.

A model of solid phase formation during titanium solution hydrolysis has been suggested by Jerman (6) based on the chemical mechanism of hydrolysis. According to this model, the polynuclear complexes of titanium with hydroxyl bridges, which is how titanium exists in acid solution, condense in contact with the nucleus surface. The chemical bonds with the substrate are produced in the course of condensation. But this can be achieved only if the complex molecule has a suitable orientation with respect to the seed surface. When the first molecule is bound to the surface nucleus, attachment of a further molecule is considerably easier and the growth of a new layer can proceed. Because more complex molecules are brought to a suitable orientation in each very short interval of time, the growth of a new layer starts from several points on the crystal surface. The growth is then polynuclear. Exactly the same mechanism of TiO_2 particle growth was found in our kinetic study.

TiO_2 GEL

The structure of the resulting TiO_2 gel was also studied. The gel consists of ellipsoidal particles approximately 2 μm in size (Fig.3). These so-called tertiary coagulates are rather loosely bound aggregates of secondary coagulates of size 5000-8000 Å. These can be easily obtained by disintegration of the hydrolysate in pyrophosphate solution using a turbine mixer.

The secondary coagulates (Fig.4), which cannot be broken into smaller pieces by mechanical forces, contain crystalline TiO_2 seeds 90 Å (determined by X-ray analysis). The formation of secondary coagulates is finished at the end of the first period of thermal hydrolysis. These particles aggregate into tertiary coagulates during the diluting period.

The filtration properties of the resulting gel may give us information about tertiary coagulates, or secondary ones, if disintigrated gel is used. The specific filtration resistance was measured in the laboratory filtration equipment shown in Fig.5. The hydrolysate maintained at 60° was filtered under constant pressure. The volume of filtrate as a function of time was measured. The general filtration equation

$$dV/dt = PA/\mu(rVA^{-1} + R)$$

can be integrated if P is constant and rewritten as

$$t/V = \mu rV/2PA^2 + \mu R/PA$$

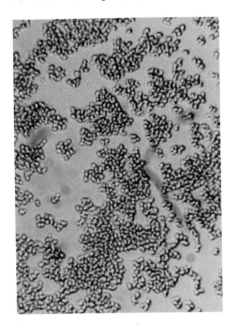
Fig. 3 TiO$_2$ gel: tertiary coagulates

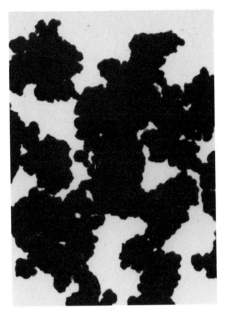
Fig. 4 TiO$_2$ gel: secondary coagulates

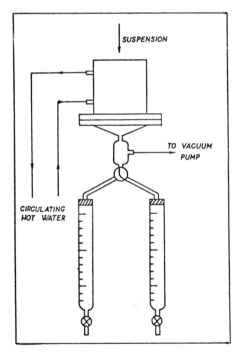
Fig. 5 Laboratory filtration apparatus

Thus a plot t/V versus V is a straight line with the slope $\mu r/2PA^2$, from which the specific resistance may be determined. According to Brownell and Katz (12)

$$r = 32 \, F_f/F_{Re} \, d^2$$

This equation enables d to be estimated if the correction factors F_f and F_{Re} are known. These can be evaluated from Brownell's plots, if the porosity of the filter cake and sphericity of the particles are estimated from the physical properties of TiO_2.

Comparison of the particle size (μm) calculated from filtration tests ($\psi = 0.7$, $\varepsilon = 0.643$ for tertiary and 0.729 for secondary coagulates respectively) and those determined from microscopic measurements is:

	Filtration	Microscopy
Tertiary:	2.4	~2
Secondary:	0.5	0.5-0.8

CONCLUSION

The kinetics of TiO_2 precipitation from titanylsulphate solution during its thermal seeded hydrolysis is controlled by polynuclear growth of the seeds with a kinetic order $p = 1$. The structure of the resulting TiO_2 gel was revealed by means of microscopy and filtration tests.

NOMENCLATURE

A	filtration area
c	actual concentration of solution
c_o	initial concentration of solution
d	particle diameter
F_f	friction factor depending on sphericity of the particles and porosity of the filter cake
F_{Re}	Reynolds number factor
I_p, I_D, I_m^+	kinetic integrals
K_p, K_D, K_m	constants
m	number of molecules forming a critical surface nucleus
p	kinetic order of polynuclear growth
P	pressure
r	specific filtration resistance
R	resistance of filter medium
t	time

V	total filtrate
X	volume of cake per unit volume of filtrate
α	degree of hydrolysis
ψ	sphericity of particles
ε	porosity of filter cake
μ	viscosity of filtrate

REFERENCES

1. Becker, H., Klein, E. and Rechman, H., Farbe Lack, 70 (1964) 779.
2. Dolmatov, Yu.D., Bobyrenko, Yu.Ya. and Sheinkman, A.I., Zh. Vses.Khim.Obshch., 11 (1966) 351.
3. Bobyrenko, Yu.Ya and Dolmatov, Yu.D., Zh.Prikl.Khim., 44 (1971) 996.
4. Bobyrenko, Yu.Ya., Dolmatov, Yu.A. and Sheinkman, A.I., Zh. Prikl.Khim, 40 (1967) 1003.
5. Jerman, Z., Blechta, V. and Blechta, Z., Chem.Prům., 16 (1966) 410.
6. Jerman, Z., Coll.Czech,Chem.Commun., 31 (1966) 3280.
7. de Rohden, C., Chem.et Ind., 75 (1956) 287.
8. Bowman, A., J.Oil.Colour Chem.Assoc., 35 (1952) 314.
9. Latty, J.E., J.Appl.Chem., 8, (1958) 96.
10. Nielsen, A.E.,'Kinetics of Precipitation', p.52, Pergamon Press, Oxford, (1964).
11. Söhnel, O., Coll.Czech.Chem.Commun., 40 (1975) 2560.
12. Brown, G.G.,'Unit Operations', p.210, John Wiley, New York, (1950).

INVESTIGATIONS ON THE OSTWALD RIPENING OF WELL SOLUBLE SALTS BY MEANS OF RADIOACTIVE ISOTOPES

ACHIM WINZER AND HANS-HEINZ EMONS

Technical University 'Carl Schorlemmer'

Leuna-Merseburg, 42 Merseburg, DDR

INTRODUCTION.

A new crystalline phase may be formed by the interaction of nucleation, crystal growth, Ostwald ripening, aggregation and re-crystallization. The individual steps may overlap, and although mathematical relations have been derived for the individual steps, an exact combination, e.g. of the nucleation kinetics with the kinetics of crystal growth, has not yet been realised. One of the main difficulties is the different crystal size distributions from which transformations during Ostwald ripening result.

We have investigated in detail, using radioactive isotopes, the mass transfer processes taking place during Ostwald ripening using the salts sodium nitrate, sodium chloride and potassium chloride. Here the idea also was to re-dissolve small crystals (< 63 μm), which appear during crystallization and create dust in technical products.

A system consisting of crystals of different size is not stable thermodynamically. By the reduction of the surface area during Ostwald ripening the free energy approaches a minimum and the final state would be one large monocrystal. During Ostwald ripening the smaller crystals dissolve, because of their thermodynamic potential, and the corresponding ions are precipitated on the larger crystals.

The theory of Ostwald ripening was developed by Lifshitz and Slyozov (1) and independently by Wagner (2). Recently Kahlweit et al.(3) have studied this problem. For the case of diffusion-controlled mass transfer the theory predicts that the increase of $\bar{r}^3$ gradually assumes a constant value. It was found, however, that

Ostwald ripening of crystalline materials proceeds more slowly than predicted. Another reason may be that the kinetics of crystal growth in solution is often not diffusion-controlled, but proceeds according to a second-order reaction. For mass transfer processes of that kind Hanitzsch and Kahlweit (4) have derived relations for Ostwald ripening and by comparing the equations for diffusion-controlled mass transfer with those for reaction-controlled mass transfer the velocity constant can be calculated.

EXPERIMENTAL

Using tracers we have studied mass transfer during Ostwald ripening between small radioactively marked sodium nitrate crystals suspended in their saturated solution together with large crystals of the same salt.

For a constant size of the radioactively marked $NaNO_3$ crystals we studied the kinetics of mass transfer as a function of the size of the inactive $NaNO_3$ crystals. The results are shown in Fig.1. The dissolution rate of the small, marked $^{22}NaNO_3$ crystals in saturated $NaNO_3$ solution is increased at the beginning of the experiment when the size of the inactive crystals is increased. After an ageing time of ~ 40 min there is a direct proportionality between the mass having gone into solution, the small $^{22}NaNO_3$ crystals and the time. At constant suspension the slope of the curves is nearly independent of the size of the non-radioactive large crystals.

In these experiments $NaNO_3$ crystals which had already been stressed mechanically (by grinding and sieving) were used so that the adequate forms of equilibrium can also form during the Ostwald ripening, whereby crystal fractions preferably go into solution. For the dissolution of the small $^{22}NaNO_3$ crystals in saturated solution the constants given in Table 1 were obtained.

We also radiochemically measured the second step of the Ostwald ripening, i.e. the growth of the large $NaNO_3$ crystals. For this purpose the saturated $NaNO_3$ solution was marked with $^{22}Na^+$ isotope and at a relatively high suspension density the mass transfer from the solution to the solid phase was determined using 50 g $NaNO_3$ (63 - 80 µm) and 50 g $NaNO_3$ (160 - 200 µm) in 200 ml by measuring the radioactivity of the solution over a period of time.

The results for the slow partial step of Ostwald ripening, the growth of the large $NaNO_3$ crystals at 25°C were

Crystal growth rate: 4.66×10^{-9} g/cm^2 sec
Mass transfer number: 0.69×10^{-8} cm/sec
Time constant: 1.8×10^{-14} cm^3/sec

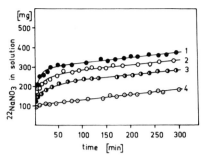

Fig. 1 Influence of the size of large inactive NaNO₃ crystals on the dissolution of small ^{22}NaNO₃ crystals at 25°C.

Curve 1: 0.5 g ^{22}NaNO₃ (< 50μm) + 9.5 g NaNO₃(160-200μm)
2: 0.5 g ^{22}NaNO₃ (< 50μm) + 9.5 g NaNO₃(125-160μm)
3: 0.5 g ^{22}NaNO₃ (< 50μm) + 9.5 g NaNO₃(80 -100μm)
4: 0.5 g ^{22}NaNO₃ (< 50μm) + 9.5 g NaNO₃(<   50μm)

^{22}NaNO₃ crystal fraction		NaNO₃ crystal fraction		Dissolution rate		Mass transfer number	Time constant per NaNO₃ particle	
mass	size	mass	size	10^{-3}	10^{-9}	10^{-8}	10^{-14}	10^{-13}
(g)	(μm)	(g)	(μm)	(mg/sec)	(g/cm²sec)	(cm/sec)	(cm³/sec)	(g/sec)
0.5	< 50	9.5	160-200	4.54	7.06	1.06	2.9	2.73
0.5	< 50	9.5	125-160	4.3	6.67	1.0	2.7	2.59
0.5	< 50	9.5	80-100	4.3	6.72	1.01	2.67	2.61
0.5	< 50	9.5	< 50	4.5	7.09	1.06	2.89	2.71

Table 1. Constants for the dissolution kinetics of small NaNO₃ crystals during Ostwald ripening at 25°C for the slow partial step.

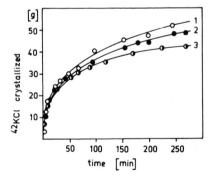

Fig. 2 Influence of stirring speed on mass transfer from solution to crystals during the Ostwald ripening of KCl at 25°C: 50 g KCl(80-100μm) + 50 g KCl(160-200μm) in 200 ml saturated solution. 1: 2700 rev/min., 2: 1600 rev/min., 3: 720 rev/min.

The time constant for the growth of $NaNO_3$ crystals during Ostwald ripening found experimentally by Hohmann and Kahlweit by means of cuvette microscopy (4) was 1.8×10^{-14} cm^3/sec. The mass transfer numbers and time constants obtained by us are generally higher, which is most likely due to the fact that we used mechanically stressed crystals.

The mass transfer rate of isothermal Ostwald ripening of well soluble salts depends mainly on the following parameters:
Temperature;
Suspension density;
Stirring rate;
Crystal size distribution;
Presence of surface-active materials;
Presence of specifically acting foreign ions (admixtures).
The influence of these parameters will be discussed using the Ostwald ripening of KCl as an example.

Studying the isothermal Ostwald ripening rate of NaCl, $NaNO_3$ and KCl we have found that the mass transfer increases with increasing temperature. With $NaNO_3$ an activation energy of 3.6 kcal/mol was measured for mass transfer from solution to the solid phase during Ostwald ripening.

Mass transfer from solution to the solid phase during Ostwald ripening also increases with increasing suspension density. We also found that mass transfer from the smaller crystals to the solution and from the solution to the bigger crystals increases with increasing stirring rate when the solution is radioactively marked (Fig.2). This effect can be observed particularly under long ageing times (stirring rate 620 - 2700 rev/min).

The influence of the crystal size distribution on the growth of the large crystals as well as on the dissolution of the small crystals was investigated. For the same size of large crystals (50 g KCl, 160 - 200 µm) the size of the small crystals was varied (<50 µm, 80 - 100 µm, 100 - 125 µm; 50 g each) and after radioactive marking the solution the mass transfer from solution to crystals was measured.

As shown in Fig.3, the mass transfer for the first partial process increases as the size of the smallest fraction decreases.

The investigations of the influence of crystal size distribution on the dissolution kinetics were carried out at a suspension density of 500 g KCl/ℓ, where 10% consisted of radioactively marked KCl crystals of the fraction <63 µm and the sizes of the inactive big crystals was 100 - 125 µm, 125-160 µm and 160 - 200 µm, respectively.

As shown in Fig.4, the dissolution rate of the small crystals increases with an increase in the size of the inactive fraction.

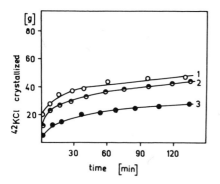

Fig. 3 Influence of crystal size distribution on mass transfer during the Ostwald ripening of KCl at 25°C. 1: 50 g KCl(160-200μm) + 50 g KCl (<50μm), 2: 50 g KCl(160-200μm) + 50 g KCl(63-80μm), 3: 50 g KCl(160-200μm) + 50 g KCl(100-125μm).

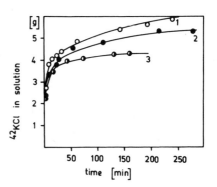

Fig. 4 Dissolution of small ^{42}KCl crystals at high suspension density at 25°C. 1: 10 g ^{42}KCl(<63μm) + 50 g KCl(160-200μm), 2: 10 g ^{42}KCl (<63μm) + 50 g KCl(125-160μm), 3: 10 g ^{42}KCl(<63μm) + 50 g KCl(100-125μm).

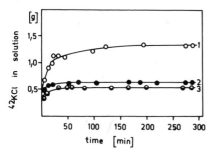

Fig. 5 Influence of alkylammonium hydrochloride on the dissolution of small ^{42}KCl crystals - 2 g ^{42}KCl (<63μm) + 18 g KCl(160-200μm) in 200 ml. 1: pure solution, 2: 1 mg alkylammonium hydrochloride, 3: 10 mg.

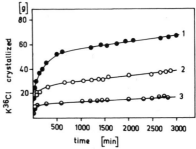

Fig. 6 Influence of alkylammonium hydrochoride on mass transfer from solution to crystals during the Ostwald ripening of KCl at 25°C - 50 g KCl(80-100μm) + 50 g KCl (160-200μm) in 200 ml. 1: pure solution, 2: 1 mg alkylammonium hydrochloride, 3: 30 mg.

In the isothermal Ostwald ripening of KCl and NaCl the influence of KCl-specific (alkylammonium hydrochloride, c_{12} - c_{18}) and NaCl-specific (stearylmorpholin) on the mass transfer was investigated. As seen in Fig.5, the presence of alkylammonium hydrochloride in concentrations of a few ppm exercises a retarding influence on the dissolution rate of the small crystals. The speed of the phase interface processes during the transition from the solution to the crystal is also considerably retarded in the early stages (Fig.6). This can be explained by adsorption of the surfaceactive materials at growth centres on the crystals. Foreign ions, which may be present in minute concentrations in the saturated salt solution, can also have a retarding influence on the rate of Ostwald ripening. Fig.7 shows that mass transfer during the isothermal Ostwald ripening of KCl in the presence of 20 mg $PbCl_2$ is completely blocked. The small KCl crystals cease to dissolve under these conditions. Mass transfer from the saturated solution to the crystal surface is also retarded or completely stopped at higher concentrations ($\simeq$ 20 mg in 200 ml).

OSTWALD RIPENING UNDER THE INFLUENCE OF PERIODIC TEMPERATURE CHANGES

Mass transfer from small radioactively marked KCl crystals to KCl solution under the influence of periodic temperature changes was also measured. In this case dissolution and crystallization proceed with a periodic rhythm due to concentration gradients.

Fig.8 shows the dissolution kinetics of small ^{42}KCl crystals during Ostwald ripening under periodic temperature changes (2°C fluctuations at a frequency of 2 per hour). With increasing crystal growth of the inactive large KCl crystals the dissolution rate of the small KCl crystals also increases compared with that for isothermal Ostwald ripening.

For the Ostwald ripening of KCl with the above periodic temperature changes the following mass transfer constants for the dissolution kinetics of small ^{42}KCl crystals were found:

Dissolution rate:	3.4×10^{-4} g/sec
Specific dissolution rate:	5.3×10^{-7} g/sec cm^2
Mass transfer number:	1.4×10^{-6} cm/sec
Time constant per KCl particle:	1.2×10^{-12} cm^3/sec

Ostwald ripening under the influence of periodic temperature changes is more suitable for increasing the mean size of crystalline products than is isothermal Ostwald ripening because the mass transfer constants for the dissolution of small KCl crystals are greater by a factor of 20.

INVESTIGATION OF OSTWALD RIPENING BY AUTORADIOGRAPHY

The study of the Ostwald ripening of potassium chloride by means of microautoradiography allowed local distributions of the ^{42}K$^+$ isotope to be determined in the KCl crystal aggregates during the crystallization process. The stripping film technique requires the KCl crystals to be surrounded by a thin waterproof protective film to prevent dissolution as well as the formation of artefacts.

To trace the mass transfer between the small and large KCl crystals, proceeding by way of the solvent phase, we labelled the small KCl crystals with the $^{42}K^+$ isotope and performed the Ostwald ripening in saturated KCl solution. The microautoradiogram of cut KCl crystals, which was obtained by the stripping film technique during the Ostwald ripening, is shown in Fig.9. KCl crystals of size $\sim$ 250 µm show a blackened edge while the small crystals of $\sim$ 50 µm are thoroughly blackened. This result can be interpreted through the mass transfer process during Ostwald ripening.

The small radioactively marked KCl crystals dissolve, due to their higher thermodynamic potential, and cause a radioactive marking of the solution. During the growth of the large KCl crystals radioactive KCl is incorporated into the crystal lattice. This is shown by the microautoradiogram. It can also be seen that, under the given experimental conditions of isothermal Ostwald ripening, the supersaturation created by the small crystals is mainly reduced by the growth of the large crystals, but agglomeration between the large and small crystals also occurs.

The mass transfer between small and large KCl crystals under the conditions of Ostwald ripening is much greater with periodic temperature changes.

INVESTIGATIONS OF OSTWALD RIPENING BY SCANNING ELECTRON MICROSCOPY

In order to observe the habit of KCl crystals during and after Ostwald ripening, the scanning electron microscope was used. As a starting material for Ostwald ripening KCl crystals of 200 - 250 µm together with crystals of <63 µm were used.

The large starting crystals shown in Fig.10 indicate clearly that they have voids which were formed by the preceding crystallization process (5).

Fig.11 shows KCl crystals after isothermal Ostwald ripening. Agglomerates characteristic of KCl can be seen. KCl crystals formed after Ostwald ripening under periodic temperature changes are shown in Fig.12; crystals with a rounded habit have been formed.

We may conclude that the method of radioactive marking is suitable for observing the mass transfer processes taking place during Ostwald ripening of well-soluble salts. However, the mass exchange processes (heterogeneous isotope exchange) must also be taken into condideration: these proceed relatively rapidly at the beginning of the experiments.

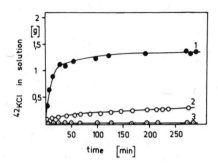

Fig. 7 Influence of PbCl$_2$ on the dissolution of small ^{42}KCl crystals at 25°C - 2 g ^{42}KCl(<63μm) + 18 g KCl(160-200μm) in 200 ml. 1: pure solution, 2: 2 mg PbCl$_2$, 3: 20 mg.

Fig. 8 Dissolution kinetics of small ^{42}KCl crystals during Ostwald ripening under periodic temperature changes showing the dependence on the inactive large crystal fraction. System volume = 200 ml. 1: 1 g ^{42}KCl(<45μm) + 9 g KCl(>250μm), 2: 1 g ^{42}KCl (<45μm) + 9 g KCl(160-200μm), 3: 1 g ^{42}KCl(<45μm) + 9 g KCl (100-125μm), 4: 1 g ^{42}KCl(<45μm) + 9 g KCl(63-80μm).

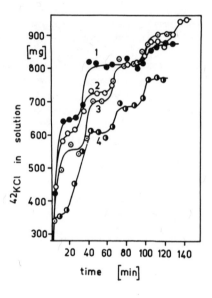

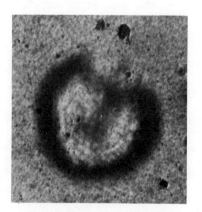

Fig. 9 Microautoradiogram of KCl after Ostwald ripening under periodic temperature changes.

Fig. 10 Scanning electron micrograph of the starting KCl crystals (250μm).

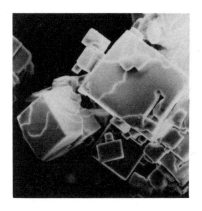

Fig. 11 Scanning electron photomicrograph of KCl after isothermal Ostwald ripening (100x).

Fig. 12 Scanning electron photomicrograph of KCl after Ostwald ripening under periodic temperature changes (100x).

These experimental results may serve to indicate that the ageing of easily soluble salts takes place according to the mechanism advanced by Ostwald.

REFERENCES

1. Lifshitz, I.M. and Slyozov, V.V., J.Phys.Chem.Solids, 19 (1961) 35.
2. Wagner, C., Z.Elektrochem., 65 (1961) 581.
3. Hanitzsch, E. and Kahlweit, M., in: Industrial Crystallization p.130, The Institution of Chem.Engineers, London (1969).
4. Hohmann, H.H. and Kahlweit, M., Ber.Bunsenges.physik.Chem., 76 (1970) 933.
5. Vetter, J., Winzer, A. and Emons, H-H., Krist.und Techn., in press.

RECRYSTALLIZATION IN SUSPENSIONS

J. SKŘIVÁNEK, S. ŽÁČEK, J. HOSTOMSKÝ AND V. VACEK

Institute of Inorganic Chemistry
Czechoslovak Academy of Sciences., Řež near Prague
Czechoslovakia

INTRODUCTION

Particles of various sizes suspended in their saturated solution constitute a system which cannot be considered stable from the viewpoint of thermodynamics. The large interfacial area of the particles represents a source of free energy which can be decreased by reducing this area. Assuming the particles do not agglomerate, the process which can take place in the system is either ripening or recrystallization, depending on the conditions. It is a common feature of both these processes that the mean particle size will increase and, simultaneously, the number of particles in the system will decrease. Given sufficient time this system may attain a state which is stable as far as the interfacial area is concerned, i.e., a condition where there is a single particle in the system, encompassing the mass of all the particles originally present.

In this paper, Ostwald ripening (1) is understood to be the growth of the mean-sized particle which occurs under isothermal conditions; its driving force is the different solubilities of small and large particles, as expressed by the Gibbs-Thomson relation (2). This process, described theoretically by Lifshitz, Slyozov (3,4) and Wagner (5), is relatively very slow and does not take place unless the particles are very small (a size of 0.5 - 50 μm has been given (7) for $CaSO_4 \cdot 2H_2O$ particles).

On the other hand, it has been ascertained (6,8-24) that if a polydispersed system is subjected to periodic temperature or concentration changes, the crystal habit is altered (16,19,24) and, in addition, the mean particle size is increased. Under these conditions, the mean growth is much faster, even with particles larger than those for which an explanation based on the Gibbs-Thomson relation can be sought. In agreement with other authors (8,9) we use the term recrystallization to describe this process. No satisfactory theoretical description of this acceleration has been published so far. The statistical model of the ripening process, applicable also to the region of larger crystals (25,26), has been extended (27-29) to allow for a description of the effect of temperature oscillation as well. However, no quantitative verification of the relationships thus obtained was made by experiment.

The process described also takes place in commercial crystallizers where the suspension of crystals is subjected to temperature oscillation by being passed at random through regions of high and low temperatures. The recrystallization process has been studied under the conditions of boiling (8,9) on the one hand, and under the conditions of periodic temperature changes in the saturated solution (6,10-19) on the other. With regard to an industrial application of the recrystallization process involving periodic temperature changes, a commercially feasible double crystallizer has been proposed (20-24). In describing the recrystallization, the retention time of the particles in regions having different temperatures (21-23) and the temperature difference between these two regions are of essential importance. However, these quantities were not defined properly in previous studies by other authors (8-19). A laboratory double crystallizer is used for systematic recrystallization studies in this work. This apparatus allows for some degree of regulation of both the retention time of particles and the temperature difference; the two quantities are clearly defined.

The description of the recrystallization process, given below, is based on balance equations set up for the model batch double crystallizer. The process description for steady-state operation in this apparatus has been described earlier (23).

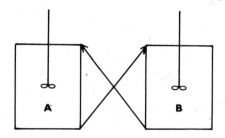

Fig. 1 Sketch of a double crystallizer

RECRYSTALLIZATION IN SUSPENSIONS

THEORETICAL

In deriving the basic balance equations, the double crystallizer schematic as shown in Fig. 1 has been considered. The double crystallizer is constituted by two perfectly mixed vessels of identical volume V. The vessels are interconnected and the volume of the tube connectors can be disregarded compared with the vessel volume. The flow rate of the suspension circulating between the vessels equals v. The temperature T_A in the vessel A is lower than the temperature T_B in vessel B. Consequently, particles grow in A and dissolve in B. Under these conditions the time development of the particle size distribution function in the two vessels is governed by the equations

$$\frac{\delta n_A(L,t)}{\delta t} = \frac{v}{V}(n_B - n_A) - \frac{\delta}{\delta L}(n_A G_A) \tag{1}$$

$$\frac{\delta n_B(L,t)}{\delta t} = \frac{v}{V}(n_A - n_B) - \frac{\delta}{\delta L}(n_B G_B) \tag{2}$$

with the initial and boundary conditions

$$n_A(L,0) = n_B(L,0) = n_o(L) \tag{3}$$

$$\lim_{L \to 0} n_A = 0 \tag{4}$$

$$\lim_{L \to \infty} n_B = 0 \tag{5}$$

The initial condition 3 indicates an identical distribution in both vessels at the beginning of the experiment. Boundary condition 4 is due to the fact that, in the cooler vessel where the particles grow, no new particles nucleate since the temperature differences used, $T_B - T_A$, are small.

Instantaneous growth and dissolution rates, G_A and G_B, must be known to solve the set of eqs 1-5. Generally, these rates depend on the solution concentrations, c_A and c_B, and on the characteristic particle size L. In this work, the kinetic relationships for G_A and G_B are approximated by the expressions

$$G_A = k_A(c_A - c_A^*)^n \tag{6}$$

$$G_B = k_B(c_B - c_B^*) \tag{7}$$

where c_A^* and c_B^* are the equilibrium concentrations (solubilities) at temperatures T_A and T_B. k_A, k_B and n are empirical constants. Hence we presume that the growth and dissolution rates are independent of particle size. The time-dependence of concentrations c_A, c_B is given by balance equations of the form

$$\frac{dc_A}{dt} = \frac{v}{V}(c_B - c_A) - \beta\rho_c \int_0^\infty n_A(L) L^2 G_A \, dL \qquad (8)$$

$$\frac{dc_B}{dt} = \frac{v}{V}(c_A - c_B) - \beta\rho_c \int_0^\infty n_B(L) L^2 G_B \, dL \qquad (9)$$

At the beginning of an experiment, i.e., at time $t = 0$ (before starting up the circulation of solutions), the concentrations c_A, c_B are identical and equal to c_o. Hence, the initial conditions for eqs 8 and 9 are:

$$t = 0 \qquad c_A = c_o \qquad (10)$$

$$c_B = c_o \qquad (11)$$

The set of eqs 1-11 represents a mathematically complete description of the processes taking place in the model double crystallizer. Using the distribution functions n_A and n_B we can also evaluate other quantities of practical importance, such as the number of particles N, the particle mean size (surface-averaged) $\bar{L}_{32}$ and the overall particle surface F.

EXPERIMENTAL

An overall diagram of the apparatus is shown in Fig. 2. The double crystallizer 1 is made of glass and composed of two jars, 130 ml capacity each, equipped with the controlled-temperature jackets 2 and interconnected by tube connectors 3. Each of the jars was provided with a vibrator 4 and an electromagnetic stirrer 5, the former ensuring circulation of the suspension from one jar to the other and vice versa, and the latter providing perfect mixing of the suspended particles in each jar. The amplitude and frequency of the vibratory stirrers could be regulated 6 within 1-10 mm and 2-14 Hz. The vibratory plates 4 were made of PTFE and had tapered openings 7 oriented so as to ensure a downward flow of the crystal suspension in each jar towards the connecting tube. The suspension flow rates and, consequently, the mean retention time of the particles in one jar, depended on the amplitude and frequency of the vibratory stirrers. The form of this dependence was established (30) by monitoring the temperature response in

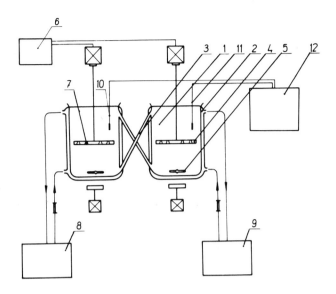

Fig. 2 The experimental double crystallizer

both jars to a temperature step function introduced to one of the jars. Taking the experimental error into account, the shape of the temperature response curve confirmed that the assumption of ideal mixing in both jars was justified. The required temperature in each jar was maintained by precision temperature controllers 8,9 to an accuracy of ± 0.01°C, measured by thermistors 10,11, and monitored by recorders 12.

With regard to earlier recrystallization studies (8-15), the system $KAl(SO_4)_2 \cdot 12H_2O$ - water was selected. Particles within the screened size fraction 0.08-0.125 mm were used in the experiments. A known weight of these particles was introduced into the double crystallizer in which an alum solution saturated at 25°C was circulating under the preset conditions. First sampling was made at time t = 0, i.e., immediately on suspending the particles, and the initial particle size distribution was determined. Sampling of 0.1 ml quantities was repeated at preset time intervals. The samples were taken with a micropipette, spread over a cover glass under a microscope, and the particle dimensions measured on magnified photographic prints. Thus the distribution f(L) was established, where f(L)dL expresses the number of particles having the sizes L to L+dL in the given series of photographs. The distribution f(L) is proportional to the distribution n(L), viz.,

$$n(L) = K f(L) \qquad (12)$$

where the proportionality constant K at the beginning of the experiment can be determined from the expression for the total mass of crystals m_{total} introduced into the system at the time $t = 0$:

$$K = m_{total} / 2 V \rho_c \alpha \sum_{i=1}^{m} f(L) L^3 \Delta L \qquad (13)$$

This formula is accurate only at the beginning of the experiment, i.e., at $t = 0$. Since the total mass of crystals is subject to little change during the experimental run, we use expressions 12 and 13 to obtain $n(L)$ from $f(L)$, even for samples taken later on in the course of the run. Samples were taken only from one jar (the cooler one), inasmuch as the distribution functions in the two jars differed from each other by an amount smaller than that representing the experimental error at the flow rates used. The shape factors for the octahedral crystals of potash alum, $\alpha = 0.471$ and $\beta = 3.46$, were taken from (31).

In the series of experimental runs, the effects of the temperature difference between the two jars (0.3 and 1.1°C) and of the mean retention time of the particles in a jar (45 and 77 sec) were studied for the suspended weight fraction of 0.038 g of crystals per gram of solution.

The experimental results were constituted by the time dependences of the particle size distribution functions determined for periods given by the duration of one experiment. From the distribution functions, the surface-averaged values of the mean particle size and the standard deviations were computed from

$$\bar{L}_{3,2} = \sum_{i=1}^{m} n(L_i) L_i^3 \Delta L / \sum_{i=1}^{m} n(L_i) L_i^2 \Delta L \qquad (14)$$

$$\sigma^2 = \frac{\sum_{1}^{m} n(L_i) L^2 \Delta L}{\sum_{1}^{m} n(L_i) \Delta L} - \left[\frac{\sum_{1}^{m} n(L_i) L \Delta L}{\sum_{1}^{m} n(L_i) \Delta L} \right]^2 \qquad (15)$$

RESULTS AND DISCUSSION

The increase with time of the mean particle size $\bar{L}_{3,2}$ is shown in Figs 3 and 4. The development of the distributions $n_A(L)$ calculated on the assumption that the mass of crystals is constant during the run is contained in Fig. 5. It is clear that both the mean particle size and the standard deviation of their distribution increase with time. Within the scope of the preset experimental conditions, the effect of the temperature difference between the jars, ΔT, is significant, whereas the effect of the mean retention time, $\bar{t} = V/v$, is less pronounced. Qualitatively, the effects of these quantities may be accounted for as follows. In the system of interconnected jars, the individual particles move stochastically (random walk). The probability that any particle considered will stay in one of the jars over the period from t to t+dt equals $(1/\bar{t})\exp(-t/\bar{t})dt$; there exists a 50% probability that the particle will leave the jar in the time $\bar{t}\cdot\ln 2$.

In the first approximation, the particle size change with time may be characterized as a random walk, viz., as the diffusion along the axis of the characteristic dimension L where the average step length is $G_{av}\bar{t}\cdot\ln 2$ and the mean frequency is $1/(\bar{t}\ln 2)$ yielding the diffusion coefficient

$$D = \tfrac{1}{2} G_{av}^2 \, \bar{t} \, \ln 2$$

Here the mean velocity G_{av} is considered to represent the mean growth rate in the cooler jar and the absolute dissolution rate in the warmer jar. For potash alum the dissolution rate is proportional to the undersaturation, whereas the growth rate is a power function of oversaturation. For the fluid-bed growth and dissolution of potash alum crystals the relationships for these rates, as obtained by Garside and Mullin (33), are as follows (at 25°C and rewritten to conform with the units used in this work):

$$G_A = 8.28 \cdot 10^{-3}(c_A - c_A^*)^{1.61} \tag{16}$$

$$G_B = -1.43 \cdot 10^{-3}(c_B - c_B^*) \tag{17}$$

The oversaturation $c_A - c_A^*$ or the undersaturation $c_B - c_B^*$ of the solution (and also the mean velocity G_A) is approximately proportional to the temperature difference between the jars ΔT. Hence, the diffusion coefficient for the motion of particles along the axis of dimension is proportional to the square of ΔT and to the mean retention time $\bar{t}$. Even though this treatment is only qualitative, disregarding the effect on the process of the time-dependent concentrations in the solution in the two jars, it does account for the overall trend of the experimental results.

As far as the quantitative description of the experimental data by the equations given in the theoretical part is considered, we are hampered by a lack of information on the kinetic equations of the type of eqs 16 and 17 describing growth and dissolution.

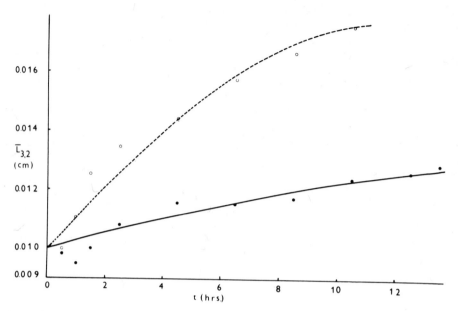

Fig. 3 Time dependence of mean particle size $\bar{L}_{3,2}$ for runs 5 and 6. Experimental data: ○ run 5, ● run 6. Calculations: broken line run 5 ($\Delta T = 1.1^\circ C$, $\bar{t} = 45s$), continuous line run 6 ($\Delta T = 0.3^\circ C$, $\bar{t} = 45s$).

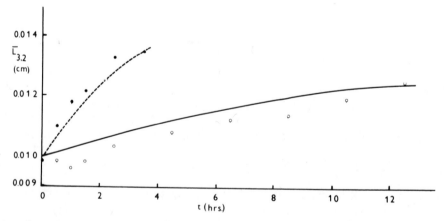

Fig. 4 Time dependence of mean particle size $\bar{L}_{3,2}$ for runs 7 and 8. Experimental data: ○ run 7, ● run 8. Calculations: continuous line run 7 ($\Delta T = 0.3^\circ C$, $\bar{t} = 77s$), broken line run 8 ($\Delta T = 1.1^\circ C$, $\bar{t} = 77s$).

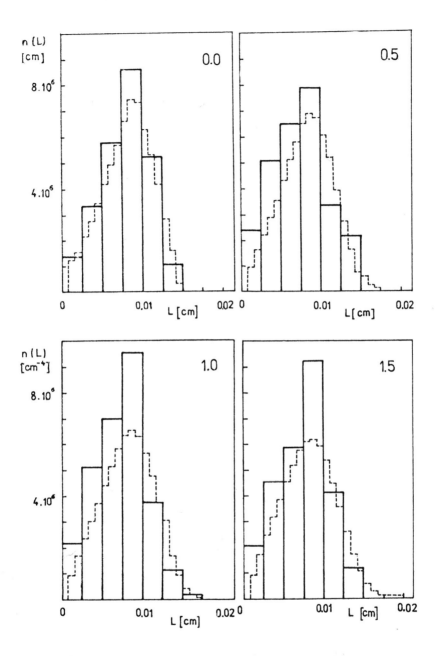

Fig. 5 Development of the particle size distribution function n(L,t) for selected times for run 7. Continuous lines: experimental. Broken lines: calculated.

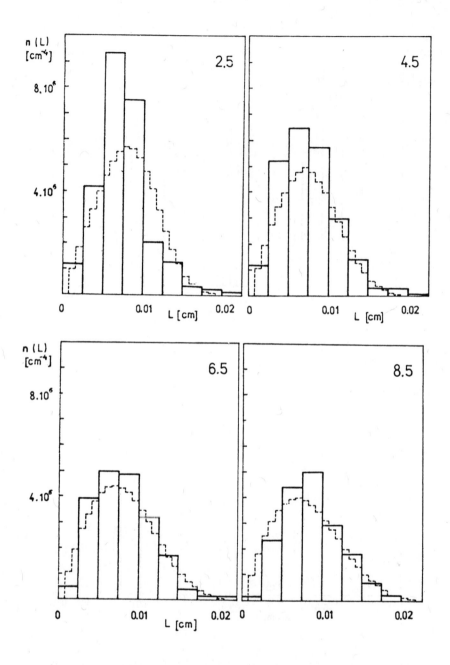

Fig. 5 (continued)

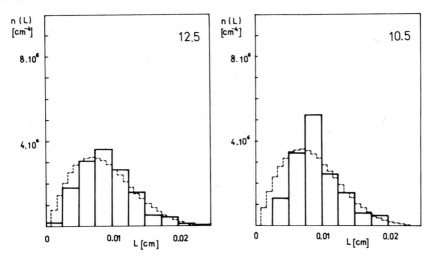

Fig. 5 (continued)

The set of equations 1-12 was solved numerically for the initial conditions corresponding to the experimental data, i.e., for the distribution function n_o evaluated by the aforementioned photographic method and for a concentration c_o equal to the initial concentration of the solution (saturated at 25°C). Eqs 1, 2, 8 and 9 were transformed to a difference form and solved using an implied operator. With regard to the integrals on the right-hand side of eqs 8 and 9 the computation was iterative, i.e., the procedure was repeated at each time step Δt until the presumed and the calculated distribution and concentration values in the time $t+\alpha t$ were identical to a preset accuracy. Eqs 16 and 17 with a multiplication factor included on their right-hand side were used as the kinetic relationships for the rates G_A and G_B. A satisfactory agreement between the experimental and the computational results of the mean particle size values $\bar{L}_{3,2}$ and of the distribution functions n_A (as shown in the Figs 3-5) was achieved for the multiplication factor value 4 (i.e., for the case where, at given oversaturation, the growth or dissolution rates were four times higher than the eqs 16 and 17 would indicate). This multiplication factor value does have a real meaning, for the mass transfer at the solid – solution interface in the double crystallizer jars where there is vibratory and rotatory agitation is higher than that at the interface in the case of crystals in the fluid bed. Further, the effect of an increased interfacial area as compared with monotonous growth or dissolution must be considered. In fact, the individual particles in the double crystallizer undergo periodic growth and dissolution so that the particle surface area exhibits a microscopical roughness (34).

In addition, it follows from the numerical calculations that the concentration of dissolved matter in the solution in both jars is subject to little change after the initial adjustment, so that the mass of crystals is practically constant during the experimental run. The growth and dissolution rates also change only little but a certain difference does exist between their absolute values which, from the standpoint of mass balance, cannot be disregarded.

NOTATION

c	concentration of solution	g/g soln.
c^*	equilibrium concentration of solution	g/g soln.
D	diffusion coefficient	cm/s
F	crystal surface area	cm^2
$f(L)$	distribution function determined from	cm^{-4}
$f(L)dL$	no. of particles having sizes L to $L+dL$	cm^{-3}
G_A	growth rate	cm/s
G_B	dissolution rate	cm/s
k_A	mass transfer coefficient for growth	$g/cm^2 s$
k_B	mass transfer coefficient for dissolution	$g/cm^2 s$
K	factor for recalculating the distribution functions in eqs 12 and 13	
L	characteristic dimension of particle	cm
m_{total}	total mass of crystals	g
n	exponent, oversaturation in eq. 6	
$n(L)$	distribution function	cm^{-4}
$n(L)dL$	no. of particles having sizes L to $L+dL$ in unit volume	cm^{-3}
N	number of particles	
t	time	s
$\bar{t}$	mean retention time, $\bar{t} = V/v$	s
T	temperature	°C
v	flow rate	cm^3/s
V	volume	cm^3

Subscripts:
A,B	denoting the two jars	
av.	average	
o	related to the time $t = 0$	
α	volumetric shape factor	
β	surface shape factor	
ρ_c	density of solid phase	g/cm^3

REFERENCES

1. Ostwald, W., Z. Physik Chem., 34 (1900) 495.
2. Mullin, J.W., Crystallization, 2nd Ed., Butterworths, London, 1972, pp. 43, 222.
3. Lifshitz, I.M. and Slyozov, V.V., Zh. eksp. teor. fiz., 35 (1958) 479.
4. Lifshitz, I.M. and Slyozov, V.V., J. Phys. Chem. Solids, 19 (1961) 35.

5. Wagner, C., Z. Elektrochemie, 65 (1961) 581.
6. Hohmann, H.H. and Kahlweit, M., Ber. Bunsenges. physik. Chem., 76 (1972) 933.
7. Jones, W.J. and Partington, J.R., J. Chem. Soc., 107 (1915) 1019.
8. Bazhal, I.G. and Kurilenko, O.D., Zh. priklad. Khim., 40 (1967) 2579.
9. Dulepov, J.N., Matusievich, L.N. and Odintsov, V.A., Zh. priklad. Khim., 45 (1972) 1922.
10. Bazhal, I.G. et al., Teoret. osnovy khim. tekhnol., 3 (1969) 931.
11. Bazhal, I.G., Kristallografija, 14 (1969) 1106.
12. Bazhal, I.G. et al., Ukr. Khim. Zh., 37 (1971) 294, 395, 508, 1075, 1174, 1294.
13. Bazhal, I.G. et al., ibid., 38 (1972) 43.
14. Bazhal, I.G. et al., Zh. prikl. Khim., 45 (1972) 1252.
15. Bazhal, I.G. et al., ibid., 46 (1973) 1973.
16. Skřivánek, J. and Žáček, S., Chem. prum., 22 (1972) 225.
17. Žáček, S. and Skřivánek, J., ibid., 23 (1973) 11.
18. Skřivánek, J. and Žáček, S., ibid., 23 (1973) 496.
19. Sutherland, D.N., Chem. Engng Sci., 24 (1969) 192.
20. Nývlt, J. and Horáček, S., Chem. prum., 14 (1964) 435.
21. Lederer, E. et al., ibid., 17 (1967) 640.
22. Skřivanek, J. and Nývlt, J., Coll. Czech. Chem. Comm., 33 (1968) 2799.
23. Žáček, S., Skřivánek, J. and Vacek, V., Krist. u. Tech., 9 (1974) 625.
24. Nývlt, J., Particle Growth in Suspensions, Academic Press, 1973, p. 131.
25. Horsák, I. and Skřivánek, J., Ber. Bunsenges. phys. Chem., 77 (1973) 336.
26. Skřivánek, J., Vacek, V. and Žáček, S., Krist. u. Techn., 10 (1975) 707.
27. Horsák, I., Vacek, V. and Žáček, S., Ber. Bunsenges. phys. Chem., in print.
28. Vacek, V., Žáček, S. and Skřivánek, J., Krist. u. Techn., in print.
29. Vacek, V., Žáček, S. and Horsák, I., Krist. u. Techn., in print.
30. Žáček, S., Skřivánek, J. and Hostomský, J., Chem. prum., 24 (1974) 275.
31. Mullin, J.W., Crystallization, 2nd Ed., Butterworths, London, 1972, pp. 397-8.
32. Reif, F., Statistical Physics (Berkeley Physics Course), Vol. 5, p. 310 (Russian ed. Nauka, Moscow 1972).
33. Garside, J. and Mullin, J.W., Trans. Inst. Chem. Engrs, 46 (1968) T11.
34. Horsák, I. and Skřivánek, J., Ber. Bunsenges. physik. Chem., 79 (1975) 433.

CRYSTAL GROWTH KINETICS OF SODIUM CHLORIDE FROM SOLUTION

RAFAEL RODRIGUEZ-CLEMENTE

Departamento de Cristalografia y Mineralogia

Facultad de Geologia, Universidad de Barcelona, Spain

INTRODUCTION

Due to the low solubility coefficient and ease of nucleation at low supersaturation, the study of the growth kinetics of NaCl crystals presents many difficulties. Previous relevant work includes that of Rumford and Bain (14) who measured crystal growth rates in fluidized beds and with a fixed seed. More recently, Sclar and Schwartz (15), Batchelder and Vaughan (1), Schichiri and Kato (16) and Sunagawa and Tsukamoto (17) have paid special attention to the influence of crystal perfection on growth rate.

However, in spite of the large amount of work done, the growth kinetics of NaCl are still unsolved because most of the previous work was done under badly controlled conditions.

The work reported here refers to the influence of supersaturation, solution velocity, temperature and experimental conditions on the growth rate of the (100) faces of NaCl crystals in pure aqueous solution. The influence of the actual crystal surface structure on the growth rate is also discussed.

EXPERIMENTAL

The apparatus is shown in Fig.1. It consists of a thermostated reactor, provided with a magnetic agitator, where a saturated solution is prepared. This solution is pumped through a flow-meter and refrigerator to the growth cell at the desired temperature and velocity.

The growth cell is shown in Fig.2. The crystal was drilled and mounted on a platinum wire without the need of an adhesive, thus eliminating one possible source of contamination. The circuit was entirely constructed from glass and Teflon tubes to avoid contamination and the temperature of the system was controlled with Chrome-Alumel thermocouples (± 0.01°C).

A cubic crystal seed, of 0.5 mm side, was mounted on the platinum wire and slightly dissolved with unsaturated solution to clean its surfaces. Immediately the seed was introduced into the cell, the solution flow was commenced through the circuit at constant temperature until thermal equilibrium between the solution and seed was reached. When cooling was started, the crystal grew flat faces and the solution temperature was fixed at the desired Temperature of Crystallization (T.C.). When conditions were steady, the experiment was started. Photographs of the upper face of the crystal were taken at regular intervals of time with a photomicroscope. Fig.3 shows some typical graphs of the face displacement as a function of time. The slope of the lines gives the rate of growth of the face.

Growth rates are calculated for faces parallel and perpendicular to the flux (Fig.2). Special attention was paid to the influence of supersaturation σ and solution velocity V_s on growth rate R_g. The same seed was employed to measure growth rates at different conditions, changing one or two of the physical variables: T.C., σ or V_s. Several series of experiments were made with different crystals and conditions were changed to try to relate those variables with the crystal growth rates.

INFLUENCE OF SOLUTION VELOCITY

In static supersaturated solutions, growing crystals of NaCl create a diffusion gradient around them; the growth rate is then controlled by mass transfer through the solution. When the solution flows, mass transfer by diffusion still operates, but it is not the rate controlling mechanism. NaCl growth rates from solution, as for many other salts (9), is solution velocity dependent. This dependence was studied by trying to find the conditions at which the growth rate became independent of solution velocity.

Series of runs were made at different values of T.C. and σ, for single seeds (Fig.4). The main problem was the tremendous scatter of the results, although the general trend of the growth rate-solution velocity dependence is asymptotic. This feature is fairly common in the literature, but up to the moment all the correlations between the two factors have been essentially empirical (8),(9). In order to obtain a theoretical correlation of the two variables, several considerations can be made:

CRYSTAL GROWTH KINETICS OF NaCl FROM SOLUTION

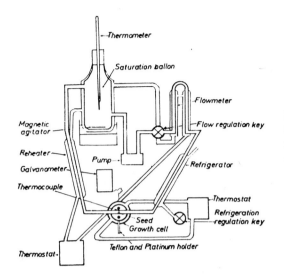

Fig. 1 Crystal growth apparatus. The growth cell is mounted on a photomicroscope.

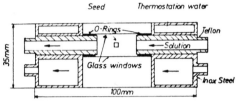

Fig. 2 Section and front view of the growth cell also showing the position of the crystal faces with respect to the solution flow.

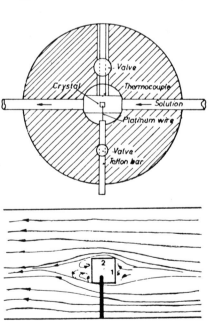

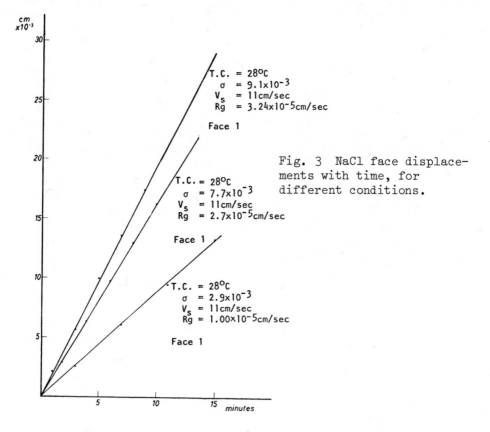

Fig. 3 NaCl face displacements with time, for different conditions.

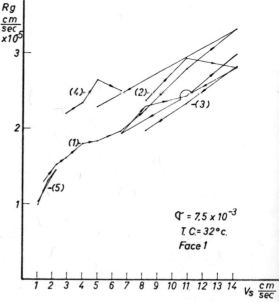

Fig. 4 Growth rate measurements, changing the solution velocity, at fixed temperature and supersaturation, for five different single crystals. The arrows indicate the path of each experiment.

CRYSTAL GROWTH KINETICS OF NaCl FROM SOLUTION

1. The increase of the growth rate with increasing solution velocity is due to the increase of the interfacial concentration.

2. The solution velocity is directly proportional to the pressure of the liquid when they are in laminar flow. This is so in the present case so the flow of solution can be considered as a pressure acting on the crystal surface that increases the concentration of solute molecules over it, up to a limit represented by the bulk concentration.

3. According to 1 and 2, the adsorption isotherm formalism can be applied, in an analogous way, to the problem of crystal growth from solution. The following equations should then hold:

$$C_i = \frac{C_\infty KV_s}{1 + KV_s} + \frac{C_{io}}{1 + KV_s} \tag{1}$$

$$R_g = K_r (C_i - C_o)^n \tag{2}$$

Therefore,

$$R_g = K_r \left[\frac{C_\infty KV_s}{1 + KV_s} + \frac{C_{io}}{1 + KV_s} - C_o \right]^n$$

$$= K_r \left[\frac{(C_\infty - C_o) KV_s}{1 + KV_s} + \frac{C_{io} - C_o}{1 + KV_s} \right]^n$$

$$= K_r C_o^n \left[\frac{KV_s \sigma}{1 + KV_s} + \frac{\sigma_{io}}{1 + KV_s} \right]^n \tag{3}$$

where C_∞ = bulk solution concentration
C_o = saturation concentration
C_{io} = concentration at the interface when $V_s = 0$
C_i = concentration at the interface when $V \neq 0$
σ = bulk supersaturation, $C_\infty - C_o/C_o$
σ_{io} = interface supersaturation when $V_s = 0$, $C_{io} - C_o/C_o$
K_r = reaction rate constant
K = Langmuir adsorption isotherm constant
R_g = rate of crystal growth.

Equation 3 holds in the limiting cases:

(a) $\sigma = 0$ $R_g = 0$
(b) $V_s = 0$ $R_g = K_r C_o^n \sigma_{io}^n$
(c) $V_s = \infty$ $R_g = K_r C_o^n \sigma^n$

The correlation curves obtained when applying equation 3, (Fig.5) shows excessive inflexions at low values of V_s, due to the narrow adjustment of K for values of $n \simeq 1$ and low values of σ.

Equation 3 indicates that the higher is σ, the higher must be the value of V_s to reach the asymptote. At low σ, saturation is obtained at lower values of V_s which coincides with reports on other systems (10, 11). Another observation is that the higher the T.C., at a given bulk supersaturation, the faster is saturation obtained (Fig.6).

The constants calculated by applying equation 4 to the results of this paper for sodium chloride, of Mullin and Garside for potash alum (10) and of McCabe and Stevens for copper sulphate (8) are given in Table I. The correlation is quite good.

INFLUENCE OF SUPERSATURATION

Growth rate measurements were made at 24, 28, 32 and 36°C and at a constant solution velocity of 11 cm/s. As described previously, runs were made using the same seed crystals for measurements at the different conditions.

The tremendous scatter of results made it impossible to establish growth rate-supersaturation curves at different crystallization temperature so the growth rate activation energy could not be calculated. However, the experimental points were correlated using the equation:

$$R_g = C \sigma \qquad (4)$$

that holds at high supersaturation (BCF) (6). The results at different temperatures, and for the two kinds of faces, are shown in Fig.7. The correlation was made using the average points of each area of dispersion.

The calculated values of constant C in equation 4, the correlation coefficient and standard deviation of R_g are given in Table II The ratio between the calculated value of C for faces 1 and 2 at the same T.C. is about 1.7. The values of C and $K_r C_o^n$ at the same temperature are expected to be the same. In fact they are of the same order of magnitude, but the scatter of the experimental points hides the true value of the constants.

SCATTER OF CRYSTAL GROWTH RATES

The literature is rich in examples where high scattering is

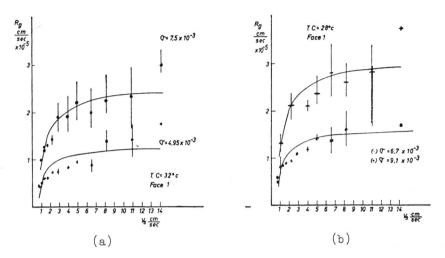

Fig. 5 Correlation of the experimental points using eq. 3 for two temperatures (a) 32°C, (b) 28°C. The vertical bars indicate the dispersion of R_g values. The points on the bars represent the average value of R_g at a given value of V_s (see Table 1).

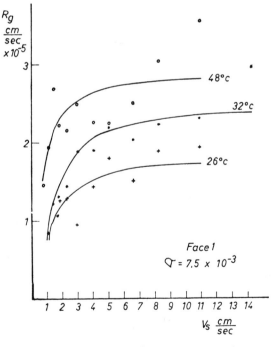

Fig. 6 Average growth rates for three temperatures at the same bulk supersaturation (see Table 1).

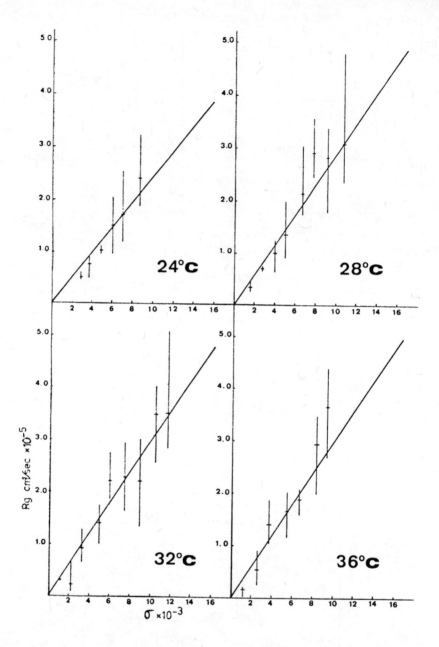

Fig. 7a Face 1 growth rates of NaCl single crystals at four temperatures and constant solution velocity (11 cm/s). Vertical bars indicate dispersion of the R_g values. Correlation made using eq. 4 (see Table 2).

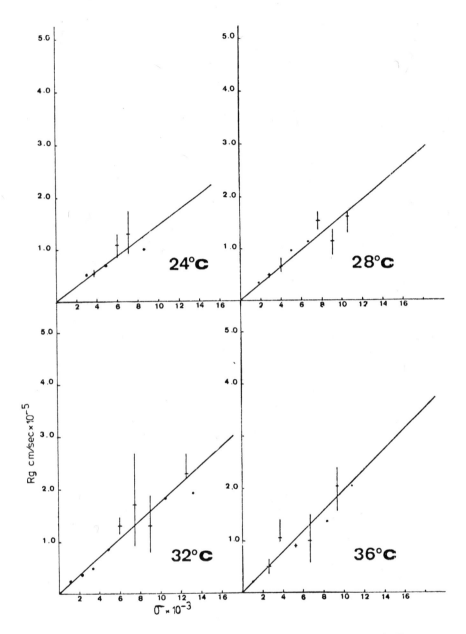

Fig. 7b Face 2 growth rates of NaCl single crystals at four temperatures and constant solution velocity (11 cm/s). Vertical bars indicate dispersion of the R_g values. Correlation made using eq. 4 (see Table 2).

Substance	Supersaturation	$K_r c_o^n \times 10^4$	K	n	Standard deviation	Reference
AlK(SO$_4$)$_2$ 12H$_2$O	0.080	32 1.56	38.8	1.011	0.127	(10)
CuSO$_4$ 5H$_2$O	0.018	27 1.76	131	1.025	0.028	(8)
"	0.018	27 1.79	650	1.015	0.020	(8)
NaCl	0.0075	26 27.7	2110	1.022	0.216	This paper
"	0.0075	32 27.7	183	1.017	0.225	"
"	0.0075	48 45.0	368	1.028	0.363	"
"	0.0067	28 25.9	257	1.011	0.216	"
"	0.0091	28 35.1	170	1.015	0.342	"
"	0.0094	32 26.9	313	1.000	0.241	"

Table 1 Constants obtained by applying equation 3 to the growth rate-solution velocity plots.

Face	T.C. °C	C cm/s × 10^4	Correlation coefficient	Standard deviation of R_g
1	24	24.40	0.985809	0.22718
1	28	29.97	0.992967	0.25276
1	32	29.38	0.992765	0.28300
1	36	34.33	0.989275	0.30385
2	24	14.74	0.984014	0.16005
2	28	15.72	0.988410	0.17272
2	32	17.48	0.986229	0.23351
2	36	19.55	0.990629	0.18338

Table 2: Constants obtained by applying equation 4 to the Growth rate-supersaturation.

shown in R_g - σ plots. Some authors attribute the phenomenon to experimental manipulations, while others (2, 3, 5) have tried to find a theoretical explanation. The following facts emerge from our experiments on NaCl.

1. Sometimes two different crystals grow with different velocities under the same environmental conditions.

2. When changes in σ and T.C. are made during the growth process and then returned to the original conditions, the final growth rate is different from the initial value (Fig.8). The ratio between the two velocities increases with the supersaturation level of the experiment.

3. The anisotropy of the environment, produced by the flow of solution, makes similar faces grow at different rates. Such a behaviour may also be produced in static solution, but in this case the concentration of dislocations is the ruling factor (12).

4. At high supersaturation, without varying the physical

CRYSTAL GROWTH KINETICS OF NaCl FROM SOLUTION

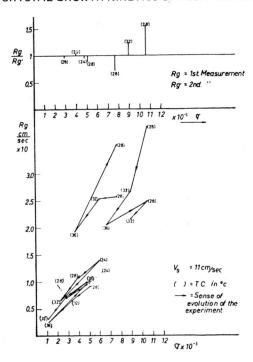

Fig. 8 Evolution of single growth rates by changing the growth conditions, the arrows indicate the path of the changes in the experiment. The relation between the initial and final growth rate, is indicated in the upper part of the diagram.

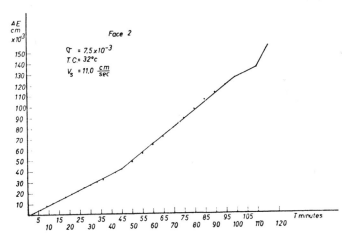

Fig. 9 Sharp change of a single crystal growth rate without varying the growth conditions. This phenomenon is probably due to the progressive increase of dislocation density on the crystal surface as a consequence of the high environment supersaturation (13). At a given moment, the distance between the emergence points of the dislocations become smaller than $2\pi r^*$, so the crystal changes its growth rate instantaneously (7).

environment a crystal sometimes changes is growth rate instantaneously (Fig.9). This phenomenon has also been recorded for the growth of urotropine from solution (7).

The following theoretical models may help to explain the scattering of growth rate data.

A. Based on the BCF theory (6), changes of R_g are produced when the concentration of dislocations in the surface of the crystal increases to the point where the distance between the centres of emergence is lower than $2\pi r^*$ (r^* is the radius of the two-dimensional critical nucleus corresponding to the given environmental supersaturation).

B. Bennema, et al.(3) suppose that at high σ, two-dimensional nucleii can be formed on the steps produced by the dislocations, increasing the concentration of kinks and hence increasing the rate of growth. Bethge et al.(4) suggested that the distance between steps, Y_o, must be higher than $4x_s$ (x_s being the main displacement of adsorbed molecules on the crystal faces) to let the above process occur.

C. Botsaris et al.(5) assumed that the concentration of impurities in the solution increased when the crystals grow due to loss of solute. They supposed that the growth rate decreases due to the build-up of impurity concentration.

These three models probably hold in different cases. From the first two it can be concluded that the growth rate is a function of the state of the crystal surface. So variations in R_g may be expected since crystal perfection is sensitive to such physical factors as the supersaturation temperature, impurity level, etc.

ACKNOWLEDGEMENTS

This research was partially supported by a grant from the "Fundacion Juan March". Helpful revision of the manuscript by Professor J.L. Amoros is gratefully acknowledged.

REFERENCES

1. Batchelder, F. and Vaughan, W.H., in: 'Crystal Growth' (ed. H.S. Peiser). Pergamon Press (1967).
2. Bennema, P., J.Crystal Growth, 1 (1967) 278.
3. Bennema, P., Kern, R. and Simon, B., Phys.Stat Solid., 19, (1967) 211.
4. Bethge, H., Zeller, K.W. and Ziegler, E., J. Crystal Growth, 3/4 (1968) 184.

5. Botsaris, G.D., Denk, F.C. and Shelder, R.A., Kristall und Technik, $\underline{8}$ (1974) 769.
6. Burton, W.K., Cabrera, N. and Frank, F.C., Phil.Trans.Roy. Soc. (1951) 299, 243.
7. Heyer, H., in: 'Crystal Growth' cited above.
8. McCabe, W.L. and Stevens, R.P., Chem.Engng.Progr., $\underline{47}$ (1951) 168.
9. Mullin, J.W., 'Crystallization', 2nd edition, Butterworths, London, (1972).
10. Mullin, J.W. and Garside, J., Trans.Instn.Chem.Engrs., $\underline{45}$ (1967) 285 and 291.
11. Mullin, J.W. and Gaska, C., Canad.J.Chem.Engng., $\underline{47}$ (1969) 483, and J.Chem.Engng.Data, $\underline{18}$ (1973) 217.
12. Mussard, F. and Goldsztaub, S., J.Crystal Growth, $\underline{13/14}$ (1972) 445.
13. Mutaftschiev, B., Crystal Growth and Dislocations in: 'Theory of Dislocations (ed. F.R.N. Nabarro) Dekker, New York (1973).
14. Rumford, F. and Bain, J., Trans.Instn.Chem.Engrs., $\underline{10}$ (1960) 38.
15. Sclar, C.B. and Schwart, C.M., in: 'Crystal Growth' cited above.
16. Shichiri, T. and Kato, N., J.Crystal Growth, $\underline{3/4}$, (1968) 384.
17. Sunagawa, I. and Tsukamoto, K., J.Crystal Growth, $\underline{15}$, (1972) 73.

Crystal Habit
Modification

SURVEY OF CRYSTAL HABIT MODIFICATION IN SOLUTION

R. BOISTELLE

Centre des Mécanismes de la Croissance Cristalline du
C.N.R.S. - Université d'Aix Marseille III - Centre de
Saint Jérome - 13397 MARSEILLE / Cédex 4 (FRANCE)

During these last years, crystal habit modification has been widely used for many purposes in Industrial Crystallization. The reader will find in certain books |1-12| and in some earlier review papers |13-18| several hundreds of references concerning habit change examples and theoretical interpretations. As in Chemical Engineering people are daily interested in obtaining different crystal habits, we shall try to summarize here the main factors involved in crystal habit modification, with the aid of a few recent and representative studies. To avoid confusion, we can just recall that, if the habit is defined as the overall shape of the crystal made up of a whole of {hkl} forms, the habit change may be understood in two ways. It is either a change in the relative development of the already existing {hkl} forms, or the appearance of some new {h'k'l'} forms.

1) EQUILIBRIUM AND GROWTH HABITS

The *equilibrium habit* of a crystal is determined by thermodynamic conditions and corresponds to the smallest convex polyhedron having the minimum surface free energy according to the well known Curie-Gibbs-Wulff relations |19-21|. In the formalism of the PBC theory |22,23| the equilibrium habit is made up of F faces which grow by lateral extension of steps. S and K faces do not grow by lateral spreading of layers and do not belong to the equilibrium habit. It could be said also |24-25| that F faces are the only ones which support stable two-dimensional layers having at least two edges with positive edge free energies, in contrast to S and K faces which have, respectively, only one or zero stable edge.

Finally, one may also notice that the equilibrium habit of a crystal is experimentally rather difficult to reach. It seems that the best method consists in creating small temperature fluctuations for a crystal placed in a droplet of its saturated solution |26-28|.

The *growth habit* of a crystal depends on the other hand on growth kinetics, and comprises solely the faces having the lowest normal growth rates. Its modification is therefore possible by slowing down selectively the growth rates of certain faces, which supposes that a strong adsorption of solvent or of certain impurities exists on these faces. Habit changes obtained by increasing the growth rates are still exceptional cases. In addition, if the new faces produced are S or K, it can be said formally that they have changed their growth mechanism into the one of the F faces. So, it is seen that a discussion of crystal habit modification returns finally to a discussion of relative growth kinetics.

2) SUPERSATURATION-SOLVENT-SOLUTE

The effect of the *supersaturation* on the crystal habit in pure solution is often not quite well understood because of the large diversity of the phenomena. We can mention here the different behaviour of solutions of highly soluble or almost insoluble species, or the increase in number of possible growth forms when the symmetry of the crystal decreases. Thus only one new form appears on the habit of the alkali halide crystals|29|, while the number of forms vary widely with the supersaturation in the case of Sr, Pb or Ba sulphate crystals |30|. Furthermore if the normal growth rates vary generally in a monotonous way with the supersaturation, one observes sometimes that they are oscillating (potash alum |31|) or that their relative values first increase, pass through a maximum and finally decrease as function of the supersaturation (rochelle salt |32|). At last, in a given solvent, some phase transformations can originate from supersaturation effects as in the case of calcium carbonate which crystallizes as calcite $|10\bar{1}0|$ at low supersaturation, and as aragonite elongated along <001>, then along <100> when the supersaturation increases |33|.

Nevertheless, in most cases, the effect of the supersaturation is related to the desolvation kinetics of the faces or of the solute. So, it has been shown that the (100) faces of KI grow faster than those of KBr and KCl because the desolvation energies of the I^-, Br^- and Cl^- ions increase in this order |34|. In a general manner it is admitted, since the equilibrium habit is composed only of F faces |35|, that there is no critical supersaturation for the appearance of a given F face |14|. The sole condition is that its desolvation kinetics is slower than those of the other faces. The habit change is progressive, as in the case of oxalic acid crystals |36| where an increase in supersaturation provokes a decrease in development of (011) and (110) and a related increase in development of (10$\bar{1}$) and (101). On the other hand, critical supersaturations are necessary in order that S or K faces appear on the crystal habit. It is assumed that the desolvation kinetics of these faces are then much lower than those of the usual F faces. This could be the case of the alkali halides |29| for which the appearance of {111} is highly favoured by more polar solvents |37,38|.

The effect of the *solvent* is directly related to the structure of the faces. The higher the number of PBC vectors piercing a face, the stronger the adsorption |39-40|. Besides the structural conditions imposed by the crystal, it is well known that the affinity of the solvents for a given face may vary considerably |41-42|. In the case of non-centrosymmetric crystals, for which the polarisabilities of the (hkl) and ($\bar{h}\bar{k}\bar{l}$) faces are different, it is the solvent (and the supersaturation) which determines the appearance of one or the other of corresponding growth forms (tetraethylammonium iodide |43|).

But strong interactions between solvent and solute are also possible and we can mention here the growth of tripalmitate and tristearate crystals |44| the growth kinetics of which is supposed to be determined by the more or less favourable orientation of the solute molecules in the crystal-solution interface. As the solvent has the same structure as the solute it is assumed that the incorporation frequency is decreased, but this hypothesis is nevertheless not corroborated in the case of paraffin crystals growing in solvents of the same structure |45-46|. From this point of view crystal habit modifications seem to be difficult to obtain.

Finally, we can mention also that according to the solvent it is possible to have solvates like $Hg(CN)_2$, CH_3OH |42| or $5Hg(CN)_2$, 4 tetrahydrofuran |47|, the habit of which depends highly on the supersaturation.

The *solute* intervenes also in different ways on the crystal habit. If it has not exactly the same composition as that of the crystal it acts sometimes like an impurity. Thus, calcite crystals have an elongated {02$\bar{2}$1} habit or a tabular {0001} habit if there is an excess, respectively, of Ca^{2+} or CO_3^{2-} ions in the solution. When the ion concentrations are the same, the habit is {10$\bar{1}$0} and {40$\bar{4}$1} |48|. If the solute is obtained from different phases, the growth kinetics of a given face may also be altered as in the case of sodium triphosphate hexahydrate crystals |49|. According to the origin of the solute ($Na_5P_3O_{10}$ of high or low temperature) there exist in the solution large $|Na_2P_3O_{10}|^{3-}$ ions, which are either isolated or bonded by Na^+ ions. In the latter case the growth rates are reduced since these bonds must at first be broken.

3) CRYSTALLIZATION TEMPERATURE

To our knowledge, if phase transformations are excluded, there are no important habit changes due to the sole effect of the crystallization temperature, even if there would be expected a simplification in the overall crystal shape when the temperature is raised. Everything happens as if the behaviour of all the crystal faces were the same in regard to the temperature. We assume here that no impurities cooperate with temperature to produce a habit change, since it is evident that adsorption depends largely on temperature.

In pure solution nevertheless, certain experiments show the manner in which intervenes the crystallization temperature. For ammonium chloride, <100> whiskers have growth kinetics highly influenced by the temperature (i.e. no effect of bulk diffusion), whereas <111> whiskers grow independently of temperature (i.e. that the rate determining step is the bulk diffusion) |50|. In the case of K or Na chlorates, of Na nitrate and nitrite, of potash alum etc.. in aqueous solution, the growth rates at low supersaturation of different faces vary very much with the crystallization temperature |51-53|. As the NMR spectra of the adsorbed water layer show some anomalies in the same ranges of temperature as do the growth rates, it is assumed that in these regions water molecules are highly orientated, and that the destruction of the adsorbed layer needs a certain time, resulting in a sharp decrease of the growth rates. So, if the adsorbed layers behave differently on different crystal faces one can imagine some important habit changes in certain well defined temperature ranges.

4) pH-SOLVENT-SOLUTE

pH modifies the nature and the concentration of ions in solution, particularly when the latter contains salts of weak acids or bases. This property is often used for the growth in gels |9|, where for instance the ionization of lead acetate increases when the pH decreases |54| resulting in an increase of the growth rate of lead dendrites.

Another example is given by ADP crystals |55| which exhibit (011) faces, the growth kinetics of which is related to the hydration of NH_4^+ ions and decreases when the pH is raised. On the other hand the (0$\bar{1}$0) faces have kinetics related to the hydration of $H_2PO_4^-$ ions and are less affected by pH. So, habit changes occur very easily by variation of the pH. The interpretation is based on the fact that addition of NH_4OH favours the formation of the complex $NH_4PO_4^-$ ions from which the free ions come out with difficulty, whereas the addition of H_3PO_4 favours another and less stable complex. But other complex ions have been suggested |56,57| to explain the growth of

ADP crystals with (001) faces well developed by increasing the growth rate of the (100) faces. When the pH is raised from 3.8 to 5 by addition of NH_4OH, the lateral extension of the steps is enhanced by a factor of 10. Here it is assumed that by increasing the pH, the concentration of hydronium ions $(H_3O)^+(H_2O)_3$ decreases in the adsorption layer on (100), which would favour the entering of solute in the steps.

5) pH-IMPURITIES-SUPERSATURATION

In most cases, the effect of pH is particularly important when some impurities are present in the growth medium, because it influences for instance the formation either of zwitterions or of complex ions, the efficiency of which is greater than that of the initial impurity.

Thus, when glycine is added to the solution, it is the $NH_3^+-CH_3-COO^-$ bipolar ions which are responsible of the appearance of {110} on the growth habit of NaCl crystals |38,58,59|. They adsorb along the <001> steps, with weak lateral interactions in the <110> direction, an interpretation which explains very well the striations of (110) in the <001> direction. At the isoelectric point (pH ~ 5.7), where the zwitterions are the most abundant in the solution, the critical supersaturation for the appearance of {110} is 10 times lower than at pH ~ 2.5 or 10 where the monopolar ions are the most abundant.

On the other hand, when NaCl crystals grow in presence of $HgCl_2$ as an impurity, the {110} growth form is due to the complex $HgCl_4^{2-}$ ions |38,60| and not to the $HgCl_2$ molecules as it has been supposed previously |59|. In this case the critical supersaturation necessary to obtain {110} decreases slightly when the pH decreases by addition of HCl, but this pH effect is almost an artifact. Indeed, in this process, the cation is not of great importance, since the impurity effect is the more pronounced when there are $4Cl^-$ ions for 1 Hg^{2+} ion in the solution. Thus, when the molar concentration ratio $NaCl/HgCl_2$ reaches the value 2, {110} is the sole growth form of the NaCl crystals, but it must be noticed also that for this value the solution has the composition of a mixed salt $Na_2HgCl_4,3H_2O$. For this reason the habit change has been attributed to the existence of a two-dimensional adsorption layer having the same composition. In such a layer the lateral bondings are strong, which explains that here the (110) faces remain perfectly smooth during the crystal growth.

6) IMPURITIES-SUPERSATURATION

Since the first quantitative studies |61|, the simultaneous influence of impurity concentration (C_i) in the solution and of supersaturation has been largely used to obtain habit changes. As the incorporation of impurities in the crystal is not a necessary condition (concave dissolution experiments |15,16|), the majority of interpretations appeals to the existence of an adsorption layer, the efficiency of which increases generally with C_i.

Certain studies, pointing out a negative adsorption |62|, i.e. an increase of the solvent molecules concentration on the crystal surface and a related removal of the impurities from the adsorption layers when C_i increases, have been later on invalidated by studies of adsorption isotherms |63,64| or epitaxies |65| with the same crystal-impurity systems (NaCl with Pb^{2+}, Urea, Cd^{2+}).

6.1) Adsorption sites.

Adsorption in *kinks* has been invoked to explain the decrease of the normal growth rates of (100) and (111) faces of $NaClO_4$ |66,67| and $NaClO_3$ |68| crystals growing in presence of Na_2SO_4. It is assumed that adsorption has an effect of passive resistance and a Langmuir type equation relates the normal growth rates of the faces observed in presence of impurity to that observed in pure solution and to C_i. In the case of $NaClO_3$ crystals, the (100) faces are quickly saturated by the impurity whereas the (111) faces are saturated more slowly, the adsorption isotherms being respectively of Langmuir and Temkin type. As has been pointed out |69,70|, if one considers the lateral advancement velocities of steps instead of the normal growth rates, it must be concluded that the adsorption in kinks has as effect to increase the mean distance between them and accordingly to decrease their density, resulting finally in a decrease of the growth rate.

Adsorption in *steps* is often difficult to distinguish from adsorption in kinks or in surface sites. Nevertheless in the case of the (100) faces of ADP crystals, growing in presence of $CrCl_3, 6H_2O$, it has been shown that the decrease of step velocities is due to a progressive increase of the adsorption in the steps when C_i increases |71|. Similar observations have been made on the (100) faces of sucrose crystals growing in presence of raffinose |72| whereas in presence of dextrose |73| the steps are higher but with the same velocities in regard to the pure solution. The interpretation given for ADP crystals |71| is based on the hypothesis |74| of a progressive coverage of the steps by the impurity. Below a critical concentration C_i of impurity, the latter has no effect, while above the critical concentration the step velocities decrease very fast. This interpretation rejoins the one proposed for sucrose, which takes into account the distance d between adsorbed impurities and the diameter $2\rho_c$ of the critical two-dimensional nucleus |72|. When $d > 2\rho_c$ the step squeezes itself between the impurities, whereas when $d < 2\rho_c$ it is stopped. But the phenomena are perhaps somewhat more complicated since the concentration of adsorbed impurities in the steps sometimes increases with the distance of the steps to their sources (Pb^{2+} on NaCl |75|).

Adsorption in *surface sites* has been suggested to explain the effect of $AlCl_3$ and $FeCl_3$ on ADP crystals |71|. The impurity is assumed to diminish the number of adsorption sites available for the solute resulting in a decrease of the flux towards the steps and the kinks. This theory, in which surface diffusion plays the most important rôle, has been elaborated |76| from the classical surface diffusion theory |77| in order to explain the experimental results of the growth of sucrose crystals in presence of raffinose and gentianose |78|.

Another general interpretation |79|, based on a higher or lower mobility of the adsorbed impurities, suggests that these impurities according to their spacings, influence not only the step velocities but also their sources. According to the growth conditions this can lead to the formation of whiskers as in the case of KCl |80| or NaCl |38,59| crystals growing in presence of $Fe(CN)_6^{4-}$ ions. It must be noticed, however, that these ions are most efficient when their concentration in the solution is such that the formation of a complete adsorption monolayer is possible |81|.

6.2) Adsorption layers

In order that S or K faces appear in the crystal habit, the adsorbed impurities must prevent the normal growth of the faces and favour on the other hand the lateral extension of growth layers |82,83|. The impurities bring thus to the S and K faces some new PBC which normally do not exist in the pure growth medium. By increasing the impurity concentration the last step of the process is the creation of an epitactic adsorption layer inside of which there are strong lateral interactions. This adsorption layer imposes then its own PBC on the crystal faces.

The composition of the epitactic layer is in most cases hypothetic, especially when two-dimensional adsorption compounds are involved. Even without direct experimental evidence for their existence, they have nevertheless been suggested long time ago and discussed in detail with the aid of a certain number of examples |14|. Recently some studies of adsorption isotherms |64| and of epitaxies |65| have shown that the number of Cd^{2+} ions adsorbed on the (100) and (111) faces of NaCl, when the isotherms are saturated, is the same as the number of Cd^{2+} ions contained in the epitaxial planes of the mixed salt $Na_2CdCl_4,3H_2O$ growing on the (100), (110) and (111) faces of NaCl. It has then be concluded that the two-dimensional adsorption compound in the crystal-solution interface has the composition of the mixed salt. Furthermore, as the structural agreement is perfect only on the (111) faces of NaCl, one can then understand that {111} is the final growth form of NaCl crystals growing in the presence of cadmium. At last, as adsorption occurs as well at low (isotherms) as at high (epitaxies) impurity concentration on all crystal faces, the idea of the *selectivity* of the adsorption must be understood in terms of bonding energies and of structural quality in the adsorption layers, but not as a discrimination of adsorption between the different crystal faces.

The assumptions, previously summarized |14|, of the first studies concerning the existence in the crystal-solution interface of mixed crystals or anomalous mixed crystals, seem to be verified also by the numerous studies of the incorporation of impurities in growing crystals. It is the case for instance |84,85| of the systems $KCl-H_2O$ or NH_4Cl-H_2O with Cu, Co, Ni, Zn, Mn, Sn as impurities. From this particular point of view, we may notice that when incorporation goes with habit change, the physical properties of the crystal may be modified according to a variation of the bonding forces in the crystalline structure. Thus |86| only few Ni^{2+} ions are incorporated in TGS crystal the habit of which remains almost the same as in pure solution, while Cu^{2+} and Fe^{3+} are incorporated to a higher degree and several growth forms disappear from the crystal habit.

6.3) Special cases

Till now, we have only considered some systems where the impurities lead to a decrease of the growth rates of certain crystal faces. Sometimes however, habit changes are more complex since an impurity can slow down the development of a face and speed up the development of another. So, the normal growth rate of the (001) and (0$\bar{1}$0) faces of sodium triphosphate hexahydrate crystals is reduced in the presence of traces of dodecylbenzenesulphonate, whereas the growth rate of (100) is enhanced at low supersaturation and later on is reduced at higher supersaturation |87|. The adsorption isotherms show that the impurity is localized preferably on (0$\bar{1}$0) resulting in a hindrance for incorporation of the solute. It would be the incorporation kinetics of the large $P_3O_{10}^{5-}$ ions which would be responsible for the growth rate changes of the faces since the latter impose different structural conditions. In the case of the action of phenol on the growth of NaCl crystals |88|, the increase of the normal growth rate of the (100) face is explained by a decrease of the interfacial crystal-solution energy. The distance between steps would be lowered resulting in a higher kink density. On the (111) face, the adsorption of phenol is higher, resulting in a normal decrease of its growth rate.

We may mention also a particular effect of some impurities on habit taking as an example the growth of spherocobaltite crystals ($CoCO_3$). If the impurity is LiCl, the crystallization leads to an anhydrous salt, while in presence of $CoCl_2$ one obtains an hydrated salt. These effects are attributed to the activity coefficient of the cation and to its hydration number. As Li^+ is better bonded to the water molecules of the solution than Co^{2+}, the crystallization of the anhydrous salt is favoured.

7) CONCLUSION

From the examples given above, it appears that most habit change interpretations are based on the existence of an adsorption layer, more or less epitaxial, in the crystal-solution interface. This layer is composed of the solvent, the impurity, or an anhydrous or hydrated mixed salt. The real composition remains however generally hypothetical.

As habit changes are due to the action of the relative growth rates of the faces, a complete theoretical interpretation must be based on growth kinetic equations. Unfortunately, experiments do not allow to have access to a large number of parameters, because of the complexity of the growth medium and of the weak investigation methods in situ of the crystalline surfaces on a molecular scale.

According to growth rate equations |77,90-92| and considering that adsorption lowers for instance the edge or surface energies, or the size of the critical two-dimensional nucleus, we see that the expected result from adsorption is an increase in the crystal growth rates. Other parameters must then act in the opposite direction in order to explain the decrease of the growth rates generally observed in habit change phenomena. We can mention here the slow down of the effective supersaturation and of the solute flux towards the steps |76| or the decrease of the lateral advancement velocity of growth layers due to step pinning |72,74,79| or to a decrease of number of kinks available for growth |69,70|. In some cases, the experiment allows to point out the competition among the different parameters which tend either to enhance or to slow down the normal growth rate of a face (hexatriacontane |93|). The reader will find in two recent papers |94,95| the references and some comments on all the growth rate theories.

As the kinetic phenomena are at the present time too difficult to describe, habit changes are mostly interpreted with the aid of crystallochemical considerations. Apart from the fact that these interpretations consider habit changes from a static point of view, it must be emphasized that the PBC theory |22,23,83| gives remarkable qualitative results when applied to adsorption problems.

REFERENCES

1) H.E. BUCKLEY, Crystal Growth, Wiley, New York (1951).
2) B. HONIGMAN, Gleichgewichts-und Wachstums-formen Von Kristallen, Steinkopf, Darmstadt (1958).
3) Growth and Perfection of Crystals, R.H. Doremus-B.W. Roberts-D. Turnbull, Wiley, New York (1958).
4) A. SMAKULA, Einkristalle, Springer Verlag, Berlin (1961).
5) Adsorption et Croissance Cristalline, Coll. Int. CNRS n° 152, CNRS, Paris (1965).
6) R.F. STRICKLAND-CONSTABLE, Kinetics and Mechanisms of Cristallization, Academic Press, London and New York (1968).
7) G. MATZ, Kristallisation, Springer Verlag, Berlin (1969).
8) R.A. LAUDISE, The Growth of Single Crystals, Prentice Hall Inc., Englewood Cliffs, New Jersey (1970).
9) H.K. HENISCH, Crystal Growth in Gels, The Pensylvania State University Press, London (1970).
10) J. NYVLT, Industrial Crystallization from Solutions, Butterworths, London (1971).
11) J.W. MULLIN, Crystallisation 2nd ed., Butterworths, London (1972).
12) M. OHARA, R.C. REID, Modeling Crystal Growth Rates from Solution, Prentice Hall Inc. Englewood Cliffs, New Jersey (1973).
13) N.N. SHEFTAL, in Growth of Crystals, A.V. Shubnikov-N.N. Sheftal, Consultants Bureau, New York 1 (1958) 5 and 3 (1962) 3.
14) P. HARTMAN, in ref. 5. p. 477.
15) R. KERN, Bull. Soc. Fr. Min. Crist. 91 (1968) 247.
16) R. KERN, in Growth of Crystals, N.N. Sheftal, Consultants Bureau, New York, 8 (1969) 3.
17) R.L. PARKER, Solid State Physics, 25 (1970) 151.
18) P. HARTMAN, in Crystal Growth - An introduction, P. Hartman, North Holland Pub. Co. Amsterdam (1973) 367.
19) P. CURIE, Bull. Soc. Fr. Miner. 8 (1885) 145.
20) J.W. GIBBS, Scientific Papers, Dover Publications Inc, New York (1961) 219.
21) G. WULFF, Z. Krist. 34 (1901) 449.
22) P. HARTMAN, W.G. PERDOK, Acta Cryst. 8 (1955) 49, 521, 525.
23) P. HARTMAN, Z. Krist. 119 (1963) 65.
24) I.N. STRANSKI, Bull. Soc. Fr. Miner. Crist. 79 (1956) 360.
25) R. LACMANN, I.N. STRANSKI, in ref. 3, p. 427.

26) G.G. LEMMLEIN, Dokl. Akad. Nauk, SSSR, 98 (1954) 973.
27) M.O. KLIYA, Dokl. Akad. Nauk, SSSR, 100 (1955) 259.
28) M. BIENFAIT, R. KERN, Bull. Soc. Fr. Miner. Crist. 87 (1964)604.
29) R. KERN, Bull. Soc. Fr. Miner. Crist. 76 (1953) 325.
30) H. ESPIG, H. NEELS, Kristall und Technik, 2, 3 (1967) 401.
31) G.D. BOTSARIS, G.E. DENK, R. A. SHELDEN, Kristall und Technik, 8, 7 (1973) 769.
32) A.V. BELYUSTIN, V.F. DVORYAKIN, in Growth of Crystals, A.V. Shubnikov-N.N. Sheftal, Consultants Bureau, New York, 1 (1958) 139.
33) G.K. KIROV, L. FILIZOVA, Kristall und Technik, 5, 3 (1970) 387.
34) G. BLIZNAKOV, E. KIRKOVA, R. NIKOLAEVA, Kristall und Technik 6, 1 (1971) 33.
35) P. HARTMAN, Acta Cryst. 11 (1958) 459.
36) S. ASLANIAN, I. KOSTOV, Kristall und Technik, 7, 5 (1972) 511.
37) M. BIENFAIT, R. BOISTELLE, R. KERN, in ref. 5, p. 515.
38) R. BOISTELLE, Thesis, University Nancy, France (1966).
39) W. KLEBER, Z. Phys. Chem. 206 (1957) 327.
40) W. KLEBER, Z. Krist. 109 (1957) 115.
41) A.F. WELLS, Phil. Mag. 37 (1946) 184.
42) M. LEDESERT, J.C. MONIER, in ref. 5, p. 537.
43) R. CADORET, J.C. MONIER, in ref. 5, p. 559.
44) W. SKODA, M. VAN DEN TEMPE L, J. of Crystal Growth, 1 (1967) 207.
45) B. SIMON, A. GRASSI, R. BOISTELLE, J. of Crystal Growth, 26 (1974) 77.
46) R. BOISTELLE, A. DOUSSOULIN, to be published in J. of Crystal Growth.
47) M. LEDESERT, P. HARTMAN, J.C. MONIER, J. of Crystal Growth, 15 (1972) 133.
48) G.K. KIROV, I. VESSELINOV, Z. CHERNOVA, Kristall und Technik, 7, 5 (1972) 497.
49) S. TROOST, J. of Crystal Growth, 13/14 (1972) 449.
50) M. KAHLWEIT, J. of Crystal Growth, 7 (1970) 74.
51) V.V. SIPYAGIN, Kristallografiya, 12 (1967) 678.
52) V.V. SIPYAGIN, A.A. CHERNOV, Kristallografiya 17 (1972) 1009.
53) A.A. CHERNOV, Annual Review of Materials Science, 3 (1973) 397.
54) H.M. LIAW, J.W. FAUST Jr., J. of Crystal Growth, 13/14 (1972)471.
55) I.M. BYTEVA, in Crystallization Processes, N.N. SIROTA-F.K. GORSKII-V.M. VARIKASH, Consultants Bureau, New York(1966)99.
56) J.W. MULLIN, A. AMATAVIVADAHANA, M. CHAKRABORTY, J. Appl. Chem. 20 (1970) 153.
57) R.J. DAVEY, J.W. MULLIN, J. of Crystal Growth, to be published.
58) M. REDOUTE, R. BOISTELLE, R. KERN, C.R. Paris 260 (1965) 2167.
59) M. BIENFAIT, R. BOISTELLE, R. KERN, in rèf. 5, p. 577.
60) M. REDOUTE, R. BOISTELLE, R. KERN, C.R. Paris, 262 (1966) 1081.
61) R. KERN, Bull. Soc. Fr. Min. Crist. 76 (1953) 365.
62) M. HILLE, Ch. JENTSCH, Z. Krist. 118 (1963) 283 and 120(1964)323.
63) H. KARGE, Kristall und Technik, 3, 4 (1968) 537.
64) R. BOISTELLE, M. MATHIEU, B. SIMON, Surf. Sci. 42 (1974) 373.

65) R. BOISTELLE, B. SIMON, J. of Crystal Growth 26 (1974)140.
66) G. BLIZNAKOV, in ref. 5, p. 291.
67) G. BLIZNAKOV, E. KIRKOVA, Z. Phys. Chem. 206 (1957) 271.
68) G. BLIZNAKOV, E. KIRKOVA, Kristall und Technik, 4, 3 (1969) 331.
69) A.A. CHERNOV, in Growth of Crystals, A.V. Shubnikov-N.N. Sheftal, Consultants Bureau, New York, 3 (1962) 31.
70) A.A. CHERNOV, in ref. 5, p. 265.
71) R.J. DAVEY, J.W. MULLIN, J. of Crystal Growth, 26 (1974) 45.
72) W.J. DUNNING, R.W. JACKSON, D.G. MEAD, in ref. 5. p. 303.
73) W.J. DUNNING, N. ALBON, in ref. 3, p. 446.
74) G.W. SEARS, J. Chem. Phys. 29 (1958) 1045.
75) L. MALIKSKO, L. JESZENSKY, J. of Crystal Growth 15 (1972) 243.
76) K.A. BURRILL, J. of Crystal Growth 12 (1972) 239.
77) W.K. BURTON, N. CABRERA, F.C. FRANK, Phil. Trans. Roy. Soc. London A 243 (1951) 299.
78) B.M. SMYTHE, Australian J. Chem. 20 (1967) 1087, 1097, 1115.
79) N. CABRERA, D.A. VERMILYEA, in ref. 3, p. 393.
80) U. STEINIKE, Z. Anorg. Allgem. Chem. 317 (1962) 186.
81) M.A. VAN DAMME-VAN WEELE, in ref. 5, p. 433.
82) G.A. WOLFF, Z. Phys. Chem. 31 (1962) 1.
83) P. HARTMAN, R. KERN, C.R. Paris 258 (1964) 4591.
84) U. STEINICKE, Kristall und Technik, 1, 1 (1966) 113.
85) U. STEINICKE, Kristall und Technik, 6, 1 (1971) 7, 17.
86) F. MORAVEC, J. NOVOTNY, Kristall und Technik, 6, 3 (1971) 335.
87) S. TROOST, J. of Crystal Growth 3/4 (1968) 340.
88) E. KIRKOVA, R. NIKOLAEVA, Kristall und Technik, 6, 6 (1971) 741.
89) N. Yu, IKORNIKOVA, in Crystal Growth, A.V. Shubnikov-N.N. Sheftal, Consultants Bureau, New York, 3 (1962) 297.
90) W.B. HILLIG, Acta Met. 14 (1966) 1868.
91) A.A. CHERNOV, Sov. Phys. Usp. 4 (1961) 116.
92) G.H. GILMER, R. GHEZ, N. CABRERA, J. of Crystal Growth 8(1971)79.
93) B. SIMON, A. GRASSI, R. BOISTELLE, J. of Crystal Growth, 26 (1974) 90.
94) A.R. KONAK, J. of Crystal Growth 19 (1973) 247.
95) P. BENNEMA, J. of Crystal Growth 24/25 (1974) 76.

SOME PROBLEMS OF CRYSTAL HABIT MODIFICATION

E.V. KHAMSKII

Institute of Rare Elements and Minerals

Academy of Sciences, Kolar, U.S.S.R.

Many problems in the chemical and other branches of process industry are closely connected with crystal habit modification, e.g. the prevention of crystal product caking, the preparation of easily filtering precipitates, and substances with optimal flow and storage characteristics and so on (1, 2). Of course, crystal habit modification is also a subject of considerable fundamental interest and this aspect of the problem must also be considered here.

Without altering their internal structure, crystals can change their external appearance due to variations in face dimensions or the appearance or disappearance of some faces (1-4). The linear growth rates of crystal faces depend on a number of crystallization conditions including the presence of impurities in the system.

Different mechanisms of impurity influence on crystal habit may be proposed (1, 3). The influence can be connected either with changes in supersaturated solution properties or with their direct interaction with the growing crystal faces. The case where a new chemical compound is formed as a result of reaction between the impurity and the substance being crystallized is not considered here because this is not habit modification in its strict meaning.

By changes in supersaturated solution properties we mean the formation of various complexes, changes in the solution ionic strength and in its structure. By interaction between the impurity and the crystal surface we mean adsorption, incorporation into the crystal lattice, concentration in the surface layers, etc. In any case, the change in solution ionic strength has an effect on the solubility of the crystallizing substance and hence on the level of supersaturation. In turn, supersaturation changes result in changes in linear growth rates of the crystal faces, these changes being different for different faces. As a result of all these effects the crystal appearance is changed.

The interaction of an impurity with the crystal surface can lead to the localization of the growth centres on the faces, resulting in habit modification. Other interactions are also possible, but in the majority of reports dealing with the properties and selection of habit modifiers the mechanism of the impurity action is rarely linked with the general theory of crystal growth. However the key to understanding the habit modification mechanism lies in the correct combination of the growth mechanism with that of impurity behaviour in the system under investigation. Hence crystal habit modification problems can be solved only if we take into consideration general conditions of precipitate formation.

In order to solve the problem of the selection of active modifiers, it is necessary to solve some co-ordinating problems such as establishing quantitative connections between crystal habit and crystallization conditions in binary systems, ascertaining the degree of interaction between the modifier and the crystal both in solution and in the solid phase, depending on their chemical and structural properties, etc.

Consider the question of the efficiency of the impurity on the crystal habit in relation to crystallization conditions. First of all it should be noted that crystal habit can also change in the absence of impurities (1, 2, 5, 6). For example, Fig.1 shows some growth rate data for potassium nitrate crystallized from aqueous solution under different degrees of supersaturation and at different temperatures. Potassium nitrate can be crystallized both in the form of oblong prisms and rounded tablets (Fig.2). Let us call the

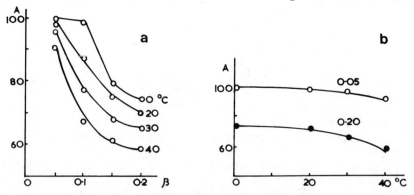

Fig. 1 Dependence of potassium nitrate crystal habit on (a) the initial relative supersaturation, β, at different temperatures and (b) the temperature of precipitation under different initial values of β. A = percentage of isometric form crystals.

PROBLEMS OF CRYSTAL HABIT MODIFICATION

oblong prisms 'non-isometric' and the rounded particles 'isometric'. Judging by the results in Fig.1 crystals of different habit are formed depending on the initial solution supersaturation. At low supersaturations isometric crystals are mainly formed, while at higher values more non-isometric crystals appear.

It should be noted that the data in Fig 1 are for agitated solutions. If the solutions are not being agitated non-isometric crystals are formed in spite of the level of supersaturation. For a given supersaturation value a temperature rise encourages non-isometric crystals to form (Fig.1b). So mixing, initial supersaturation and a temperature decrease appear to promote isometric

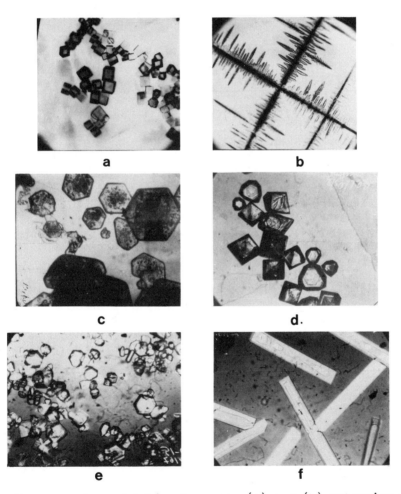

Fig.2 Some examples of habit changes: (a) and (b) potassium nitrate, (c) and (d) barium nitrate, (e) and (f) ammonium chloride.

forms. The higher the initial supersaturation the greater is the velocity of crystal formation and growth (1, 2). Therefore, with the average linear growth rate decreasing, the differences between the linear growth rates of individual faces disappear. It is important to realise that not only impurities but other conditions of crystallization can also promote habit modification. It is natural to assume that impurities influences will be different with other conditions of crystallization being different.

In Table 1 data are reported for the influence of lead nitrate on the habit of potassium nitrate crystals. β = relative supersaturation, C_{eq} and C'_{eq} = solubility in the binary system and in the presence of impurity respectively, C_{lim} = limiting concentration of the impurity. All the concentrations are expressed in g per 100 cm^3 H$_2$O. In the present case C_{lim} decides the transition from an isometric form to a non-isometric one. At impurity concentrations less than C_{lim}, non-isometric crystals are formed. Two values of the supersaturation, β, corresponding to potassium nitrate solubility in the presence and absence of impurity are presented in Table 1. In both cases, however, the basic salt concentration in the initial solution is the same. Comparing the values in Columns C_{eq} and C'_{eq} we could come to the conclusion that the impurity influence consists in changing the solubility of KNO_3 and the solution supersaturation. Confirmation of this is found by comparing the results in Table 1 and Fig.1.

Another mechanism of impurity influence is observed when methylene blue interacts with barium nitrate to change the crystal habit (8). This salt crystallizes from pure solution as octahedra (4) and from the solutions containing methylene blue as hexagonal plates (see Fig.2). Limiting methylene blue concentrations converting octahedra into lamellar crystals are given in Table 2.

Methylene blue does not change solubility of the barium nitrate to any significant extent so only one value of the initial supersaturation, β, is given in Table 2. The limiting concentration does not depend on the initial supersaturation but it increases as the temperature is increased.

However, if we take into consideration the molal ratio and the fact that barium nitrate solubility increases with temperature, it turns out that less and less methylene blue per unit mass of precipitate is required as the temperature increases and the crystallization process accelerates. In other words, the impurity modifying effect increases. We suppose we are dealing with the selective adsorption of methylene blue on the faces of barium nitrate crystals (8). This adsorption is likely to take place more easily on defects, the number of which depends on the rate of precipitation and the temperature. In this case a definite connection between the impurity influence and crystallization conditions is observed.

Table 1. The influence of lead nitrate on the habit of potassium nitrate crystals

°C	β c_{eq}	c'_{eq}	c_{lim}	Molal ratio impurity: salt
0	0.10	0.076	0.87	1 : 55
	0.15	0.81	3.05	1 : 16
	0.20	0.119	4.93	1 : 11
20	0.05	0.047	0.17	1 : 65
	0.10	0.067	3.32	1 : 34
	0.15	0.074	7.68	1 : 16
	0.20	0.081	12.21	1 : 10
30	0.05	0.028	2.40	1 : 65
	0.10	0.046	6.02	1 : 27
	0.15	0.042	12.33	1 : 14
	0.20	0.039	18.62	1 : 10
40	0.05	0.022	3.37	1 : 65
	0.10	0.023	11.30	1 : 20
	0.15	0.021	18.46	1 : 13
	0.20	0.020	26.19	1 : 10

Table 2. The influence of methylene blue on barium nitrate crystal habit.

°C	β	c_{lim}	Molal ratio impurity: salt
0	0.2	0.01	1 : 730
	0.3	0.01	1 : 790
	0.4	0.008	1 : 1060
	0.5	0.008	1 : 1140
20	0.3	0.012	1 : 1200
30	0.3	0.014	1 : 1300

Ammonium chloride belongs to the salts which are easily crystallized in the form of dendrites (Fig.23). The crystal form is gradually changed in the presence of impurity until cubic crystals appear (3, 9) (Fig.2f). Data concerning the influence of supersaturation and temperature on the limiting concentrations of impurity are presented in Figs.3 and 4. The concentrations beginning with which ammonium chloride was crystallized in the cubic form only were also assumed to be the limiting ones. As seen in the figures, the value of C_{lim} is increased with increases in the relative supersaturation and temperature. But when the temperature and supersaturation are increased, the quantity of the precipitate increases

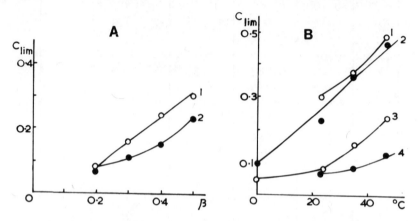

Fig. 3 Ammonium chloride. A: dependence of the modifier limiting concentration C_{lim} on the initial relative supersaturation β of the solution at 20°C (1 - $FeCl_3$, 2 - $CuCl_2$). B: dependence of C_{lim} on the crystallization temperature β = 0.05, 1 - $FeCl_3$, 2 - $CuCl_2$; β = 0.20, 3 - $CuCl_2$, 4 - $FeCl_3$.

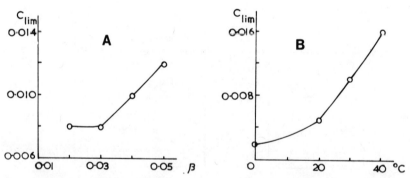

Fig. 4 Ammonium chloride. A: dependence of murexide limiting concentration C_{lim} on the solution initial relative supersaturation β at 30°C. B: dependence of C_{lim} on the crystallization temperature with β = 0.04.

as well, although a direct connection between the precipitate mass increase and the limiting concentration change was not observed (9). For instance, the precipitate mass being increased by 2.5, the limiting concentrations of different impurities increase by 1.5 to 5. At the same time insignificant changes of the precipitate mass under the influence of temperature change (by 1.5) results in changing C_{lim} by 1.5 to 4. These results tend to support the assumption that crystallization conditions greatly affect the impurity action efficiency. Most likely in this case we are dealing with an adsorption mechanism of habit modification. To a certain extent this concept included the above-mentioned observations as to both chemical and physical nature of impurities influence.

Even a small number of examples of crystal habit modification enables us to conclude that for the selection of suitable impurities it is necessary to take into consideration both the properties of the substance being crystallized and the crystallization conditions. When studying modifier properties it is necessary to indicate conditions of their application and to fix parameters of the crystallization process and the impurity behaviour. Information about the impurity should contain not only data about its nature but also about possible chemical interaction in solid and liquid phases and about impurities contained in the crystals. All this information must be taken together to enable us to work out the scientific foundations of habit modifier selection.

REFERENCES

1. Khamskii, E.V., "Crystallization from Solution", "Nauka", Leningrad, (1967).
2. Mullin, J.W., "Crystallization", 2nd Edition, Butterworths, (1972).
3. Kuznetsov, V.D., "Crystals and Crystallization", Gos.isd.technotero,lit.M., (1954).
4. Buckley, H.E., "Crystal Growth", Chapman and Hall, London, (1952).
5. Chernov, A.A., Collected articles in "Crystal Growth", Isd.AN SSSR, M., 3 (1961) 47.
6. Belustin, A.V. and Dvoryankin, V.F., Collected articles in "Crystal Growth", Isd. AN SSSR, M., (1957) 147.
7. Khamskii, E.V. and Dvegubsky, N.S., Zhur.Prinkl.Chim. 44 (1971) 468.
8. Khamskii, E.V. and Dvegubsky, N.S., Teor.Osn.Chim.Technol. 5 (1971) 823.
9. Khamskii, E.V. and Dvegubsky, N.S., Zhur.Prinkl.Chim. 46 (1973) 2307.
10. Kirkova, E. and Nikolaeva, R., Krist.und Techn. 6 (1971) 741.
11. Kern, R. and Tollimanu, M., Compt.Rend. 236 (1963) 942.
12. Belikova, M.I. and Neiman, O.V., Kolloid Zhur. 35 (1973) 1037.

SOLVENT EFFECTS IN THE GROWTH OF HEXAMETHYLENE TETRAMINE CRYSTALS

J.R. BOURNE AND R.J. DAVEY

Swiss Federal Institute of Technology

CH-8006 Zurich, Switzerland

INTRODUCTION

Gilmer and Bennema (1) have shown that the mechanism by which a crystal surface grows, is determined by a surface energy parameter (hereafter the α factor) given by:

$$\alpha = \frac{2(\Phi_{ss} + \Phi_{ff}) - 4\Phi_{sf}}{kT} \quad [1]$$

This definition results from the use of a Tempkin interface model in which the solid and fluid phases are divided into blocks Φ_{ss}, Φ_{ff} and Φ_{sf}, then being the bond energies between adjacent solid blocks, fluid blocks and solid and fluid blocks respectively. For a simple (001) face of a Kossel crystal, computer simulation of crystal growth (1,2) has shown that when $\alpha < 3.2$ the interface is molecularly rough and growth is continuous. As α rises above 3.2, the interface becomes molecularly smoother so that for $3.2 < \alpha < 5.2$ nucleation (birth and spread) growth is possible in the absence of steps, while for $\alpha > 5.2$ only step growth requiring dislocations is possible.

α as defined by equation [1] is experimentally accessible if complete wetting of the surface is assumed since then $\Phi_{sf} = \Phi_{ff}$ and it may be shown that (3):

$$\alpha = \frac{n_s}{n_t} \frac{\Delta H}{kT} \quad [2]$$

where n_s and n_t are the numbers of nearest neighbours in the surface and in the bulk respectively. ΔH for growth from solution may be taken to be the heat of solution. We note that α cannot be calculated from equation [2] when dissolution is exothermic (i.e. ΔH negative). In solution growth, crystal forms are bounded by the slowest growing faces which are generally, in the notation of Hartman and Perdok (4), F-faces and often assumed to require nucleation or dislocation centres to grow. However, it is clear from the definition of α given by equation [1] that the mechanism of growth is not determined solely by the bonding in the solid phase as assumed in the Hartman and Perdok-type analyses, but also by the solvent-solvent and solute-solvent interactions in the fluid phase. For example, if we consider the growth of a material from two different solvents A and B, we have two values of α:

$$kT\,\alpha^A = 2\,\Phi^A_{ss} + 2\,\Phi^A_{ff} - 4\,\Phi^A_{sf} \quad \text{and:}$$

$$kT\,\alpha^B = 2\,\Phi^B_{ss} + 2\,\Phi^B_{ff} - 4\,\Phi^B_{sf} \quad \text{Thus since } \Phi^B_{ss} = \Phi^A_{ss},$$

$$\alpha^A = \alpha^B + \frac{4(\Phi^B_{sf} - \Phi^A_{sf})}{kT} + \frac{(\Phi^A_{ff} - \Phi^B_{ff})}{kT}$$

Thus if, for example, $\Phi^A_{ff} \approx \Phi^B_{ff}$ it is clear that, if the solute-solvent interactions are stronger in solvent B than in solvent A, $\alpha^B < \alpha^A$. Hence the stronger the solvent-solute interactions the smaller will be the α value for the interface. It follows therefore that, since α determines the growth mechanism of an interface, solvent-solute interactions may play a role in determining the measured kinetics of crystal growth.

Following Bennema et al. (2), we shall consider two theoretical kinetic relationships relating the rate of crystal growth to the supersaturation:

a) the Burton-Cabrera-Frank (BCF) surface diffusional model

$$\dot{L} = C\,\frac{\sigma^2}{\sigma_1}\,\tanh(\sigma_1/\sigma) \qquad [3]$$

in which $C = \dfrac{kT}{h}\,\beta\,c_o\,\Lambda\,\Omega\,N_o\,\exp(-\Delta G_{des}/kT)$

and $\sigma_1 = \dfrac{9.5}{S}\,\dfrac{\gamma'}{kT}\,\dfrac{a}{\lambda_s}$ \qquad [3a]

b) the Birth and Spread (B and S) nucleation model

$$\dot{L} = A \sigma^{5/6} \exp(-B/\sigma) \qquad [4]$$

where $A = (\dfrac{2\pi C_o}{3})^{1/3} \dfrac{2 \lambda_s C}{a}$

and $B = \dfrac{\pi}{3} (\dfrac{\gamma'}{kT})^2 \qquad [4a]$

The overall magnitude of the growth rate given by equations [3] and [4] is determined by the parameters C and A, while σ_1 and B decide the shape of the curves. Further, since the factor γ'/kT which appears in the above expressions for σ_1 and B, is related to α through the equality (2):

$$\dfrac{\gamma'}{kT} = \dfrac{\alpha}{4} - \Delta S_{cn} \qquad [5]$$

It follows that solvent-solute interaction which tend to decrease α will also decrease γ'/kT. Indeed, Bennema et al. (2) find for a Kossel crystal (001) face that as α falls below 3.2 γ'/kT becomes zero. This implies that both σ_1 and B become zero, so that equations [3] and [4] reduce to:

$$\dot{L} = C \sigma \qquad [3b]$$

and $\dot{L} = A \sigma^{5/6} \qquad [4b]$

respectively as α falls below 3.2.

It may thus be concluded that through their effect on α, solute-solvent interactions can determine the form of the measured relationship between the crystal growth rate and supersaturation, a linearisation occuring when solvent-solute interactions cause α to decrease.

In this paper we shall consider the above predictions in the light of existing and newly measured data for the growth of hexamethylene tetramine (HMT) from aqueous solution (5), ethanolic solution and from the vapour phase (6), the latter being considered the limiting case of solute-solvent interactions in which $\Phi_{sf} = \Phi_{ff} = 0$.

THE NATURE OF SOLVENT SOLUTE INTERACTIONS IN AQUEOUS AND ETHANOLIC HMT SOLUTIONS

The available thermodynamic data relating to aqueous and ethanolic solutions of HMT at 25°C is summarised in Table 1.

Table 1: Thermodynamics of HMT solutions.

Solvent	ΔH_f kJ/mol	ΔH_S^m kJ/mol	ΔH_S^i kJ/mol	Activity coefficient solvent	HMT
C_2H_5OH	18.8	15.8	18.5	0.99 a)	2.61 c)
H_2O	18.8	-12.6	-0.7	0.85 b)	0.27 c)

a) obtained from ebullioscopic data.
b) obtained from Ref. (11).
c) calculated from the van't Hoff relationship.

In this Table, ΔH_S^m is the measured heat of solution, ΔH_S^i the heat of solution estimated from the available solubility data for the two systems using the van't Hoff equation, and ΔH_f the heat of fusion of HMT. For an ideal solution, all three quantities should be equal. Since the molten state of HMT is a hypothetical one (due to sublimation), the value of ΔH_f was estimated by measuring the heat of solution in the alcohol series CH_3OH, C_2H_5OH and $C_5H_{11}OH$ since with increasing molecular weight of the solvent the solution may be assumed to approach ideality with ΔH_S^m reaching an assymptotic value equal to ΔH_f. Such behaviour was observed and ΔH_f estimated as 18.8 kJ/mol (7). We note that in water the solubility of HMT ($x \approx 0.1$) decreases in the temperature range 0-70°C and that dissolution is exothermic (8), while in ethanol the solubility is an order of magnitude lower ($x \approx 0.01$) but increases with increasing temperature (9) and dissolution is endothermic (7). As seen in Table 1, the ideal criterion $\Delta H_f = \Delta H_S^m = \Delta H_S^i$ is most nearly obeyed for ethanolic solutions while for water the deviation from ideality is large.

This view is substantiated by considering the relevant activity coefficients (referred to the pure components) of saturated solutions (Table 1). The values suggest qualitatively a solvation of the HMT molecules and structuring of the water in aqueous solution, while in ethanol the reverse is true for the HMT molecules and the structure of the ethanol remains essentially unchanged from that in pure ethanol.

The enormous difference in solubility of HMT in the two solvents makes a quantitative comparison rather difficult. However, by assuming the solutions to be regular, the normalised excess free energy of mixing $G^E/(x_{HMT} \cdot x_{solvent})$ may provide a more realistic basis. For saturated HMT aqueous solution we find -7.4 kJ/mol and for saturated ethanol solution 0.02 kJ/mol, again indicating the large deviation from ideality ($G^E = 0$ for an ideal solution) due to solvent-solute interactions in aqueous solutions and the near ideality of ethanolic solutions.

In the case of water solutions, the solute-solvent interactions most probably occur through hydrogen bonding, causing an increased structuring of the water lattice in which the HMT molecules occupy specific sites. This view is consistent with the exothermic heat of solution and the decrease in solubility with increasing temperature. The form of the solution structure may be expected to resemble that of the crystalline clathrate hexahydrate phase which forms below 13.5°C and in which each HMT molecule is hydrogen bonded to three water molecules (10) and occupies a cavity in an ice-like water lattice. Viscosity data of Crescenzi et al. (11) substantiate that in solutions also an average of 3 water molecules are bonded to each HMT molecule.

On the other hand if a continuous chain model is assumed to describe the structure of ethanol (12), then it must be concluded that breakage of the chains is necessary to accomodate the HMT molecules in the liquid phase and that the resulting solution is structurally similar to pure ethanol with the HMT molecules occupying random positions. This is consistent with the positive dependence of the solubility on temperature and the endothermic heat of solution.

These considerations indicate clearly that the solute-solvent interactions in aqueous solutions are much stronger than in ethanolic solutions. We may thus expect that the growth kinetics measured from the two solvents will differ due to the effect of the interactions on α, and will also differ from vapour growth in which no solvent is present. In order to predict roughly the form of the kinetic relationships in each case, we now discuss the methods, by which the factor α may be estimated.

EVALUATION OF THE α FACTOR

The habit form of HMT crystals is bounded almost entirely by {110} faces (an F-face according to Hartman and Perdok (4)) so that the following discussion relates only to these faces. The following estimation methods were used for α:

1) Direct evaluation of α.
   The use of equation [2] avoids the difficulties associated with the definition of a fluid phase block model required for equation [1] and gives useful results for growth from ethanol and

from the vapour phase when the heats of solution and sublimation respectively may be used, n_s/n_t being equal to one half for the {110} faces. This gives for growth from ethanol solution $\alpha = 3.2$ and from the vapour phase $\alpha = 15$. The negative heat of solution of HMT in water precludes the use of equation [2] in this case.

2) By estimating the pure edge energy.
According to Bennema et al. (2) for a Kossel crystal (001) face the edge energy γ'' is related to α via:

$$\gamma''/kT = \alpha/4 \qquad [5]$$

The edge energy may be calculated from the surface energy (F) using the relationship:

$\gamma'' = F \times$ step height $\times$ length of step occupied by one molecule

which for a monomolecular step on a {110} face of an HMT crystal reduces to:

$$\gamma'' = 34.8 \times 10^{-16} \, F \quad \text{(erg/molecule)} \qquad [6]$$

The surface energy may be calculated from the heat of solution assuming this to be the maximum value of the energy required to form a new phase, when:

$$F \approx \Delta H_s^m / Nf$$

For ethanolic solutions this gives $F \approx 22.8 \text{ erg/cm}^2$, but may not be used for aqueous solutions since the negative ΔH_s^m gives a meaningless result. γ'' is then calculated from equation [6] and α from equation [5] giving $\alpha = 7.7$ for growth from ethanol solution. Just as the value of F is a maximal value, this value of α should also be considered as maximal.

3) If the change in free energy (ΔF), occuring upon the creation of solid HMT can be estimated, then the graphical relationship between ΔF and α given by van Leeuwen (13) may be used to estimate α. In the present work we have estimated ΔF from the measured metastable zone widths of supersaturated aqueous and ethanolic solutions to be respectively -0.24 erg/cm^2 and -0.40 erg/cm^2 giving α values of 2 and 4.8.

The estimates are summarised in Table 2 and it is emphasized that methods 2 and 3, which rely on the results of computer simulation experiments, are only strictly valid for a simple cubic lattice, while HMT is face centred cubic. The resulting values of α are therefore considered only as a guide offering some basis on which to make predictions regarding the growth mechanism of HMT in the systems considered.

GROWTH OF HEXAMETHYLENE TETRAMINE CRYSTALS

Table 2: The estimation of α.

Solvent	Estimation method	F (erg/cm^2)	ΔF (erg/cm^2)	α
Vapour	Equation [2]			15
C_2H_5OH	Equation [2]			3.2
C_2H_5OH	$F_{max} = \Delta H/(N_o f_o)$	22.8		7.7
C_2H_5OH	Measured metastable zone width		-0.40	4.8
H_2O	Measured metastable zone width		-0.24	~ 2

We thus see that as the solvent-solute interactions decrease (aqueous solution → ethanolic solution → vapor), so the value of α increases. Using the computer simulation results of Gilmer and Bennema (1), it may be concluded that growth will be continuous from aqueous solution since $\alpha < 3.2$. From ethanolic solution $3.2 < \alpha < 8$ so that either nucleation or step growth may occur, while from the vapour since $\alpha \approx 15$ only step growth will be possible.

It follows qualitatively that the measured $\dot{L}$ (σ) relationships should show a linearisation as the growth medium is changed from vapour to ethanolic to aqueous solution.

CRYSTAL GROWTH RATE DATA

In order to test the above predictions we make use of the following data:

1) Growth from aqueous solution.

Bomio et al. (5) have reported previously that the growth of the {110} faces of HMT from aqueous solution is strongly influenced by mass transfer (volume diffusion) effects, and using this data, measured at 30°C on polycrystalline spheres at relative solution-particle velocities of 84 and 210 mm/s, the interfacial supersaturation (σ_i) during growth has been estimated via the relationship:

$$\dot{L} = \frac{k_d}{\rho} \frac{(W - W_i)}{(1 - W_i)} \qquad [7]$$

in which k_d is the mass transfer coefficient assumed to be identical in growth and dissolution, and W_i the interfacial concentration of solute. In addition to this data, the recent measurements of Hungerbühler (14), reported for growth onto a rotating polycrystalline disc of HMT have been treated in the same way, and in Fig.1 the resulting experimental points relating the {110} face growth rate to the interfacial supersaturation are shown. The agreement between measurements made on spheres and discs is excellent.

2) Growth from ethanolic solution.

New data has been measured for the growth HMT crystals from ethanolic solution. Measurements were made at 25°C in both a fluidised bed system and a simple flow system in which a single {110} face was suspended perpendicular to a solution flow stream. The results of the fluidised bed experiments showed that in the range of relative crystal-solution velocities $32 < u_r < 45$ mm/s (crystal size range 0.6 - 1.0 mm) the growth rates were independent of solution velocity. Face growth rates of the single crystals of ~ 3 mm diameter showed no dependence on solution velocity in the range 4 - 10 mm/s, and the activation energy for growth was found to be 58.8 kJ/mol. It is therefore concluded that mass transfer effects are unimportant in growth from ethanol. The measured data, shown in Fig.2, are therefore taken as representing the true surface kinetics. It should be noted that the agreement between data measured in the two different experimental situations is good.

3) Growth from the vapour.

In order to obtain growth rate supersaturation relationship for vapour growth, the data of Heyer (6) for the growth of undisturbed {110} faces of HMT measured at 50, 70 and 90°C has been used. To facilitate comparison with the solution data, the measurements have been correlated with BCF curves (equation [3]) at each temperature and the values of C and σ_1 at 25°C found by extrapolation. A growth curve corresponding to 25°C was then constructed. This is shown in Fig.3, together with the measured data for ethanol solutions and aqueous solutions, the latter being corrected to 25°C using the previously reported activation energy of 71.4 kJ/mol (5).

COMPARISON BETWEEN THEORY AND EXPERIMENT

Fig.3 shows clearly the linearisation of the growth rate supersaturation relationships which occurs when the growth medium is changed from vapour to ethanol to aqueous solution. For a given supersaturation there is also a marked increase in the rate of the surface kinetics suggesting that the number of growth sites (kink

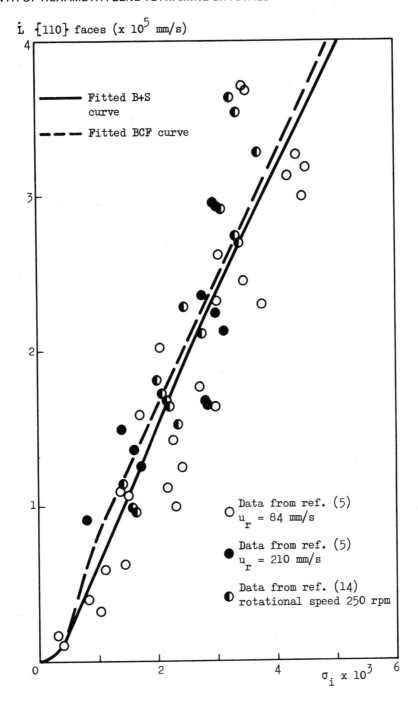

Fig. 1: The growth kinetics of HMT from aqueous solution.

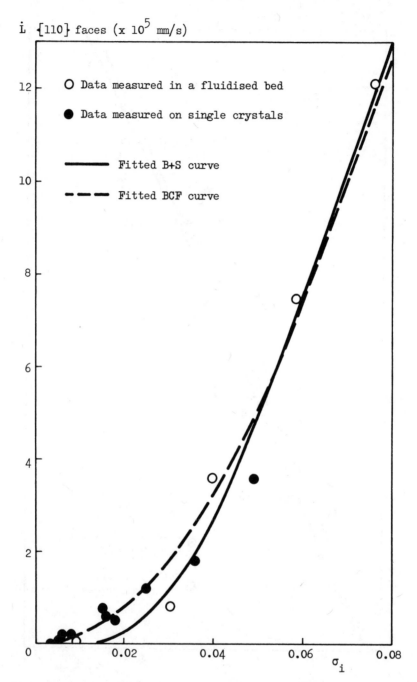

Fig. 2: The growth kinetics of HMT from ethanolic solution.

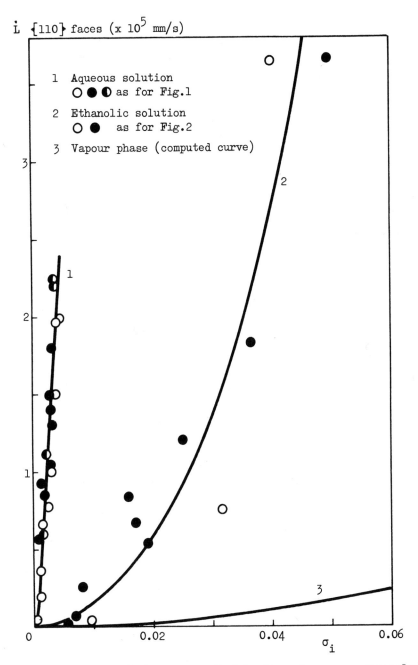

Fig. 3: Comparison between growth kinetics from aqueous solution, ethanolic solution and the vapour phase.

sites) also increases (i.e. an increase in surface roughness occurs). These observations are qualitatively in agreement with the estimated decrease in the α values occuring in the different growth mediums (Table 2). A quantitative comparison between the data and theory requires that the data be fitted to theoretical curves of the form of equations [3] and [4]. This was carried out using a non-linear regression programme based on the method of steepest descent. For each set of data, the values of C, σ_1 A and B in equations [3] and [4] best correlating the data are given in Table 3.

Table 3 : Best values of parameters.

Model	HMT-vapour	HMT-C_2H_5OH	HMT-H_2O
BCF	$C = 26.7 \times 10^{-5}$	$C = 1770 \times 10^{-5}$	$C = 815.7 \times 10^{-5}$ mm/s
	$\sigma_1 = 0.425$	$\sigma_1 = 0.886$	$\sigma_1 = 0.914 \times 10^{-3}$
B + S		$A = 2830 \times 10^{-5}$	$A = 370.2 \times 10^{-5}$ mm/s
		$B = 0.0778$	$B = 0.613 \times 10^{-3}$

The data for vapour growth was well fitted by BCF type curves in agreement with the calculated value of α = 15. At 25°C σ_1 was estimated to be 0.43. Since α = 15 is roughly equivalent to $\gamma'/kT = 3.0$ (2) and taking $1 \leq S \leq 5$ (2), we find using equation [3a] that $10a \leq \lambda_S \leq 50a$ which is reasonable.

The best BCF and B + S curves fitting the data measured in ethanolic solutions are shown in Fig.2. Unique definition of C and σ_1 was not possible since all the data points lie below σ_1. However, taking σ_1 = 0.89 and $\gamma'/kT \approx 1.0$ (this corresponds to α = 5.0)(?) we find from equation [3a] for $1 \leq S \leq 5$ that $2a \leq \lambda_S \leq 10a$. This indicates a smaller mean surface diffusion distance in growth from ethanol than in growth from the vapour and may be understood since when ethanol is present, the diffusing HMT molecule must carry with it a solvation jacket which would tend to increase the relaxation time for surface diffusion. The best value of B from the B + S model gives, from equation [4a], $\gamma'/kT = 0.3$ which is roughly equivalent to α = 3.5 and certainly within the range of estimated α values given in Table 2.

For HMT growth in aqueous solution, the fitted BCF and B + S curves are shown in Fig.1. The rather large scatter of data points does not allow precise definition of the relevant parameters. For the BCF equation we find $\sigma_1 = 0.9 \times 10^{-3}$ which implies either that γ'/kT is very low ($\approx$ zero) or that λ_S is very high. The latter possibility is rather remote, since λ_S in aqueous solution is not likely to differ too much from its value in ethanol solution. It must be concluded that $\gamma'/kT \approx 0$ implying that $\alpha < 3.2$. The B + S model gives $\gamma'/kT = 0.024$ which also implies that $\alpha < 3.2$. Both models are thus consistent with the prediction of $\alpha \approx 2$ given in Table 2 and imply that the surface should, during growth, be rough and growth continuous.

In neither growth from ethanolic nor aqueous solution does the curve fitting indicate one theoretical growth model to give a better fit than the other. It can only be concluded that in both cases the models give satisfactory descriptions of the data resulting in reasonable estimates for λ_S and values of α which are in good agreement with those given in Table 2.

The conversion of γ'/kT values to α values is based on the computer simulation data of Bennema et al. (2).

CONCLUSIONS

It has been shown that through their effect on the α factor solute-solvent interactions can determine the growth mechanism of a crystal face from solution. For the growth of HMT from the vapour, ethanolic solution and aqueous solution estimates of α have confirmed this possibility and indicate that for vapour growth a BCF mechanism should operate, for growth from ethanol a dislocation or nucleation mechanism may occur, and for growth from aqueous solution growth may be continuous.

The comparison of theoretical BCF and B + S models with available growth rate supersaturation data for the three systems has been found to confirm these predictions.

ACKNOWLEDGEMENT

The authors wish to acknowledge the help of Mr. F.Thoresen who made growth rate measurements in the fluidised bed and assisted in the evaluation of results.

NOMENCLATURE

a	shortest distance between molecules in the crystal
A, B, C_o	parameters in the B + S nucleation equation
C, c_o	parameters in the BCF surface diffusion model
f	area of molecule in the crystal
F	surface energy
G^E	excess free energy of mixing
h	Planck's constant
k	Boltzman's constant
k_d	mass transfer coefficient
L	linear crystal growth rate
n_s	number of nearest neighbours per layer
n_t	total number of nearest neighbours
N_o	number of growth units per unit volume of saturated solution
N	Avogadro's number
S	number of collaborating spirals
T	absolute temperature
u_r	relative velocity between crystal and solution
W	mass fraction of solute in solution
W_i	interfacial concentration of solute (mass fraction)
x_i	mole fraction of component i in solution
α	characteristic energy parameter for a crystal surface
Φ_{ss}	energy of the solid-solid first neighbour bond
Φ_{tt}	energy of the fluid-fluid first neighbour bond
Φ_{st}	energy of the solid-fluid first neighbour bond
γ'	edge free energy of a molecule in the edge of a critical nucleus
γ''	edge energy of a growth unit in a straight step
ρ	solution density
β	retardation factor in BCF equation
Λ	thickness of adsorbed layer
Ω	volume of a molecule
λ_S	mean surface diffusion distance
ΔF	change in surface free energy upon creation of solid phase
ΔG_{des}	activation free energy for desolvation and entry in surface phase
ΔH_f	heat of fusion
ΔH_s^i	heat of solution obtained from solubility data
ΔH_s^m	measured heat of solution
ΔS_s^i	entropy of solution obtained from solubility data
ΔS_{cn}	configurational edge entropy of a critical nucleus
σ_1	parameter in BCF equation
σ_i	interfacial supersaturation defined by $\sigma_i = x_i - x_S/x_S$
BCF	abbreviation for Burton, Cabrera and Frank growth model
B + S	abbreviation for the Birth and Spread growth model

REFERENCES

1. Gilmer, G.H. and Bennema, P., J.Cryst.Growth, 13/14 (1972) 148.
2. Bennema, P., Boon, J., van Leeuwen, C. and Gilmer, G.H., Krist. und Tech., 8 (1973) 659.
3. Bennema, P. and Gilmer, G.H. in "Crystal Growth: An Introduction", ed. P. Hartman, North Holland/American Elsevier (1973) 274.
4. Hartman, P. and Perdok, W.G., Acta Cryst., 8 (1955) 49.
5. Bomio, P., Bourne, J.R. and Davey, R.J., J.Cryst.Growth, (in press).
6. Heyer, H., J.Phys.Chem.Solids Suppl. No.1 (1967) 265.
7. Olbrecht, M., Diss. ETH Zurich No. 5367 (1974).
8. White, E.T., J.Chem.Eng.Data, 12 (1967) 285.
9. Bourne, J.R. and Davey, R.J., J.Chem.Eng.Data, 20 (1975) 15.
10. Mak, T.C.W., J.Chem.Phys., 43 (1965) 2799.
11. Crescnezi, V., Quadrifoglio, F. and Vitagliano, V., J.Phys. Chem., 71 (1967) 2313.
12. Sassa, Y. and Katayama, T., J.Chem.Eng.Japan, 6 (1973) 31.
13. van Leeuwen, C., J.Cryst.Growth, 19 (1973) 133.
14. Hungerbühler, K., Diplomarbeit Sommersemester 1975, ETH Zurich.

EFFECT OF ISODIMORPHOUSLY INCLUDED Co(II), Fe(II) AND Cu(II) IONS ON THE CRYSTAL STRUCTURES OF $ZnSO_4 \cdot 7H_2O$ AND $MgSO_4 \cdot 7H_2O$

C. BALAREW AND V. KARAIVANOVA

Academy of Sciences of Bulgaria, Sofia 13

Institute of General and Inorganic Chemistry

In a previous paper it was shown (Balarew, Karaivanova and Aslanian) that the electron configurations of Mg(II), Ni(II) and Zn(II) ions favour a comparatively regular octahedral arrangement of the six water molecules in $Me(H_2O)_6^{2+}$ presumably existing in the crystal structures of $MgSO_4 \cdot 7H_2O$, $NiSO_4 \cdot 7H_2O$ and $ZnSO_4 \cdot 7H_2O$. The salts (sulphate heptahydrates) of these ions crystallize in the orthorhombic crystal system. The electron configurations of the ions Cu(II), Fe(II) and Co(II), however, due to the Jahn-Teller effect, favour a distorted octahedral arrangement of the water molecules in $Me(H_2O)_6^{2+}$ presumably existing in the crystal structures of $CuSO_4 \cdot 7H_2O$, $FeSO_4 \cdot 7H_2O$ and $CoSO_4 \cdot 7H_2O$. This distortion is ultimately responsible for the crystallization of the Cu(II), Fe(II) and Co(II) sulphate heptahydrates in the monoclinic crystal system.

If Zn(II), Mg(II) or Ni(II) ions in their sulphate heptahydrates are isodimorphously substituted by Cu(II), Fe(II) or Co(II), a deformation stress is expected to appear in the crystal structure of the host crystal. For a specific degree of substitution this stress will lead to an abrupt conversion of the orthorhombic crystals into monoclinic crystals. There are literature data (Retgers, Stortenbecker and Hollmann) indicating that "mixed crystals containing Fe(II) or Cu(II) ions are obtained from aqueous solutions, part of which are based on the labile monoclinic $ZnSO_4 \cdot 7H_2O$". It was reported also (Retgers) that monoclinic mixed crystals of $MgSO_4 \cdot 7H_2O$ were prepared from solutions containing Cu(II) ions. The distortion effect of the admixed ions should be reflected in the solubility diagrams: a crystallization field of monoclinic mixed crystals based on $ZnSO_4 \cdot 7H_2O$ or $MgSO_4 \cdot 7H_2O$ should be present. In the case of the system $ZnSO_4$-$CuSO_4$-H_2O this field should be separated from all the other fields, whereas for the systems $ZnSO_4$-$FeSO_4$-H_2O and $ZnSO_4$-$CoSO_4$-H_2O and $MgSO_4$-$CoSO_4$-H_2O this field will be separated by means of an eutonic point only from the crystallization field of the orthorhombic mixed crystals based on $FeSO_4 \cdot 7H_2O$ or $CoSO_4 \cdot 7H_2O$ because they are isotypic. In this case the single eutonic point should separate the comparatively narrow crystallization field of the host orthorhombic $ZnSO_4 \cdot 7H_2O$ or $MgSO_4 \cdot 7H_2O$ from a comparatively wider crystallization field of the monoclinic mixed crystals. The solubility diagrams of the systems $ZnSO_4$-$CoSO_4$-H_2O (Balarew, Karaivanova and Oikova), $MgSO_4$-$FeSO_4$-H_2O and $MgSO_4$-$CoSO_4$-H_2O (Balarew, Dobreva and Oikova) at 25°C studied previously, have exactly these features.

In the system $ZnSO_4$-$FeSO_4$-H_2O studied by Gorshtein at 0°C the crystallization field of $ZnSO_4 \cdot 7H_2O$ is represented only by three experimental points which are due to phases of similar compositions. The system $ZnSO_4$-$CuSO_4$-H_2O was studied by Jangg and Gregori. The eutonic point which should separate the crystallization field of the mixed crystals based on orthorhombic $ZnSO_4 \cdot 7H_2O$ from those based on monoclinic $ZnSO_4 \cdot 7H_2O$ was not found by these authors. Thus the data are useless with respect to the assessment of the effect of the FE(II) or Cu(II) ions on the crystal structure change of $ZnSO_4 \cdot 7H_2O$. For the system $MgSO_4$-$CuSO_4$-H_2O there are data at 25°C (Averina and Shevchuk) and at 30°C (Inguzzi). These data however are contradictory. This is the reason why we have chosen to study again the systems $ZnSO_4$-$CuSO_4$-H_2O, $ZnSO_4$-$FeSO_4$-H_2O and $MgSO_4$-$CuSO_4$-H_2O at 25°C.

The above systems were studied using Khlopin's method of isothermal recording of the saturation. The method has been described in detail previously (Balarew, Karaivanova and Oikova). The reagents used were of A.R. grade.

The sum of the Zn(II) and Cu(II) concentrations as well as that of the Zn(II) and Fe(II) concentrations were found by complexometric back titration at pH 4.5-5 using xylenol orange (Fe(II) was oxidized in advance by warming with concentrated nitric acid). Mg(II) is determined by direct complexometric titration at pH 9 using eriochrom black T as indicator after prior precipitation of Cu(II) with ammonium sulphide in the presence of NH_4Cl. Fe(II) was separately determined permanganometrically and Cu(II) iodometrically. The composition of the ideally filtered solid phase was determined by a modification of the Schreinemakers' method - algebraic indirect identification of the solid phase composition (Trendafelov and Balarew).

To find the time needed to reach equilibrium between the liquid and solid phases kinetic curves were plotted. The results show that the systems are in equilibrium after 10 hours of vigorous stirring.

The experimental results for the system $ZnSO_4-CuSO_4-H_2O$ at 25°C are plotted (Roseboom's method) in Fig. 1. Two eutonic points are observed in the solubility diagram. These points separate three crystallization fields: a very narrow crystallization field of mixed crystals based on orthorhombic $ZnSO_4 \cdot 7H_2O$ (distribution coefficient D appreciably lower than 1), a comparatively wider crystallization field of mixed crystals based on the monoclinic $ZnSO_4 \cdot 7H_2O$ ($D_{av.}$ = 2.6) and finally a crystallization field of practically pure $CuSO_4 \cdot 5H_2O$. The three different crystal phases were also detected in the X-ray diffraction patterns.

The mixed crystals of orthorhombic $ZnSO_4 \cdot 7H_2O$ as host and those of monoclinic $ZnSO_4 \cdot 7H_2O$ as host differ also in habit. The orthorhombic mixed crystals are of a prismatic habit and the monoclinic of tabular habit. The optical studies have shown that the mixed crystals with a tabular habit have oblique extinction in contrast to the mixed crystals with a prismatic habit which display parallel extinction.

The experimental data on the system $ZnSO_4-FeSO_4-H_2O$ at 25°C are plotted in Fig. 2. The results obtained show the existence of two crystallization fields: a comparatively narrow crystallization field of mixed crystals of orthorhombic $ZnSO_4 \cdot 7H_2O$ as host (D ca. 1.15) and a wide crystallization field of the monoclinic mixed crystals (D ca. 0.22). The mixed crystals from both fields of crystallization were studied by X-ray diffraction and microscopically.

The isothermal diagram of the system $MgSO_4-CuSO_4-H_2O$ at 25°C is shown in Fig. 3. It shows the existence of three crystallization fields: a crystallization field of mixed crystals based on orthorhombic $MgSO_4 \cdot 7H_2O$ ($D_{av.}$ = 0.2), a comparatively narrow crystallization field of mixed crystals based on the monoclinic $MgSO_4 \cdot 7H_2O$ ($D_{av.}$ = 2.2) and a crystallization field of mixed crystals based on the triclinic $CuSO_4 \cdot 5H_2O$ ($D_{av.}$ = 0.08). The crystal phases differing in structure are proved also by the X-ray diffraction patterns. The microscopic pictures of the mixed crystals from the two crystallization fields of basic component $MgSO_4 \cdot 7H_2O$ show the existence of two crystal phases differing in habit: prismatic and tabular. The optical studies of these crystals have shown a parallel extinction when the mixed crystals have a prismatic habit and oblique extinction when the mixed crystals are tabular.

The experimental data on the $ZnSO_4-CuSO_4-H_2O$, $ZnSO_4-FeSO_4-H_2O$ and $MgSO_4-CuSO_4-H_2O$ systems at 25°C as well as those on the systems $ZnSO_4-CoSO_4-H_2O$ (Balarew, Karaivanova and Oikova), $MgSO_4-FeSO_4-H_2O$ and $MgSO_4-CoSO_4-H_2O$ (Balarew, Dobreva and Oikova) at 25°C published elsewhere illustrate the effect of the isodimorphously included admixtures on the crystal structure of the host salt. Furthermore, the position of the eutonic point, or the minimum

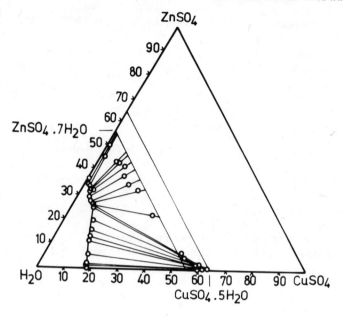

Fig.1 Solubility diagram of the system $ZnSO_4-CuSO_4-H_2O$ at $25°C$

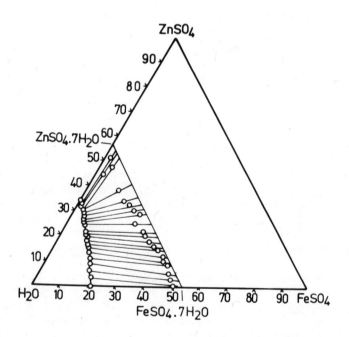

Fig.2 Solubility diagram of the system $ZnSO_4-FeSO_4-H_2O$ at $25°C$

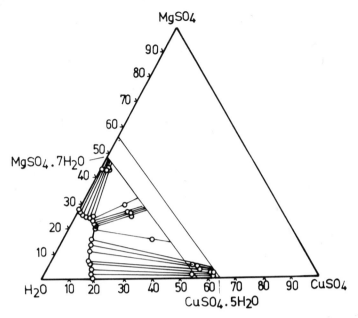

Fig. 3 Solubility diagram of the system $MgSO_4$ $CuSO_4-H_2O$ at $25°C$

concentration of the admixture which may cause the crystal structure conversion of orthorhombic $ZnSO_4 \cdot 7H_2O$ or $MgSO_4 7H_2O$ into monoclinic can be used as a measure of the deforming effect of the admixed ions on the structure of the orthorhombic crystals. In this sense the strongest deforming effect is that of the Cu(II) ions, followed by the Fe(II) ions, the weakest deforming effect being that of the Co(II) ions.

REFERENCES

1. Averina, P.A. and Shevchuk, V.G., J.neorgan.khimii, 12 (1967) 3138.
2. Balarew, C., Dobreva, P. and Oikova, T., Jahrbuch Chem.-Technolog. Hochschule Burgas, 10 (1973) 523.
3. Balarew, C., Karaivanova, V. and Aslanian, S., Kristall und Technik, 8 (1973) 115.
4. Balarew, C., Karaivanova, V. and Oikova, T., Commun.Departm. Chem.Bulg.Acad.Sci., 3 (1970) 637.
5. Gorshtein, G.I., Tr.Inst.Chim.reaktivov IREA, 22 (1953).
6. Hollmann, R., Z.phys.Chem., 54 (1905) 98.
7. Inguzzi, C.M., Atti Soc.Toscana Sci.Nat., 55 (1948) 2684.
8. Jangg, G. and Gregori, H., Z.anorg.allg.Chem., 351 (1967) 81.
9. Retgers, J.W., Z.phys.Chem., 15 (1894) 529.
10. Retgers, J.W., Z.phys.Chem., 16 (1895) 596.
11. Stortenbecker, W., Z.phys.Chem., 22 (1897) 60.
12. Trendafelov, D. and Bałarew, C., Commun.Departm.Chem.Bulg.Acad. Sci., 1 (1968) 73.

THE EFFECT OF IONIC IMPURITIES ON THE GROWTH OF AMMONIUM DIHYDROGEN PHOSPHATE CRYSTALS

R.J. DAVEY AND J.W. MULLIN

Department of Chemical Engineering

University College London, Torrington Place
London WC1E 7JE, England

INTRODUCTION

Additions of trace amounts, often only a few parts per million, of specific substances (generally termed impurities or poisons) to a crystallizing system are known to cause dramatic changes in crystal face growth rates and habit of the crystallizing component. Such impurities are widely utilized comercially to yield crystals with both desirable physical and physio-chemical properties[1].

Theoretical approaches to crystal growth in the presence of impurities have assumed that impurity adsorption occurs at specific sites on the growing crystal surface[2] resulting in a decrease in the face growth rate[3,4,5]. This assumption is supported experimentally by the specificity of impurities in their effect on face growth rates and by the way in which measured relationships between growth rates and impurity concentration in solution have been successfully compared with various theoretical models[2,3,4].

We present here data on ammonium dihydrogen phosphate (ADP) crystals grown in aqueous solution to which had been added trace quantities of $CrCl_3 \cdot 6H_2O$, $FeCl_3$ and $AlCl_3$. The effects on the {100} surface growth features are discussed.

EXPERIMENTAL

The crystals ($\sim$ 0.5 mm long) were grown on the lower, non-reflecting surface of a small observation cell (Figure 1). The central portion (20 mm diameter and 4 mm deep) in which the crystals

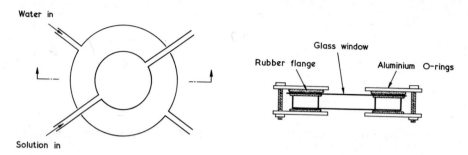

Fig. 1 Observation cell used for studying crystal surfaces

were grown, was enclosed in a water jacket which controlled the cell temperature to within ± 0.05°C of a preset value. Aqueous solutions of ADP were circulated through the cell to provide a constant relative supersaturation σ = 0.052, at a temperature of 23.9$_6$°C and known impurity levels. The crystal surfaces were illuminated with a highly collimated, intense light beam from a 24 V, 150 W tungstenhalogen lamp. Angular adjustment of the cell in the horizontal and vertical planes allowed the light reflected from the crystal surfaces to be diverted into a low-power microscope. Photographs were taken through the microscope using 50 ASA Panatonic X film and Promicrol fine grain developer, to give negatives of high contrast.

As described previously[2], the growth of the {100} faces of ADP crystals in pure solution is characterised by the creation of elliptical layers at well-defined growth centres and their lateral movement across the crystal face.

RESULTS

The effect of $CrCl_3.6H_2O$

Figures 2a-d show a typical sequence of photographs demonstrating the effect of adding $CrCl_3.6H_2O$ to the growth solution. Figure 2a shows a (100) surface growing from pure solution, with growth layers moving outwards from the centre A in the $[00\bar{1}]$ direction with a velocity of 3.5×10^{-7} m/s and a mean separation of 33 μm. On the addition of 15 ppm $CrCl_3.6H_2O$ to the growth solution the velocity of the growth layers was reduced to 1.2×10^{-7} m/s and their separation increased to 100 μm due to a decrease in the rate of creation of layers at the growth centre A. Figure 2c shows the same surface growing in a solution which contained 20 ppm $CrCl_3.6H_2O$ when the centre at A ceased emitting layers entirely while the existing layers became stationary.

AMMONIUM DIHYDROGEN PHOSPHATE CRYSTALS 247

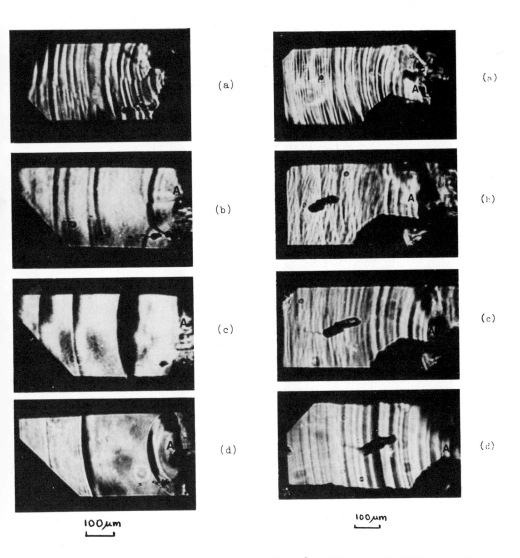

Fig. 2 Effect of increasing concentration of added $CrCl_3 \cdot 6H_2O$ on the growth layers on the {100} face of an ADP crystal. (a) pure solution, (b) 15 ppm $CrCl_3 \cdot 6H_2O$, t = 65 min, (c) 20 ppm $CrCl_3 \cdot 6H_2O$, t = 90 min, (d) the effect of recirculating pure solution.

Fig. 4 Effect of $AlCl_3$ on the growth layers on a {100} face of an ADP crystal. (a) in pure solution, (b) 80 ppm $AlCl_3$, t = 0 min, (c) 80 ppm $AlCl_3$, t = 70 min, (d) 80 ppm $AlCl_3$, t = 140 min.

Figure 2d shows the effect of recirculating pure solution over the crystal surface. The growth centre A once again became active and it was observed that the existing layers gradually speeded up until they reattained a velocity characteristic of that in pure solution.

Figure 3 shows the relation between time and layer velocity for one such layer. Thus at t < 0 the crystal grew in a solution containing 20 ppm $CrCl_3 \cdot 6H_2O$. At a time t = 0 pure solution was circulated over the crystal and the velocity of the layer can be seen to increase linearly with time until after six minutes the velocity characteristic of that in pure solution was reattained.

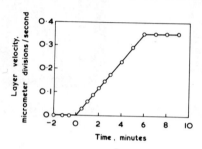

Fig. 3   Dependence of layer velocity on time (t < 0 surface growing in presence of 20 ppm $CrCl_3 \cdot 6H_2O$, t > 0 pure solution recirculated). 1 micrometer reading = 8.8×10^{-7} m.

Effect of $AlCl_3$ and $FeCl_3$

The growth of the {100} faces of ADP crystals in the presence of $AlCl_3$ and $FeCl_3$ was characterised by slightly different changes in surface features from those observed with $CrCl_3 \cdot 6H_2O$. Figure 4 shows typical surface changes occurring when either $AlCl_3$ or $FeCl_3$ was added to the growth solution.

In Figure 4a the surface is seen growing from pure solution with elliptical layers originating at the centre A moving across the surface in the [001] direction with a velocity of 5.8×10^{-7} m/s and mean layer separation of 20 µm. Figure 4b shows the same surface immediately after the addition of 80 ppm of $AlCl_3$ to the growth solution. The regular layer structure characteristic of growth from pure solution is replaced by smaller layer-like features moving roughly in the [001] direction with a velocity of 0.3×10^{-7} m/s. 70 minutes later the surface appeared as in Figure 4c with layers beginning to reform at the centre A. After 140 minutes the surface had returned to normal with elliptical growth layers moving across the surface from the centre A as shown in Figure 4d. Their movement was steady state in its nature and their velocity reduced from the value in pure solution to 0.73×10^{-7} m/s. The mean layer separation increased slightly to 25 μm indicating a much smaller change in the rate of layer creation than that in the case of $CrCl_3 \cdot 6H_2O$.

The steady-state nature of layer movement in the presence of $FeCl_3$ and $AlCl_3$ is well illustrated by Figure 5 which is a time-distance plot for a layer moving across a {100} face of ADP in the [001] direction in the presence of 20 ppm $AlCl_3$. The portion AB of this plot shows the movement of the layer near the growth centre, while the region BC corresponds to its movement near the intersection of (100) and (101) faces.

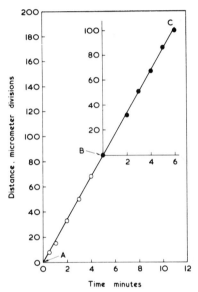

Fig. 5 Time-distance plot for a layer moving in the [001] direction on a {100} face of an ADP crystal during growth in the presence of 20 ppm of $AlCl_3$. o - near the growth centre ● - near the (100)/(101) intersection. 1 micrometer reading = 8.8×10^{-7} m.

A further general property of both $FeCl_3$ and $AlCl_3$ was the promotion of new growth centres, giving simpler layer patterns, occasionally in the form of single or double spirals. Over the course of hundreds of observations made over a period of more than 1 year, no spirals were observed during growth from pure solution. During the same time 3 cases of spiral growth were recorded for growth occurring in the presence of $FeCl_3$ and 2 cases when $AlCl_3$ was present in the growth solution.

Figure 6 demonstrates this feature of the growth process. Figure 6a shows a part of a (100) surface growing in the presence

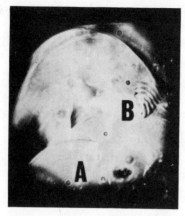

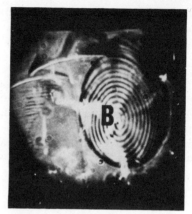

Fig. 6 Spiral growth on a {100} face of an ADP crystal in the presence of 40 ppm AlCl$_3$ showing the promotion of new growth centres - photograph (b) was taken 20 min after (a).

of 40 ppm AlCl$_3$. In pure solution the dominant growth centre was situated in the region A, but during growth in the presence of 40 ppm AlCl$_3$ a new centre B became active as shown in Figure 6a After 20 minutes this centre had developed into a double spiral as shown in Figure 6b.

DISCUSSION

The Reduction in Growth Centre Activity by CrCl$_3$.6H$_2$O

The nature of the growth centres observed on the {100} faces of ADP has been discussed previously[2,6] where it was concluded, in agreement with earlier studies by Dunning, Jackson and Mead[7], that the growth centres were most probably the result of co-operation between groups of dislocations in such a way that the individual spiral steps from each dislocation in the group were able to bunch together creating a visible growth layer. The criterion for such co-operation is that the separation, d, of dislocations within the groups should obey the relationship[8]

$$d < 2 \pi r^* \qquad (1)$$

where r^* is the radius of the critical nucleus defined by

$$r^* = \frac{\gamma a}{\sigma k T} \qquad (2)$$

where γ is the edge free energy of a growth unit in a step, σ is the relative supersaturation and a is the distance between neighbouring growth units in the crystal.

If the impurity species were adsorbed at a dislocation growth step, the edge energy would decrease and the diameter of the critical nucleus would be reduced. Thus for a given supersaturation a group of dislocations with $d < 2\pi\gamma a/kT\sigma$ would co-operate to form step bunches in pure solution, while in the presence of adsorbed impurity the value of $2\pi\gamma a/kT\sigma$ would decrease and the distance, d, between neighbouring dislocations in the group could become greater than the diameter of the critical nucleus. This would promote the formation of single monolayer steps[8] rather than step bunches. Experimentally this would lead to a reduction in the observed rate of layer creation as reported here for growth in the presence of $CrCl_3 \cdot 6H_2O$. This conclusion is also consistent with previously reported kinetic data[2] which indicated that for growth in the presence of $CrCl_3 \cdot 6H_2O$ adsorption of impurity species at the step was important. (Note: the impurity is referred to as the solid form added; its nature in solution is not anticipated here - see references 6 and 9).

The reversibility of the adsorption process is clearly shown by Figure 3 which shows the effect of recirculation of pure solution over the surface poisoned during growth in the presence of $CrCl_3 \cdot 6H_2O$.

The Effect of $FeCl_3$ and $AlCl_3$ on the Surface Structure of the {100} Faces of ADP

As illustrated above the initial effect of both $AlCl_3$ and $FeCl_3$ added to the growth solution was to cause a breakdown of the existing layer structure of the {100} faces of ADP crystals giving a completely disordered surface. This may result from the disruption of steady-state conditions when the impurity species initially enter the adsorbed layer. However, the surface eventually regains its ordered nature and grows again in a steady-state manner, presumably due to the attainment of an adsorption equilibrium between impurity in solution and at the surface.

That such an equilibrium is attained is illustrated by Figure 5 which shows a time-distance plot for a layer moving in the [001] direction across a {100} face. It is clear that the layer velocity does not depend on the position of the layer on the surface. A non-steady-state adsorption may be expected to give local build-ups of impurity at certain points on the surface resulting in fluctuations in layer velocity. No fluctuations were observed.

It should also be noted that, unlike growth in the presence of $CrCl_3 \cdot 6H_2O$, very little change in growth centre activity occurred with $FeCl_3$ and $AlCl_3$. This observation is consistent with ledge adsorption, as reported previously[2], since this would have little effect on co-operation between dislocations.

The Creation of New Growth Centres by $FeCl_3$ and $AlCl_3$

We have reported previously[9] that when $CrCl_3.6H_2O$ is added to the growth solution, some incorporation of impurity species occurs in ADP crystals. Although no analyses were made, it seems feasible[10] that similar incorporation of $AlCl_3$ and $FeCl_3$ could occur

This incorporation may well lead to the situation in which impurity species are incorporated into the crystal and yet are accommodated imperfectly (i.e. at interstitial sites) causing the introduction of new dislocations and hence new growth centres, as observed in the present study. Indeed X-ray studies made by Deslattes et al.[11] have indicated that when ADP crystals are grown in the presence of Cr^{3+} such an increase in dislocation density does occur.

CONCLUSIONS

Previously reported kinetic data[2] for the growth of the {100} faces of ADP in the presence of added $CrCl_3.6H_2O$ led to the conclusions that the impurity species were adsorbed at growth steps. This view is supported here by qualitative observations of reduced rates of layer creation in the presence of $CrCl_3.6H_2O$ which may also be explained by adsorption at the step and the subsequent reduction of the edge free energy.

For growth in the presence of $FeCl_3$ and $AlCl_3$ kinetic studies[2] indicated that adsorption occurred on the ledges between the growth steps which would not lead to substantial changes in edge free energy. This view is supported by the observation that little change in growth centre activity occurs when $AlCl_3$ and $FeCl_3$ are present in solution. However, random incorporation into the crystal lattice could lead to the creation of new dislocations and hence the observed increase in growth centres.

REFERENCES

1. Mullin, J.W., Crystallization, Butterworths, London, 2nd Ed., 1972.
2. Davey, R.J. and Mullin, J.W., J. Crystal Growth, 26 (1974) 45.
3. Bliznakov, G., Forsch. Mineral., 36 (1958) 149.
4. Burrill, K.A., J. Crystal Growth, 12 (1972) 239-244.
5. Albon, N. and Dunning, W.J., Acta. Cryst., 15 (1962) 474.
6. Davey, R.J., Ph.D. Thesis, University of London, 1973.
7. Dunning, W.J., Jackson, R.W. and Mead, D.G., Adsorption et Croissance Crystalline, p 303 Centre Nationale de la Recherche Scientifique, Paris, 1965.
8. Burton, W.K., Cabrera, N. and Frank, F.C., Phil. Trans. Roy. Soc., 243 (1951) 299.
9. Davey, R.J. and Mullin, J.W., J. Crystal Growth, 23 (1974) 89.

INFLUENCE OF Mn(II), Cu(II) and Al(III) IONS ON SIZE AND HABIT OF AMMONIUM SULPHATE CRYSTALS

MIROSLAV BROUL

Research Institute of Inorganic Chemistry

400 60 Ústí nad Labem, Czechoslovakia

INTRODUCTION

The effects of various additives (Mn^{2+}, Cu^{2+}, Al^{3+}) on the process of ammonium sulphate crystallization were investigated at varying rates of supersaturation of its solutions and at different intensities of agitation. Since ammonium sulphate represents a bulk quantity product, a considerable interest has been given to the possibilities of improving its quality. As it is chemically indifferent to most admixtures, it is possible to study the influence of a wide range of additives on its crystallization and properties. We selected bivalent cations Mn^{2+} and Cu^{2+}, whose favourable effects have been known and qualitatively described in literature, and Al^{3+} ion, which is considered to be one of the generally most effective additives.

The shape of crystals is controlled primarily by the rates of growth of the individual crystal faces. According to the principle of crystal faces overlapping, the area of slowly growing faces is relatively increasing and, ultimately, these faces become predominant in the final habit of the crystal, whereas the fast growing faces may even disappear. The ratio of the growth rates of individual faces is determined by the simultaneous action of a number of factors (1), such as the nature of the crystallizing substance proper, type of solvent, crystallization temperature, rate of supersaturation and degree of supersaturation, hydrodynamics of the system pH of the solution, presence of impurities and additives, and so on.

In most cases it is not possible to assess the mechanism of

effects brought about by one particular variable, since the existing inter-relations among the individual factors are rather complex and cannot be generalized explicitly. This is true especially as far as the effects of admixtures are concerned in spite of the fact that both the practical aspects and theory of these problems have been studied extensively. Any generalization of experimental results is very difficult due to the inconsistency of the available published data. Experimental procedures vary widely in scale and apparatus: from microscopic observations of crystallization in a few drops of a saturated solution containing the examined additive, to commercial-scale testing and verification of laboratory data. It is obvious that the other parameters of crystallization, which are usually not specified precisely, may affect the final result considerably, depending on the experimental method used. In certain cases the differences in hydrodynamic conditions may play a particularly important role.

The final effect, which is attributed to the presence of the examined additive may be, in fact, the result of a simultaneous action of several factors. A further trouble lies in the fact that there is not a generally accepted criterion for purity of the treated substances. A change in the crystallization parameters is often brought about by extremely small amounts of admixtures (2,3) and therefore one should not be surprised when two authors, dealing independently with an "identical" system, report results which are markedly different, or even antagonistic.

The problems become even more serious when using additives in plant-scale crystallization processes. In addition to several essential substances, the solution contains other dissolved materials impurities, originating from raw materials, undesired chemical reactions and corrosion of the equipment. These impurities may react with the studied substances, to promote or inhibit their effects, and their concentration may vary considerably with time. Further factors that must be taken into consideration are the differences in experimental conditions, especially in the hydrodynamics of the systems, which affect the shape of crystals substantially and, consequently, it is quite natural that the results of plant-scale measurements will almost always differ more or less fr those obtained in labroatory tests with synthetic model solutions.

As stated earlier, the shape of a crystal is given by the rate of growth of its individual faces. The size of a crystal, too, depends on the rate of growth of its faces; a quantitative description of the dependence, however, is a very complex problem. In the common industrial practice the size of crystals is evaluated on the basis of product sieve analyses, which enables also to assess the mean size $\overline{L}$ of product crystals. In most cases the results thus obtained make it possible to conclude that a certain additive affects the crystal size favourably. Only in a limited number of cases one may succeed in developing a more or less complex empirical relation between $\bar{L}$ and concentration of an additive but such a relation usually holds only within certain specified ranges of parameters of the particular crystallization process.

Extensive efforts have been devoted to studies of effects of various admixtures on crystallization of ammonium sulphate. For example, more than a hundred references to literature on the subject are cited in (4). A favourable effect of manganese has been described (5,6,7); large tabular crystals are formed in the presence of Mn^{2+} (8). It is claimed (7) that Cu^{2+}, in combination with Mn^{2+}, in combination with Mn^{2+}, improves the crystal shape, too. According to (9) copper ions inhibit the adverse effect of ferric ions. The presence of Al^{3+}, even at extremely low concentrations (0.005%), brings about a distinct extension of crystal length, at higher concentrations fissures appear in the crystals and they are brittle and milky (5,10). The addition of small amounts of Al^{3+}, however, is recommended by some authors (5,11-14), especially when the so-called rice grain shape of product crystals is required.

The presented paper deals with our studies of the effects of the rate of supersaturation and intensity of agitation on the shape and size of ammonium sulphate crystals, grown from aqueous solutions in the presence of manganese (II), copper (II) and aluminum (III) sulphates.

The experiments were carried out using a laboratory-scale discontinuous isothermal vacuum crystallizer equipped with a programmed evaporation. A schematic diagram of the apparatus is given in Fig.1. The crystallizer proper (a 2 ℓ glass sulphonation flask) provided with a stirrer, is situated in a water bath, fitted with a heater. Temperature in the crystallizer is measured by a resistance thermometer and registered by a recorder. Water vapour from the boiling solution, in which crystallization is taking place, enters a water cooler and the condensate then flows down through an electromagnetically controlled ball valve either into a condensate receiver or back into the crystallizer. The source of the input signal for the relay, controlling the ball valve coil, is a mobile indicator of liquid level in the condensate receiver. The indicator is a modified contact thermometer: it is coupled with an adapted laboratory device for programmed temperature control and this arrangement enables to control the rate of evaporation in accordance with a chosen programme, recorded on a punched tape. The agitator speed is measured by an electronic revolution counter. A constant pressure within the system is maintained by a contact pressure gauge, an electromagnetically controlled air valve and a water pump. The process is carried out under isobaric conditions at the vapor pressure of saturated solution of the examined substance, corresponding to the selected temperature, which ensures that temperature in the crystallizer will remain constant throughout the experiment.

The factorial planning method was adopted for conducting and evaluating of experiments (15):
Concentration (expressed as molality) of admixtures (5 levels, including zero, for each additive):

Mn^{2+} : 17.0×10^{-6}, 34.0×10^{-6}, 170.0×10^{-6} and 386.0×10^{-6}
Cu^{2+} : 2.96×10^{-6}, $14.8 \cdot 10^{-6}$, 29.5×10^{-6} and 145.7×10^{-6}
Al^{3+} : 0.222×10^{-6}, 0.554×10^{-6}, 1.66×10^{-6} and 3.29×10^{-6}

Rate of agitation - 2 levels: 600 and 1200 r.p.m.
Rate of supersaturation - 2 levels, corresponding to two constant evaporation rates of 470 g H_2O over 80 or 320 min.

The effects of the respective additive and its concentration, rate of agitation and rate of supersaturation on crystal shape were evaluated by assessing the length/width ratios of a total of fifty crystals on microphotographs. The effect of the above variables on crystal size was determined on the basis of sieve analysis data and by their subsequent treatment according to a method suggested by our laboratory (17). The mean crystal size $\overline{L}$ was then calculated using a standardized computer programme.

EFFECT OF ADMIXTURES ON CRYSTAL SHAPE

The influence of the examined variables upon the final shape of crystals may be seen in Fig.2. The individual drawings always relate

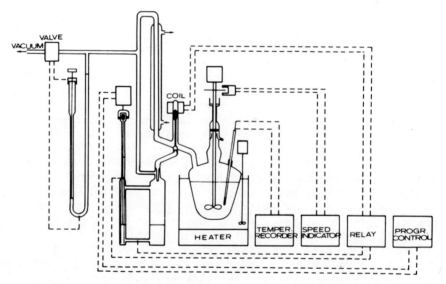

Fig. 1 Schematic diagram of experimental apparatus

to crystals formed in solutions at the highest concentration level
of the respective admixture.

A statistical evaluation of the factorial experiments reveals
clearly and unambiguously that the effect of the rate of
supersaturation exceeds many times that of the presence of the
admixtures: whereas the additive concentrations were varied within
one to two orders, the higher value of the supersaturation rate was
only 4 times the lower level value. The rate of supersaturation
affects the influence of the admixtures as well.

At the high supersaturation rate relatively very long rod-like
crystals are formed in pure solution but immediately after the
first addition of the admixtures a considerable linear contraction
occurs. A further increasing of the additives concentration,
however, has no marked effect on the relative length of crystals.

At the low supersaturation rate short and thick crystals are
formed in the pure solution, whose length increases slightly after
the first addition of the admixture; as its content is further increasing, however, the length of crystals decreases.

Slow saturation facilitates the formation of better developed
crystals, in which less imperfections and intergrowths occur.

The effect of the examined ions is demonstrated in Fig.2.
Mn^{2+} has a comparatively distinct effect; its presence brings about
a progressive rounding off of crystal corners, partial shortening
of prisms and formation of pyramidal ends. It is possible to
prepare, under optimum conditions (low supersaturation rate, slow
agitation and 0.5-1.0% Mn^{2+}) rice grain shaped crystals of ammonium
sulphate.

The influence of copper is not so pronounced as that of
manganese. A typical phenomenon is the formation of thick prisms.
The rounding off of corners occurs but only at the highest
concentrations of the admixture.

The additions of aluminum bring about a favourable effect only
at its lowest concentrations. Higher contents of Al^{3+} lead to
distinct imperfections on crystal faces and to the occurrence of
intergrowth, whose number increases rapidly as the Al^{3+} concentration in the solution rises. Milky aggregates are formed at the two
highest admixture concentration levels, which is in agreement with
published data.

Of all the variables examined, intensity of agitation exhibits
the least pronounced effect on the crystallization process. This
may be attributed to the fact that both the two levels of stirring
intensity are such that the relative velocities of the solid and
liquid phases are sufficiently high and, therefore, crystallization

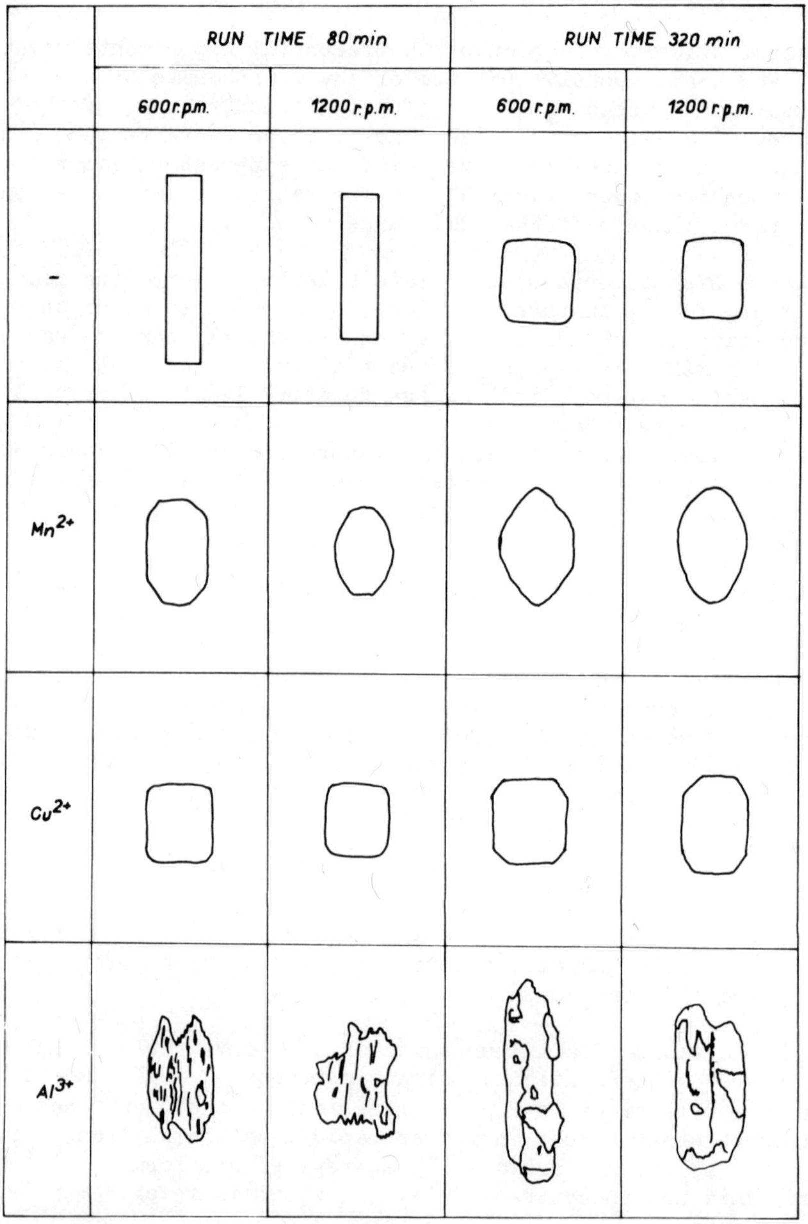

Fig. 2 Influence of certain process variables on the shape of ammonium sulphate crystals

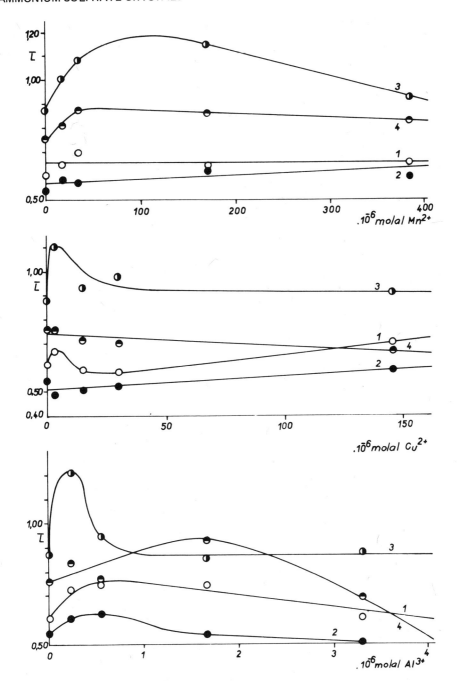

Fig. 3 Influence of certain process variables on the mean size of ammonium sulphate crystals. 1:80 min, 600 rpm; 2:80 min, 1200 rpm; 3:320 min, 600 rpm; 4:320 min, 1200 rpm.

is not controlled by the diffusion processes between the bulk solution and crystal surface. Slightly more regular and round-shaped crystals are formed at the higher speed of stirring, but this is most probably due to abrasion.

EFFECT OF ADMIXTURES ON CRYSTAL SIZE

The effect of the type and concentration of the additive on the mean size of crystals at different supersaturation rates and intensities of agitation is shown in Fig.3. It follows from both the diagrammatical representation and statistical evaluation that the rate of supersaturation affects the crystal size most profoundly.

At the low supersaturation rate (curves 3 and 4), when obviously an intensive nucleation does not take place in the solution, comparatively large crystals can be formed. The dependences of $\bar{L}$ on admixture concentration exhibit relatively distinct maxima, i.e. for Mn^{2+} at 50×10^{-6}, Cu^{2+} at 5×10^{-6} and Al^{3+} at 0.5×10^{-6} molar.

The maxima on curve 3 are approximately equal ($\bar{L} \simeq 1.1$-1.2 mm) for all the examined additives. When defining the "relative efficiency" of the individual admixtures as the inverse value of those concentrations which are needed to attain the same effect, one obtains for the efficiency of the studied ions: $Mn^{2+} \sim 1$, $Cu^{2+} \sim 10$ and $Al^{3+} \sim 100$.

However, efficiency thus defined, is limited to specific systems in which the values of the other process variables are accurately specified.

At the high supersaturation rate (curves 1 and 2) the dependence of mean crystal size on additive concentrations is only slight. On the other hand, the intensity of agitation has a distinct effect. The mean size of product crystals is smaller at the higher stirrer speed, probably due to disintegration of crystals and more intensive secondary nucleation taking place in the system.

KINETICS OF CRYSTALLIZATION OF AMMONIUM SULPHATE IN THE PRESENCE OF ADMIXTURES

The kinetics of a crystallization process carried out in a MSMPR crystallizer can be described by equations derived on the basis of the population balance (16). Modifications of this method for batch-type crystallizers have resulted in rather complex relations, whose practical application is difficult. In most cases the crystal size distributions from discontinuous crystallizers exhibit linear dependences when plotted on linearised co-ordinates, as suggested for the MSMPR crystallizer (17). Therefore, we have made an attempt to apply formally the method for evaluating the kinetic parameters of crystallization, based on the population density concept, to an analogical treatment of our results.

Using the least squares method, a straight line was drawn through the experimental points $n = f(L)$ in such a way that two conditions were satisfied simultaneously, i.e. the relation for particle size distribution

$$\log n = \log n^o - (L/2.3 \, \dot{L} t_c)$$

and that for suspension density

$$m_c = 6 \, \alpha \, \rho_c \, n^o \, (\dot{L} t_c)^4$$

The dominant particle size $\bar{L}$ could then be calculated from values of $\dot{L} t_c$ according to

$$\bar{L} = 3 \, \dot{L} t_c$$

Thus determined values of $\bar{L}$ were compared with those of mean sizes of crystals read off diagrams of linearised dependences of oversize fractions on the size of crystals (17).

The comparison was made with 109 experimental results. A statistical evaluation makes it possible to conclude that the results obtained on the basis of the formally applied method of population density are lower by only 5-10% than those based on the linearisation of the product size distribution. Such a surprisingly good agreement justifies the application of the suggested approach. The use of simple relations, valid for a continuous crystallizer for interpreting experimental data obtained on a batch-crystallizer, gives acceptable results.

It is, therefore, possible to conclude that the described method is suitable at least for a comparative appraisal of the influence of individual additives on the crystallization process parameters.

LIST OF SYMBOLS

L crystal size
$\bar{L}$ mean (dominant) crystal size
$\dot{L}$ linear crystal growth rate
m_c mass of crystals
n^o nuclei population density
t_c overall time of batch crystallization
α volume shape factor
ρ_c density of crystals

REFERENCES

1. Nývlt, J., Industrial Crystallisation from Solutions, Butterworths, London, (1971).
2. Buckley, H.E., Crystal Growth, Wiley, New York (1951).
3. Chernov, A.A., Sov. Phys. Usp. $\underline{4}$, (1961), 116.
4. Broul, M., Ph.D. Thesis, Praha (1975).
5. Nývlt, J., Netuka V., Václavů V., Miček F., Research Report RIIC - No. 259 (1960).
6. Glund W., Klempt, W., Ritter, H., Ber. Ges. Kohlentechn. $\underline{3}$ (1931), 371.
7. Fr. Pat. 917 528 (1945).
8. Ger. Pat. 666 546 (1938).
9. Ger. Pat. 627 511 (1936).
10. Dal V.I., Shapiro, M.D., Gubergric, M.J.: Koks i Khim. $\underline{1957}$, 38.
11. U.S. Pat. 2 782 097 (1957)
12. Nývlt, J., Chem. průmysl $\underline{12}$ (1962), 170.
13. Ger. Pat. 636 057 (1936).
14. Messing, T., Chem. Ing. Techn. $\underline{42}$, (1970), 1141.
15. Davies, O.L., The Design and Analysis of Industrial Experiments, Oliver & Boyd, London 1954.
16. Randolph, A.D., Larson, M.A., The Theory of Particulate Processes, Academic Press, New York 1971.
17. Nývlt, J., Kožmín B., Chem. průmysl $\underline{23}$, (1973), 173.

THE INFLUENCE OF SURFACE-ACTIVE AGENTS ON GYPSUM CRYSTALLIZATION IN PHOSPHORIC ACID SOLUTIONS

JERZY SCHROEDER, WIKTORIA SKUDLARSKA, ANNA SZCZEPANIK, EWA SIKORSKA AND STEFAN ZIELINSKI

Technical University of Wroclaw

Wybrzeze Wyspianskiego 27, Wroclaw, Poland 50370

It is well known that surface-active agents like alkylbenzene sulphonic acid salts affect the filterability of gypsum crystals formed as a by-product in the wet-phosphoric acid process. The object of our experiments was to explain the influence of the sodium salts of alkylbenzene-sulphonic acid, C_{10}-C_{14}, hereinafter referred to as "sulphapol", on the yield of gypsum crystallized in H_3PO_4. Moreover we intend to explain some problems concerning the mechanism of crystallization in the above mentioned system.

EXPERIMENTAL

Gypsum was crystallized in phosphoric acid solutions at $70^\circ C$. The H_3PO_4 solution (30% P_2O_5) of Ca^{2+} ions (200 cm^3) and H_3PO_4 solution of SO_4^{2-} ions (200 cm^3) containing 30% P_2O_5, 1% Al_2O_3, 1.5% Fe_2O_3 and 1.5% F were added to a glass beaker filled with 200 cm^3 of H_3PO_4 solution containing 30% P_2O_5, 2% H_2SO_4, 0.5% Al_2O_3, 0.75% Fe_2O_3 and 0.75% F (weight percentage). The final volume was 600 cm^3. The addition lasted for 4 hours, after which the crystallizing solution was kept for 16 hours at $70^\circ C$. During the addition and crystallization periods the mixture was stirred with a flat paddle (1 cm) glass agitator rotating at 500 rev/min. Most of the experiments were carried out in the presence of gypsum seed crystals. The sulphapol concentration was varied from 0 to 3 g/dm^3.

We considered the gypsum yield, i.e. the ratio of the weight of crystallized gypsum to the stochiometric amount of gypsum, to be the most important result of the experiments. The solution

composition was similar to the crude phosphoric acid analysis.
The calcium ion concentration was kept similar to that brought into
an industrial reacting vessel to crystallize. In most experiments
the concentration was 10 g CaO/dm^3, while in some 20 g CaO/dm^3 was
used.

DISCUSSION

The highest gypsum yield, about 85%, appeared in clear H_3PO_4
solutions, to which neither ions of Fe^{3+}, Al^{3+}, F^- and SiF_6^{2-} nor
seed crystals were added. In these experiments needle-like crystals
were formed. The gypsum yield from solutions containing Fe^{+3},
Al^{+3}, F^-, SiF_6^{2-} ions and surfactant was only about 10% when no
seed crystals were present. However, by introducing seed crystals
we got closer to the industrial process, with a two-phase mixture
in the reaction vessel. In order to examine the influence of seed
crystal surface on the crystallization yield we added different
amounts. The ratio of the highest amount of seed (50 g) to the
Ca^{2+} and SO_4^{2-} contents was the same as in the industrial system.

It appeared that the shape of the seed crystals affected the
final shape of the gypsum precipitate as well as the crystallization
yield. The mean size of plate-like seed crystals was 60x20μm.
There were often small crystals attached to the plates, so that
most of the seed crystal faces were not smooth (Fig.1). The mean
size of needle-like seed crystals was about 190x30μm (Fig.2).

According to the general rules of precipitation we assume that
gypsum crystallization may start either by nucleation in the liquid
phase, or by gypsum deposition on the seed crystal surface. It
seems that in the investigated system the presence of the solid
phase at the beginning of crystallization is not the essential
factor governing the process. It is important to mention that we
use the term "nucleation in the liquid phase" (or bulk nucleation)
to distinguish the situation when the crystal forms and grows
irrespective of the seed, from the situation when the crystallization
process is highly affected by the seed crystal surface (surface
nucleation).

From our experiments we concluded that the mechanism of the
process differs when the Ca^{2+} concentration is smaller than
10 g CaO/dm^3 and higher than 20 g CaO/dm^3.

At the higher level of concentration the crystallized gypsum
yield was nearly theoretical in spite of the different amounts of
sulphapol added, as well as different amounts and shape of seed
crystals. The gypsum crystals precipitated in these experiments
were small, and it seemed that even when the seed quantity was
about 40% (weight ratio of seed to gypsum precipitate) the nuclei

formed mainly in the liquid phase (Figs 3 and 4).

The mechanism observed in solutions where the CaO content was 10 g/dm^3 seemed to be different. The problem became more complicated since the course of crystallization also depended on the shape of the seed crystals. Needle-like seed crystals favoured a high yield of a needle-like precipitate (Fig. 5). Plate-like seed crystals caused another crystallization course, governed by the quantitative relations of the calcium and sulphapol concentrations to the seed crystal surface. The crystallization yield from solutions containing sulphapol and no more than 10 g CaO/dm^3 considerably decreased when the sulphapol to calcium ratio increased, especially when the seed quantity was not high. The correlation factors are given at the bottom of Table 1.

We explain these differences as follows. The surface area of the plate-like seed crystals seems to be insufficient to allow the deposition of the total gypsum yield without nucleation occurring in the liquid phase (without bulk nucleation). Sulphapol suppresses this kind of nucleation. We did not notice an effect like this in solutions containing more than 20 g CaO/dm^3. It is probable that due to the higher concentration of Ca^{2+} and SO_4^{2-} the bulk nucleation process runs more smoothly. It is worth noting that the discrepances in experimental results are higher when, due to the lack of seed surface, surface nucleation proceeds with more difficulty. When the amount of seed crystals is increased, the effect of the sulphapol to calcium ratio becomes less important. The crystallization yield is higher as there is enough surface for the surface nucleation process. It is worth mentioning again that in experiments with needle-like seeds we did not notice any dependence of the crystallization yield on the sulphapol to seed crystal ratio. More than 70% of gypsum was precipitated in every experiment and we assume that even small amounts of needle-like crystals favour surface nucleation.

Our suggestions concerning the mechanism seem to be confirmed by the gypsum crystal shape and size. When the amount of sulphapol increases and the calcium amount remains constant the probability of bulk nucleation decreases. In consequence, the number of small crystals diminishes and the seed crystals grow to bigger sizes. Fig. 6 shows many small crystals obtained when there was only 5 g of seed crystals and the sulphapol concentration was low. Increasing the sulphapol content suppresses bulk nucleation and small crystals form only close to the seed crystals (Figs 7 and 8). Crystallization is finally limited only to gypsum deposition on the seed crystals (Fig. 9) and simultaneously the precipitation yield drops. The same tendency in crystal shape change was noted in experiments with 10 g of seed.

The influence of sulphapol on the shape and size of gypsum

Surfactant concentration g/dm³	Amount of gypsum precipitated (g)		
	5 g seeds	10 g seeds	50 g seeds
0	11,12,7,12,8	14,14	13,13
0.01	12		
0.016	13,13		12,13
0.033	12,13,13,13		
0.066	12,12		
0.1	12.3,12		
0.08	11,8.6	11	11
0.2	7.7,7,7,10,9,4,11.8,3	8.3,10	12.4,13
0.4	7.5,11.5,8,14,16,15.4,13	11	14,16
0.16		10.2	
0.04		12	
0.6		11	
0.8	3	12,8	3.5
1.0	6.5,7,7		
1.2	4,3.5	4	10,12
1.6	3,3,6	6	11.6,4
2.0	3,2,8,7.8,8.0,2	5,7,7	9,3.5,10,10.4
3.0			9,3.5,10,10.4
8.5			10.2,10

Table 1 Gypsum crystallization efficiency using plate-like seed crystals

crystals is less distinct when as much as 50 g of seed is added. This is in accordance with the higher crystallization yield with a large amount of seed. It is interesting to compare the crystals shown in Figs 7, 9, 6 and 10-13. In each case the sulphapol to seed ratio is constant, but the surface area available for gypsum deposition is increasing. It seems that nicely shaped gypsum crystals can only be obtained when there is enough surface not covered with sulphapol.

CONCLUSIONS

The influence of sulphapol on the gypsum crystallization process has been studied. The distinctly marked dependence on the crystallization yield, as well as the shape and size of

GYPSUM CRYSTALLIZATION IN PHOSPHORIC ACID SOLUTIONS 267

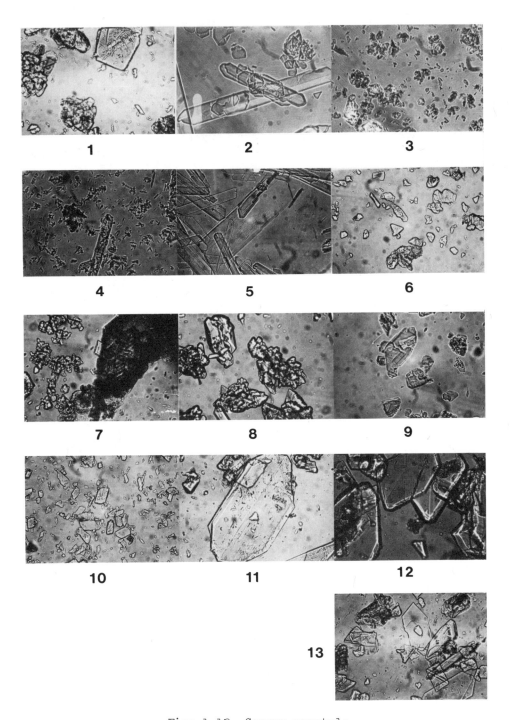

Figs 1-13 Gypsum crystals

gypsum precipitated on a plate-like seed, on the sulphapol content may be a consequence of (a) the change of interfacial free energy or (b) mass transport disturbances due to surfactant adsorption on the crystal surface.

On the other hand, the fact that the crystallization yield in the presence of needle-like crystals does not depend on the sulphapol concentration may be accounted for by assuming that the sulphapol adsorbs selectively on some crystal faces. Different crystal faces may show different affinity to sulphapol and gypsum adsorption.

The above mechanism may explain some of the difficulties noticed when surface active substances are used in the phosphoric acid process. High surfactant concentration makes it possible to obtain nice large gypsum crystals at first, but the decrease in the crystallization yield causes an increase in Ca^{2+} and SO_4^{2-} concentration. As a result the probability of bulk nucleation increases and the final gypsum precipitate often contains large numbers of small crystals.

ISOMORPHOUS SERIES IN CRYSTALS OF $MeSO_4 \cdot nH_2O$ TYPE

S. ASLANIAN[*] AND C. BALAREW[**]

Bulgarian Academy of Sciences, Sofia 13.

Institute of Geology[*]
Institute of General and Inorganic Chemistry[**]

Historically, the scope of the term isomorphism has been extended several times so that the meaning imparted by the various authors to it is not always the same. For this reason it is necessary to specify what is meant by the term when using it. For example, when isovalent isomorphism is discussed, cases in which the ion-for-ion substitution leads to changes in the symmetry of ligand field are also included (Vinokurov). Obviously, the two compounds participating in the process must have different structures, i.e. this is a case of the so-called forced isomorphism or isodimorphism. Although isodimorphism phenomena show great variety, in all of them a discontinuity in the series of mixed crystals must occur. Most often a single discontinuity is observed. Thus, two isomorphous series of mixed crystals occur in the system, each of which has the crystal structure of the respective end member - pure salt.

In studying the isomorphous relationships in mixed crystals of salts of the $MeSO_4 \cdot nH_2O$ type (where $Me = Mg^{2+}$, Ni^{2+}, Zn^{2+}, Fe^{2+}, Co^{2+}, Cu^{2+} and $n = 0 - 7$) it became necessary to consider in some detail the crystal chemistry of the sulphates in question. For the purpose we had data available in the literature on 24 sulphate structures belonging to this type (see Table 1) built up of (MeO_6) octahedra and (SO_4) tetrahedra, separate water molecules being also present in some of the structures, e.g. $MeSO_4 \cdot 7H_2O$ and $MeSO_4 \cdot 5H_2O$. The various structures differ in the octahedral coordination around the metal cation and in the manner in which the octahedra are bound between each other.

In water-free sulphates the octahedral coordination is donated entirely by oxygens belonging to the sulphate tetrahedra (Bokij and

	Mg^{2+}	Fe^{2+}	Co^{2+}	Ni^{2+}	Cu^{2+}	Zn^{2+}
$MeSO_4 \cdot 7H_2O$	(R=4,8%) $Mg-O_{w1}$ 2,057 $Mg-O_{w2}$ 2,109 $Mg-O_{w3}$ 2,050 $Mg-O_{w4}$ 2,051 $Mg-O_{w5}$ 2,096 $Mg-O_{w6}$ 2,070 $Mg-O(m)$ 2,072	(R=4,5%) Fe_1-O_{w1} 2,068 Fe_1-O_{w2} 2,144 Fe_1-O_{w3} 2,136 Fe_2-O_{w4} 2,096 Fe_2-O_{w5} 2,109 Fe_2-O_{w6} 2,188 $Fe-O(m)$ 2,124		$Ni-O_{w1}$ 2,03 $Ni-O_{w2}$ 2,02 $Ni-O_{w3}$ 1,96 $Ni-O_{w4}$ 1,97 $Ni-O_{w5}$ 2,13 $Ni-O_{w6}$ 2,08 $Ni-O(m)$ 2,03		
$MeSO_4 \cdot 6H_2O$	Mg_1-2O_{w1} 2,046 Mg_1-2O_{w2} 2,044 Mg_1-2O_{w3} 2,080 Mg_2-2O_{w4} 2,083 Mg_2-2O_{w5} 2,054 Mg_2-2O_{w6} 2,059 $Mg-O(m)$ 2,061		(R=10,9%) Co_1-2O_{w1} 2,11 Co_1-2O_{w2} 2,05 Co_1-2O_{w3} 2,14 Co_2-2O_{w4} 2,13 Co_2-2O_{w5} 2,12 Co_2-2O_{w6} 2,11 $Co-O(m)$ 2,11			
$MeSO_4 \cdot 5H_2O$	(R=5,7%) Mg_1-2O_{w1} 2,049 Mg_1-2O_{w2} 2,059 Mg_1-2O_1 2,091 Mg_2-2O_{w3} 2,038 Mg_2-2O_{w4} 2,046 Mg_2-2O_2 2,097 $Mg-O(m)$ 2,063				(R=5,7%) Cu_1-2O_{w1} 1,98 Cu_1-2O_{w2} 1,99 Cu_1-2O_1 2,38 Cu_2-2O_{w3} 1,95 Cu_2-2O_{w4} 1,98 Cu_2-2O_2 2,43 $Cu-O(m)$ 2,12	
$MeSO_4 \cdot 4H_2O$	(R=7,1%) $Mg-O_1$ 2,082 $Mg-O_2$ 2,083 $Mg-O_2$ 2,053 $Mg-O_{w1}$ 2,087 $Mg-O_{w2}$ 2,072 $Mg-O_{w3}$ 2,072 $Mg-O(m)$ 2,077	(R=6,8%) $Fe-O_1$ 2,120 $Fe-O_2$ 2,120 $Fe-O_2$ 2,099 $Fe-O_{w1}$ 2,129 $Fe-O_{w2}$ 2,127 $Fe-O_{w3}$ 2,126 $Fe-O(m)$ 2,122				
$MeSO_4 \cdot 3H_2O$					$Cu-O_1$ 1,943 $Cu-O_2$ 2,448 $Cu-O^2$ 2,399 $Cu-O^4$ 1,976 $Cu-O_{w1}$ 1,968 $Cu-O_{w3}$ 1,955 $Cu-O(m)$ 2,115	
$MeSO_4 \cdot H_2O$	(R=4,8%) $Mg-2O_1$ 2,024 $Mg-2O_1^2$ 2,045 $Mg-2O_{w1}$ 2,180 $Mg-O(m)$ 2,083	$Fe-2O_1$ 2,03 $Fe-2O_1^2$ 2,10 $Fe-2O_{w1}$ 2,20 $Fe-O(m)$ 2,11	$Co-2O_1$ 2,04 $Co-2O_1^2$ 2,07 $Co-2O_{w1}$ 2,17 $Co-O(m)$ 2,09	$Ni-2O_1$ 2,02 $Ni-2O_2$ 2,07 $Ni-2O_{w1}$ 2,06 $Ni-O(m)$ 2,05		$Zn-2O_1$ 2,05 $Zn-2O_1^2$ 2,08 $Zn-2O_{w1}$ 2,15 $Zn-O(m)$ 2,09
$MeSO_4$ Cmcm	$Mg-2O_1$ 2,01 $Mg-4O_2$ 2,09 $Mg-O(m)$ 2,06	(R=5,9%) $Fe-2O_1$ 2,023 $Fe-4O_2$ 2,205 $Fe-O(m)$ 2,144	$Co-2O_1$ 2,01 $Co-4O_2$ 2,10 $Co-O(m)$ 2,07	$Ni-2O_1$ 1,99 $Ni-4O_2$ 2,06 $Ni-O(m)$ 2,04		
$MeSO_4$ Pbnm	(R=14,4%) $Mg-2O_1$ 2,29 $Mg-2O_2$ 2,08 $Mg-2O_3$ 1,99 $Mg-O(m)$ 2,12		$Co-2O_1$ 2,34 $Co-2O_2$ 2,14 $Co-2O_3$ 1,93 $Co-O(m)$ 2,14		(R=19%) $Cu-2O_1$ 2,37 $Cu-2O_2$ 2,00 $Cu-2O_3$ 1,89 $Cu-O(m)$ 2,09	(R=18,6%) $Zn-2O_1$ 2,35 $Zn-2O_2$ 2,13 $Zn-2O_3$ 1,95 $Zn-O(m)$ 2,14

Table 1. Interatomic distances Me-O in $MeSO_4 \cdot nH_2O$.

Gorogockaja 1967). The structure consists of zig-zag chains of metal octahedra oriented along the c-axis and linked by the sulphate tetrahedra, two of the sulphate tetrahedron corners belonging to a single chain while the other two are shared between two neighbouring chains. The low-temperature modifications of $MgSO_4$, $FeSO_4$, $CoSO_4$ as well as $CuSO_4$ and $ZnSO_4$ formed at high pressure belong to the structural type of $NiSO_4$ (Cmcm). In the $CuSO_4$ type structure (Pbnm) the basic pattern of zig-zag chains is preserved, but the relative lengthening of two of the six Me-O distances in the octahedron leads to certain changes in the structure lowering its symmetry. Besides, the transition from the $NiSO_4$ type of structure into the $CuSO_4$ type of structure leads to changes in the unit cell parameters. A relatively slight shortening of $\underline{a}$ and lengthening of $\underline{b}$ take place, while $\underline{c}$ remains almost the same. The high-temperature modifications of $MgSO_4$, $FeSO_4$, $CoSO_4$, as well as $ZnSO_4$ crystallize in the $CuSO_4$ type of structure. The differences in the two structural types are expected to produce a discontinuity in the isomorphous series.

In monohydrates, the octahedral co-ordination of the metal ion involves four oxygens donated by sulphate tetrahedra and two water molecules common to two neighbouring octahedra. Here again as in the water-free salts there are chains of metal octahedra between which the sulphate tetrahedra are located. The chains are connected by hydrogen bonds and by oxygens belonging to the sulphate tetrahedra (Bregeault, Herpin and Coing-Boyat; Le Fur, Coing-Boyat and Bassi). The isomorphous series of monohydrates will be continuous.

Dihydrates probably do not exist, because no stable arrangement of octahedra and tetrahedra forming the proper pattern is possible. In the sulphate dihydrates known, e.g. $CaSO_4.2H_2O$, the bivalent metal ion is not in an octahedral co-ordination, but has a co-ordination number of 8.

In trihydrates the corners of the octahedra are occupied by three water molecules and three oxygen atoms donated by three different sulphate groups. Thus, three oxygen atoms of each sulphate tetrahedron participate in the octahedral co-ordination of three different metal ions, while the fourth takes part in the formation of hydrogen bonds only (Zahrobsky and Baur). It should be noted that copper sulphate only forms a stable trihydrate. The square-planar co-ordination around Cu contains three oxygen atoms belonging to water molecules and one oxygen from the sulphate group. This is the basic portion of the structure because it includes all short Cu-O distances. The other two long Cu-O distances combined with the square-planar co-ordination form a distorted-lengthened-octahedron. There are no chains here. Neighbouring units are connected by means of hydrogen bonds. The presence of a single type of lengthened octahedra in this structure accounts for the fact that the single representative of this type of structure known so far is $CuSO_4 \cdot 3H_2O$, since copper and chromium are the only bivalent metal ions in the first row of transition metals that show a marked affinity towards formation of strongly lengthened octahedra.

In tetrahydrates, each octahedron is formed by oxygens of all four water molecules as well as by one oxygen of two neighbouring sulphate groups each (Baur). The interesting point in this structure is that two metal octahedra and two sulphate tetrahedra share common corners, thus forming a closed "molecular" ring. The ring is connected by hydrogen bonds. The identical structures will determine a continuous series of mixed crystals (Bokij and Gorogockaja 1973).

In pentahydrates, only four of the water molecules donate oxygens to each octahedron, the other two oxygens coming from two different sulphate tetrahedra. The fifth water molecule does not take part in the co-ordination polyhedron. Alternating metal octahedra and sulphate tetrahedra share common oxygen corners to form chains parallel to [110]. The remaining two oxygen atoms of the sulphate tetrahedron participate in the formation of hydrogen bonds, which connect the chains with the fifth water molecules and between each other. The arrangement of the building units in this type of structure permits a comparatively regular octahedral co-ordination around the metal ion in some cases, as in $MgSO_4 \cdot 5H_2O$, while in others the octahedra may be distorted-lengthened, as in the case of $CuSO_4 \cdot 5H_2O$ (Baur and Rolin). As a result, a discontinuity of the isomorphous series could be expected in some cases.

In hexahydrates, the octahedral co-ordination is donated by water molecules only. The number of water molecules here equals the cation co-ordination number, as a result of which the sulphate tetrahedra are separated from the metal octahedra. The bonding between them is by hydrogen bonds. In the structure of the monoclinic hexahydrates (where belong all known hexahydrates with the exception of the low-temperature modification of $NiSO_4 \cdot 6H_2O$ which is tetragonal) there are two kinds of octahedra. The water molecules in one of them are connected by hydrogen bonds to the oxygens of the sulphate tetrahedra only. Water molecules of the other sort of octahedra are connected by hydrogen bonds not only to the oxygens of the sulphate tetrahedra but also to water molecules belonging to neighbouring octahedra of the same sort (Zalkin, Ruben and Templeton). The hexahydrates discussed will form continuous isomorphous series. Discontinuities in the isomorphous series will be observed in those series only where the tetragonal $NiSO_4 \cdot 6H_2O$ is an end member.

In heptahydrates, the metal octahedra and sulphate tetrahedra remain separated as in the case of hexahydrates. The seventh water molecule is not co-ordinated to the metal ion but is connected by hydrogen bonds to a metal octahedron and a sulphate tetrahedron. Two different structures, rhombic in $MgSO_4.7H_2O$ (Fig. 1), $NiSO_4.7H_2O$, $ZnSO_4.7H_2O$, and monoclinic in $FeSO_4.7H_2O$ (Fig. 2), $CoSO_4.7H_2O$ and $CuSO_4.7H_2O$, occur in this case. In the rhombic structure the octahedra are comparatively regular. Every metal octahedron is connected by hydrogen bonds to a water molecule not co-ordinated to the metal atom (Baur; Ferraris, Jones and Yerkess). In the monoclinic structure there are two different metal octahedra, lengthened and shortened respectively (Baur). The one sort of octahedra, the shortened ones, are bound by hydrogen bonds to the sulphate tetrahedra only, while the other sort, the lengthened octahedra, are bound by hydrogen bonds both to sulphate tetrahedra and to two of the water molecules not belonging to the metal atoms co-ordination. The occurrence of two different structures in the group of heptahydrates is the result of the electron configurations of the metal ions, which in Fe^{2+}, Co^{2+} and Cu^{2+} are split in an

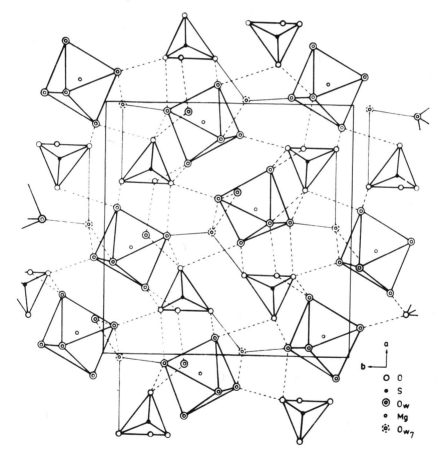

Fig.1 Projection of the $MgSO_4.7H_2O$ structure on (001); dashed lines represent hydrogen bonds.

octahedral crystalline field according to the Jahn-Teller effect producing distorted metal octahedra. In Mg^{2+}, Ni^{2+}, Zn^{2+}, the Jahn-Teller effect is not operative, and the co-ordination around the metal ions is comparatively regular (Balarew, Karaivanova and Aslanian). This determines the difference in the symmetries of the heptahydrates crystal structures, monoclinic and rhombic. The different arrangements of the building units in the two structures leads in turn to different habits of their crystals - an axial habit (Kostov) along the c-axis in sulphates of Mg^{2+}, Ni^{2+}, Zn^{2+} and a pseudoisometric habit along the same axis in the sulphates of Fe^{2+}, Co^{2+} and Cu^{2+}. For this reason isomorphous series whose end members are two heptahydrates of different type of structure are discontinuous, forming actually two isomorphous series. The mixed crystals have the habit and the crystal forms of the respective pure salt, end member of the series. Isomorphous series whose end members are two heptahydrates of one and the same structural type should be continuous. The experiments carried out with sulphate heptahydrates of Mg^{2+}, Ni^{2+}, Zn^{2+}, Fe^{2+}, Co^{2+} and in all possible combination confirm the conclusion reached here (Aslanian, Balarew and Oikova; Balarew, Karaivanova and Aslanian).

The isomorphous series whose end members are any combination of salts of different crystalline-water content will be discontinuous

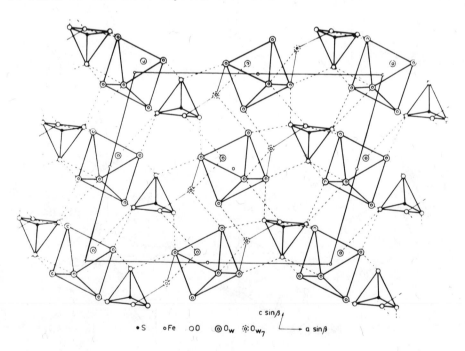

Fig.2 Projection of the $FeSO_4 \cdot 7H_2O$ structure on (010); dashed lines represent hydrogen bonds.

in all cases.

Comparing the distances in sulphate groups of all different structures it was confirmed that the mean S-O distance remains constant in the tetrahedra of all salts of the type discussed (Baur; Larson). It varies from 1.44 to 1.56, its mean value being 1.47 Å.

In comparing the values of the Me-O distances in the octahedra, they were found to vary in fairly wide limits not only in the various structures but even in one and the same octahedron, which can be seen in Table 1. It was found, however, that the mean value of the Me-O interatomic distances within the octahedra of all different crystal hydrates is practically constant for a given metal ion. Thus, the differences between the values of the Me-O distance in a given octahedron are mutually balanced: for example, lengthening of a bond in a given direction leads to the shortening of another bond in the same octahedron.

The differences in the structure of crystal hydrates will be reflected in the isomorphous series formed by them. Thus, for instance, a continuous series of mixed crystals will exist if the end members of the isomorphous series crystallize in identical structures. If the end members of the isomorphous series have different structures, a discontinuous series of mixed crystals will be formed, i.e. there is an isodimorphism between the two end members. The review made of the structures of the $MeSO_4 \cdot nH_2O$ type salts provides means for predicting whether a discontinuous or continuous series of mixed crystals will be formed.

REFERENCES

1. Aslanian, S., Balarew, Chr. and Oikova, T., Kristall und Technik, 7 (1972) 525.
2. Balarew, Chr., Karaivanova, V. and Aslanian, S., Kristall und Technik, 8 (1973) 115.
3. Baur, W., Acta Cryst., 17 (1964) 863, 1167 and 1361.
4. Baur, W. and Rolin, J., Acta Cryst., B28 (1972) 1448.
5. Bokij, G. and Gorogockaja, L., J. strukt. himij, 8 (1967) 662 and 14 (1973) 313.
6. Bregeault, J-M., Herpin, P. and Coing-Boyat, J., Bull. soc. chim. France, 6 (1972) 2247.
7. Ferraris, G., Jones, D. and Yerkess, Y., J.chem.Soc. Dalton Trans., 8 (1973) 816.
8. Kostov, I., Miner.sb.L'vovsk.geol.obštestva, 16 (1962) 75.
9. Larson, A., Acta Cryst., 18 (1965) 717.
10. Le Fur, Y., Coing-Boyat, J. and Bassi, G., CR Acad. Sci., 262C (1966) 632.
11. Vinokurov, V., sb. "Problema isomorphnich samešenij atomov v kristallah", M. Nauka 172 (1971).

12. Zahrobsky, F. and Baur, W., Acta Cryst., B24 (1968) 508.
13. Zalkin, A., Ruben, H. and Templeton, D., Acta Cryst., 15 (1962) 1219 and 17 (1964) 235.

THE TRANSFORMATION OF AMORPHOUS CALCIUM PHOSPHATE INTO

CRYSTALLINE HYDROXYAPATITE

LJ. BREČEVIĆ AND H. FÜREDI-MILHOFER

"Rudjer Bošković" Institute

Zagreb, Croatia, Yugoslavia

INTRODUCTION

It has been shown that at medium and high supersaturations in the neutral and basic pH range the formation of hydroxyapatite (1-3) and octacalcium phosphate (4,5) is usually preceded by the precipitation of a cryptocrystalline solid phase (amorphous calcium phosphate, ACP) which is metastable for a given length of time. Amorphous calcium phosphates have been recognized (6) as a class of highly hydrated (7) salts, their chemical composition depending on the solution environment prevailing during precipitation. The mechanism of the conversion of ACP into crystalline calcium phosphate is not yet clear. It has been suggested (2,8,9) that the principal mechanism of conversion is the dissolution of ACP before the onset of hydroxyapatite precipitation. However, kinetic pH and turbidity curves (4,5) and electron micrographs (4,10-12) showing that ACP spheres are intimately associated with the formation of crystalline material are at variance with the preceding conception. An alternative conception (11), suggesting that ACP is transformed during the period of metastability and serves as a template for the nucleation of the secondary, crystalline precipitate, is being examined in our laboratory. In the present paper some new experimental evidence is presented, which, together with earlier observations (4,5,11) should contribute to the understanding of the processes involved in the formation and transformation of ACP.

METHODS

All experiments were conducted at 25°. Samples were prepared as previously described (4,13) by direct mixing of equal volumes of

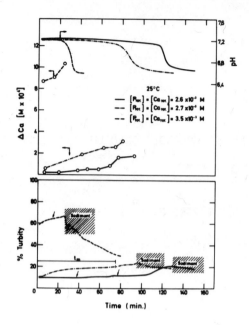

Fig. 1 Changes of pH, turbidity and Ca concentration as function of time at 25°C. Zero time initial reactant concentrations of Ca and phosphate were as indicated. Changes in Ca concentration are expressed as ΔCa (difference between initial Ca concentration and that at any given time. t_m is the lifetime of ACP. Arrows point to the position on the turbidity curves at which samples were taken for electron microscopy (Fig. 2).

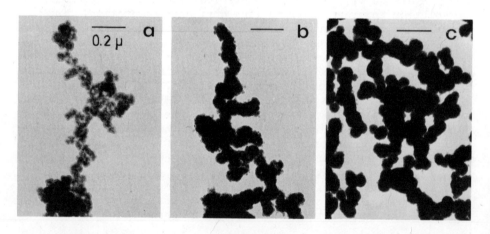

Fig. 2 Electron micrographs of ACP precipitates: a and b, $(P_{tot}) = (Ca_{tot}) = 2.6 \times 10^{-3}$M, 38 and 77 min after sample preparation. c, $(P_{tot}) = (Ca_{tot}) = 3.5 \times 10^{-3}$M, 15 min. Time intervals are marked on respective curves in Fig. 1.

calcium chloride solutions with solutions of sodium phosphate. The later solutions were preadjusted to pH 7.4-7.5 in order to obtain a constant pH 7.2 + 0.05 at time zero after sample preparation. Changes of the pH and the relative turbidity were continuously recorded as a function of time (4). Turbidities were measured at 540 nm against a 6.28×10^{-5} g/ml suspension of latex LS-54-2 ($R_{90}0 = 0.040$ cm^{-1} at 546 nm). Precipitates were separated at given time intervals and the concentrations of calcium in the supernatants were determined by atomic absorption spectrophotometry.

If not otherwise stated citrate was added to the phosphate solutions prior to the adjustment of the pH to 7.4. In most experiments the initial calcium concentration (3×10^{-3}M) was exceeded the concentrations of citrate ions by approximately 10 times.

Another set of experiments was carried out at constant pH 7.2+0.02 by means of a pH-stat device (Radiometer Co.). The reaction slurry was stirred at constant rate throughout the experiment and the consumption of sodium hydroxide was automatically recorded. In these experiments the citrate solution was added either prior to or after sample preparation. Samples containing the same concentrations of calcium chloride and sodium phosphate, but without citrate ions were used as controls.

The particle charge was determined by microelectrophoresis.

The number of particles was estimated on the basis of the particle volume and the total volume of the precipitate. Particles were assumed to be spherical (12) and their radius was obtained by electron micrography. The total volume of the precipitate was determined from the results of solution calcium analysis on the basis of a molar Ca/P ratio of 1.4, assuming that the density of the solid phase was between 1 and 2.

RESULTS

The following results show the influence of changes in reactant concentrations and the addition of citrate ions on the properties and lifetime of ACP. In figure 1 three sets of curves show changes of the pH, the turbidity and calcium concentration with time for different initial reactant concentrations. In accordance with previous results (4,5) the turbidity and pH curves show that precipitation occured in two steps, the first step being the formation of ACP (Fig.2). The lifetime of ACP decreases with increasing initial reactant concentrations. The solution calcium concentration decreased continuously during the reaction (ΔCa increases in Fig.1) signifying continuous precipitation. As a consequence the size of the ACP spherules increased as shown in Figs 2a,b. At the time just before the onset of secondary

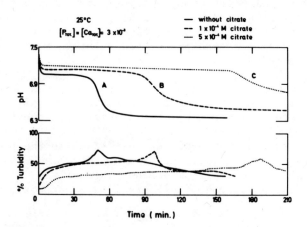

Fig. 3 Turbidity and pH versus time curves recorded at 25°. Zero time total concentrations: $(P_{tot}) = (Ca_{tot}) = 3 \times 10^{-3}M$, citrate A. -, B. $1 \times 10^{-4}M$, C. $5 \times 10^{-4}M$.

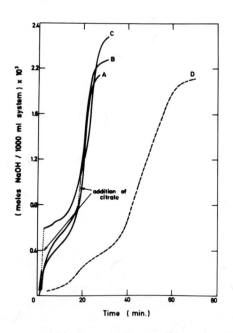

Fig. 4 Consumption of NaOH as a function of time at 25°C. Zero time total concentrations: $(P_{tot}) = (Ca_{tot}) = 3 \times 10^{-3}M$, citrate A. -, B. $1 \times 10^{-4}M$ added 2 min after sample preparation, C. $1 \times 10^{-4}M$ added 18 min after, D. $1 \times 10^{-4}M$ added prior to sample preparation.

precipitation the particle sizes were comparable (r = 30 - 32 nm; Figs 2b,c). The number of particles was between 4×10^{13} (ρ=1) and 8×10^{13} (ρ=2) for $P_{tot.} = Ca_{tot.} = 2.6 \times 10^{-3}$M and 4×10^{14} (ρ=1) and 1×10^{15} (ρ=2) for $P_{tot.} = Ca_{tot.} = 3.5 \times 10^{-3}$M. Particle numbers estimated in one particular system at different time intervals were approximately constant. The electrophoretic mobility of ACP particles was approx. zero. In the presence of citrate ions the surface charge changed to negative. Thus for concentration of citrate ions 1×10^{-4}M and higher the electrophoretic mobility was of the order of magnitude of $\omega_- \sim 10^{-4}$ cm^2 sec^{-1} V^{-1}.

Figs 3 and 4 show the effect of different concentrations of citrate ions on the lifetime of ACP. In the system shown in curves B and C in Fig.3 and curve D in Fig.4 citrate was present "in statu nascendi" i.e. added to the phosphate solution previous to sample preparation. In these experiments secondary precipitation was appreciably delayed as compared to the controls (curves A in Figs 3 and 4). However, if citrate was added at any given time after sample preparation no effect on the lifetime of ACP could be detected (curves B and C in Fig.4).

DISCUSSION

The above results are in accordance with earlier work (5,9,13) showing that the stability of ACP is most sensitive to variations in the conditions of precipitation. At constant pH stabilization of ACP, i.e. a delay in secondary precipitation is achieved by
- decreasing reactant concentrations (Fig.1) and/or increasing ionic strength (5,14)
- decreasing temperature (5,9)
- the addition of certain foreign substances (Fig.3, refs. 5,14).
Stirring of the reactant solutions increases the rate of ACP transformation (5). All these experimental facts might be explained if it is possible to show what is happening to ACP during the period of relative metastability (t_m in Fig.1).

The actual number of particles ($10^{13} - 10^{15}$) and its increase with increasing initial reactant concentrations, indicate that the precipitates were formed by homogeneous nucleation. Most probably agglomeration of primary particles leads to the formation of the spherulites, which are so typical for ACP (5). These then grow (as evidenced by electron microscopy (Figs 2a, b, ref. 12) and the decrease in solution calcium (Fig.1) and agglomerate into chain-like structures (Fig.2, ref. 4). It is interesting, that before secondary precipitation sets in, the size of the individual spherules is comparable in all experiments, regardless of the initial reactant concentrations (Figs 2b,c).

The sensitivity of the stability of ACP to stirring (5) suggests that solution mediated processes, such as the transport of new material to the ACP spherules and/or their agglomeration are partly responsible for the onset of secondary precipitation. A highly effective stabilizer, pyrophosphate has been shown to inhibit the growth process. However, the observation, that citrate ions can effectively stabilize ACP only if present "in statu nascendi", i.e. in the very beginning of the formation of the amorphous material (Fig.4) shows that occlusion of a foreign substance may be required to achieve this effect. This result is in support of the idea (11) that rearrangements within the amorphous particles, such as dehydration and crystal ordering are also to a large degree responsible for the nucleation of the crystalline material. Such rearrangements could be severely hindered by the coprecipitated foreign substance, the consequence being stabilization of ACP. Recently published electron micrographs (12) also indicate that crystallization within ACP particles can take place.

CONCLUSIONS

It is suggested that under conditions where ACP is the initial precipitate, the amorphous particles are intimately associated with the nucleation of the secondary, crystalline phase. The mechanism of convertion of ACP to a suitable template for secondary nucleation is very complex, involving both external, solution mediated processes and internal rearrangements.

REFERENCES

1. Eanes, E.D., Gillesen, I.H. and Posner, A.S., Nature, $\underline{208}$ (1965) 365-7.
2. Eanes, E.D. and Posner, A.S., Trans. N.Y. Acad. Sci., $\underline{28}$ (1965) 233-41.
3. Termine, J.D. and Posner, A.S., Arch. Biochem., $\underline{140}$ (1970) 307-17.
4. Brečvić, LJ. and Füredi-Milhofer, H., Calc. Tiss. Res., $\underline{10}$ (1972) 82-90.
5. Füredi-Milhofer, H., Brečević, LJ., Oljica, E., Purgarić, B., Gass, Z., and Perović, G., S.C.I. Monogr., $\underline{38}$ (1973) 109-129.
6. Termine, J.D. and Eanes, E.D., Calc. Tiss. Res., $\underline{10}$ (1972) 171-97.
7. Holmes, J.M. and Beebe, R.A., Calc. Tiss. Res., $\underline{7}$ (1971) 163-74.
8. Blumenthal, N.C. and Posner, A.S., Calc. Tiss. Res., $\underline{13}$ (1973)

235-43.
9. Boskey, A.K. and Posner, A.S., J. Phys. Chem., 77 (1973) 2313-17.
10. Eanes, E.D., Colloques internationaux C.N.R.S. No.230 (1975) 295-310.
11. Füredi-Milhofer, H., Bilinski, H., Brečević, LJ., Despotović, H., Filipović-Vinceković, N., Oljica, E. and Purgarić, B., ibid 303-10.
12. Eanes, E.D., Termine, J.D. and Nylen, M.U., Calc. Tiss. Res., 12 (1973) 143-58.
13. Füredi-Milhofer, H., Purgarić, B., Brečević, LJ. and Pavković, N., Calc. Tiss. Res., 8 (1971) 142-53.
14. Termine, J.D., Peckauskas, R.A. and Posner, A.S., Arch. Biochem. Biophys., 140 (1970) 318-25.

THE EFFECT OF Fe(III) IONS ON THE GROWTH OF POTASSIUM DIHYDROGEN PHOSPHATE MONOCRYSTALS

J. KARNIEWICZ, P. POSMYKIEWICZ AND B. WOJCIECHOWSKI

Institute of Physics, Technical University of Lodz

ul. Zwirki 36, 90-924 Lodz, Poland

INTRODUCTION

The mechanism of crystallization from solution is described by the interaction of ions or molecules of the solute with those of the solvent and the growing crystal. However, there are still a number of unsolved problems in this area. In particular this is true of the structure of supersaturated solution, the type of nucleation, the effect of additives on the properties of solution and their incorporation into the growing crystal. Additives in a solution can influence the growth of the crystal in various ways as well as its properties. The effects of additives on the shape of the crystal, the speed of growth and the type of nucleation are currently under investigation, in particular the effect of trivalent metal ions M^{3+} where M = Fe, Al, Cr.

In saturated acid solution Fe^{3+} ions adsorbed by the crystal significantly change the crystal habit (1,2). However these ions have no effect on the process of crystallization or the habit so long as concentration does not exceed 0.1 wt. % (3).

According to Mullin et al. (4) the concentration of impurities in excess of 10^{-3} g-ion Fe(III)/litre of solution causes tapering of ammonium dihydrogen phosphate (ADP) crystals. The growth rate of ADP is also significantly influenced by Fe^{3+} ions (5,6). The growth rate of the {100} face is strongly influenced and tapering occurs, but the authors do not give the amount of M^{3+} ions adsorbed into the crystals. This amount depends on the type of face, type of impurities and properties of the solution. Measurements of the segregation coefficient for Fe^{3+} ions ($K_c = C_c/C_s$) show that for KDP (7,8), DKDP (8) and LiCl (9) the segregation coefficient for

{100} face is greater than one. This means that the concentration of Fe(III) in the crystal, C_c, is larger than that in the solution, C_s.

EXPERIMENTAL

KDP monocrystals have been grown simultaneously in six crystallizers in a common water-bath, thus maintaining the same temperature in all crystallizers. The temperature was controlled by a thyrystor device to ±0.01 K. Two different methods of growth were used, namely evaporation of the solvent and lowering of the solution temperature. In the first method the crystals were grown on seeds introduced to the solution after a supercooling of 0.5 K had been achieved. In the second case crystallization was preceded by spontaneous nucleation. The growth of the single series lasted for 20 days.

The KDP solutions were purified in an ionite column, and afterwards doped with controlled quantity of the ferric citrate $C_3H_4(OH)(COO)_3$ Fe·$3H_2O$ or ferric ammonium sulphate $Fe(NH_4)(SO_4)_2 \cdot 12H_2O$, in such a way that the concentration of Fe^{3+} in the solutions contained amounts ranging from 0.1 to 0.0001 wt. %. One crystallizer contained undoped solution. The chosen dopants were easily soluble in water and did not affect the pH significantly. The concentration of Fe^{3+} in solution as well as in the crystals was determined polarographically. Recommended buffer electrolytes (10-12) such as sodium citrate and potassium tartarate were not suitable for the determination of trivalent iron in the KDP solutions. However, the use of sodium acetate enabled us to obtain a suitable shape of the polarographic wave of Fe^{3+} so that it was possible to detect concentrations of the order of 10^{-5} wt. % (Fig. 1).

In the process of constant increase of potential, Fe^{3+} in sodium acetate solution changed into Fe^{2+} when the semi-wave potential was 0.42 V and later Fe^{2+} changed into Fe atoms at 1.55 V. The measurements were made with polarograph LP-60 in which the dependence of the diffusion current, J_d, on the concentration of Fe^{3+} and is described by

$$J_d = k C_c$$

where k is a proportionality constant.

RESULTS AND DISCUSSION

In previous investigations the amount of impurities have been determined spectroscopically, colorimetrically and neutronographically. However, the polarographic method is equally precise and easier to use.

Table 1. Segregation coefficient K_c for Fe^{3+} ions.

Crystal face	K_c	Ref.
{100}	3.1	(6)
{101}	0.6	
{100}	1.9	(8)
{101}	0.6	
{100}	2.0	this work
{101}	0.7	

Table 2. Concentration, C_c, of Fe^{3+} in KDP crystallizing from solutions containing concentration C_s of Fe^{3+}.

g Fe(III)/g KDP

$10^3 C_s \pm 0.02$	$10^3 C_c \pm 0.02$	K_c
0.15	0.32	2.1
0.23	0.46	2.0
0.44	0.88	2.0
0.72	1.45	2.0
2.00	4.08	2.0
4.21	4.20	-
9.20	4.19	-
14.20	4.22	-

Table 1 gives concentrations of Fe^{3+} in solutions and corresponding crystals of KDP. The segregation coefficient K_c for the {100} face is in general agreement with that quoted in references (7) and (8). Both methods of crystal growth with the same concentration of dopant in solution give the same segregation coefficient. The relation between Fe^{3+} in the crystal and in the

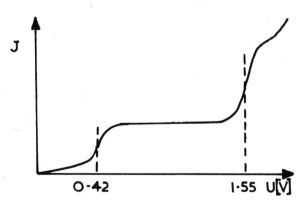

Fig. 1 Polarogram for Fe^{3+} and Fe^{2+} ions in sodium acetate

solution is given in Table 2. It is interesting to note that above a certain concentration C_s of Fe^{3+} ions in solution the Fe^{3+} concentration C_c in the crystal reaches a limiting value. Similar results have been obtained for NaCl crystals doped with $(Fe(CN)_6)^{4-}$ (13).

The effect of Fe^{3+} ion concentration on the shape of KDP crystals is well-known: the tapering of the crystal is proportional to the concentration of Fe^{3+} in solution and we find that at concentrations <10^{-4} wt. % tapering ceases. The effect of Fe^{3+} concentration on the speed of nucleation and growth has also been noted but it requires further consideration.

CONCLUSIONS

Fe^{3+} ions cause tapering of KDP crystals.

With small concentrations of Fe^{3+} in solution the segregation coefficient is greater than unity. With large concentrations of Fe^{3+} in solution the Fe^{3+} concentration in the crystal attains a limiting value.

Information relating the mechanism of the interaction between Fe^{3+} and the face of the crystal can only be obtained by EPR, NMR or X-ray methods which allow determination of the symmetry and nature of the bonds (14,15).

REFERENCES

1. Kolb, H.J. and Comer, J.J., J.Am.Chem.Soc., 67 (1945) 894.
2. Byteva, I.M., Rostkristallov, Nauka, Moskva, V (1955).
3. Mareček, V., Dobiasova, L. and Novak, J., Kristall und Technik, 4 (1969) 39.
4. Mullin, J.W., Amatavivadhana, A. and Chakraborty, M., J.Appl. Chem., 20 (1970) 153.
5. Davey, R.J. and Mullin, J.W., J. Crystal Growth, 23 (1974) 89.
6. Davey, R.J. and Mullin, J.W., J. Crystal Growth, 26 (1974) 45.
7. Czerednik, L.A., Portnov, W.N. and Belustin, A.W., Izv.vyzsz. ucz.zav.Fizika, 129 No. 2 (1973).
8. Belouet, C., Dunia, E. and Petroff, J.F., J. Crystal Growth, 23 (1974).
9. Hinks, D. and Susman, S., Mat.Res.Bull., 9 (1974) 53.
10. Milner, G.W.C., Polarografia, WNT, 1969.
11. Kolthoff, I.M. and Lingane, J.J., Polarography, 1952, New York.
12. Zagórski, Z., Metodapolarograficzna w analizie chem., 1956, PWN.
13. Glasner, A. and Zidon, M., J. Crystal Growth, 21 (1974) 294.
14. Torgesen, J.L. and Jackson, R.W., Science, 148 (1965) 952.
15. Stjern, D.C., DuVarney, R.C. and Unruh, W.P., Phys.Rev.B, 10 No. 3 (1974).

**Crystallizer
Design**

CRYSTALLIZER DESIGN AND OPERATION

J.W. MULLIN

Department of Chemical Engineering

University College London, Torrington Place
London WC1E 7JE, England

INTRODUCTION

Crystallization has attracted a considerable amount of attention from chemical engineers in the past decade or so, and several notable advances have been made in the analysis of crystallization processes. We now have a deeper understanding of the kinetics and mechanisms involved, the importance of secondary (as opposed to classical primary) nucleation, the interaction of the various operating parameters and the crystal size distribution, the influence of equipment geometry and system hydrodynamics. Useful models have been developed for the cycling and instability of crystallizers, cascade operation, optimum cooling rates, and so on.

As a result of all this intensive effort the crystallizer designer and operator have undoubtedly gained a better insight into the complexities of the crystallization process. But are industrial crystallizers now designed from first principles? Regrettably the answer is 'no'. Why is this?

Crystallization is known to be a very complex operation and some of the reasons for this complexity are fairly obvious. For example, the growth of crystals in a crystallizer involves the simultaneous processes of heat and mass transfer in a multi-phase, multi-component system. These conditions alone present complications enough, but the crystallization process is also strongly dependent on fluid and particle mechanics in a system where the size and size distribution of the particulate solids, neither property being capable of unique definition, can vary with time.

Furthermore, the solution in which the solids are suspended is thermodynamically unstable, frequently fluctuating between so-called metastable and labile states, and sometimes entering the unsaturated condition. Traces of impurity, sometimes a few parts per million, can profoundly affect the nucleation and crystal growth kinetics. It is perhaps understandable why crystallization has been slow to submit to simple analytical procedures.

On the other hand, the many advances made in recent years cannot be ignored. The concept of the population balance and the techniques of mathematical modelling, together with the intelligent use of the computer, have led to some spectacular successes.

DESIGN REQUIREMENTS

It is worth recalling how the foundations of the unit operation of distillation were laid down in the 1930s. Chemical engineers, working with the relatively simple concepts of the heat and mass balance, coupled with a knowledge of phase equilibria, established the basic theories of stage-wise contact and developed design procedures that have basically changed little over the past 40 years. Subsequent mathematical sophistications have embellished these procedures, and computers have greatly facilitated their implementation and significantly cut down the work load. But can distillation columns now be designed from first principles? Has there been developed one supreme, universally accepted method of designing distillation columns? The answer is 'no' to both questions. And I think there is a lesson to be learned here.

A distillation column designer frequently uses at least two or more different methods to arrive at the various basic equipment dimensions, especially if he is dealing with a new working system. If the values roughly agree, he gains confidence. If they do not, he begins a thorough re-examination of his basic data. He may even initiate fresh laboratory work.

Crystallizer designers should do the same. Crystallization is infinitely more complex than distillation. No single design technique can possibly reign supreme for all cases. Neither will any given design method give a unique answer. It is all a matter of compromise and what can be considered to be reasonable under the prevailing circumstances.

Table 1 gives an example of the sort of tabular calculations a designer might make. These data, incidentally, refer to the design of a crystallizer to produce 1000 kg/h of 1 mm potassium sulphate crystals in a fluidized bed classifying crystallizer. Initially it is necessary to determine by experiment the maximum allowable working supersaturation (0.01 kg/kg), the crystal

Table 1 Calculations of the dimensions of a classifying crystallizer for potassium sulphate (production rate = 1000 kg/h, working supersaturation = 0.01 kg/kg, 20°C)

Desupersaturation, γ	1.0	0.9	0.5	0.1
Circulation rate, Q, (m³/h)	103	115	208	1030
Maximum growth rate, g_p, (m/s) x 10⁸	5.6	5.6	5.6	5.6
Crystal residence time, τ, (h)	408	25.2	14.4	3.5
Weight of crystals, W, (kg)	145000	9000	5100	1250
Suspension volume, V, (m³)	364	22.5	12.8	3.15
Up-flow velocity, u, (m/h)	144	144	144	144
Cross-sectional area, A, (m²)	0.72	0.80	1.45	7.2
Crystallizer height, H, (m)	505	28	8.8	0.44
Crystallizer diameter, D, (m)	0,96	1.01	1.36	3.02
H/D	525	28	6.5	0.15
Separation intensity, SI	3	45	78	320
Economically possible?	no	no	yes	no

growth rate at the maximum supersaturation (5.6 x 10^{-8} m/s) and the upflow velocity to suspend 1 mm crystals (144 m/h). Then the calculations can be commenced.

The relative desupersaturation γ is an important quantity. Complete desupersaturation, $\gamma = 1$, is difficult to achieve: only a small quantity of solution would be allowed to flow through the crystallizer (103 m³/h), a residence time of 408h and a very deep bed of crystals (505 m) would be needed. As the desupersaturation is reduced the various quantities such as the height:diameter ratio, solution circulation rate, suspension volume, and so on become more reasonable.

Although one of the cases in Table 1 seems to be an economic possibility it does not mean, of course, that this is the optimum condition of operation. Further calculations would be necessary to determine this. Nor should the entirely empirical aids be neglected. A useful check on the feasibility of such calculations is the Separation Intensity (SI) factor which expresses the production rate in terms of kg of 1 mm equivalent crystals per m³ of crystallizer volume per hour. For inorganic salts in aqueous solution at ambient temperature a value of SI in the range 100 - 150 might be expected.

It is very encouraging to see different design routes being devised, but there is a long way to go before crystallizer design reaches maturity. We should stop arguing about which design method is best. No design equation is any more reliable than the data fed into it, and crystallization data are notoriously unreliable. We should concentrate, therefore, on devising methods of measuring reliable data.

Significant advances have been made in recent years on various aspects of crystallizer design, but real progress from now on can only be peripheral until we have a better understanding of the fundamental processes. We cannot go on building on unsure foundations.

Admittedly the areas where more attention should be devoted may sound less glamorous than the current academic exercises, but they are still very important.

On the fundamental side work should continue in the areas of growth and nucleation kinetics (particularly secondary nucleation) and the effect of impurities. Studies of solution-phase phenomena and non-equilibrium processes would also be profitable.

On the practical side there are some surprising gaps. Heat exchange in two-phase crystallizing systems is not at all well understood. Neither is encrustation on internal surfaces despite its notoriously detrimental effect on crystallizer performance. Problems associated with the suspension and removal of solids from agitated vessels need attention, and the design of agitators has been neglected. The effect of boiling conditions on the kinetics of nucleation and growth have not been studied to any extent. Yet this information would be of considerable importance in the choice between cooling and evaporation modes of operation. Problems of energy conservation, alternative operating schemes and optimum methods of operation are all potentially fruitful fields.

SOME BASIC TYPES OF CRYSTALLIZER

It has been said that there are almost as many different crystallizers as there are products being crystallized, but most fall into a few well-defined categories. Batch-operated cooling crystallizers, for example, are often nothing more than simple agitated vessels, but the continuously operated units are a little more complex. Figure 1 shows three basic types. The heat exchanger in each case may be either a heater or a cooler, i.e. the crystallizers can be of the evaporative or cooling types. Alternatively, they may be operated adiabatically, i.e. as vacuum crystallizers, which act as both evaporators and coolers.

The high rate of recirculation through the external heat exchanger in the forced circulation crystallizer (Figure 1a) improves heat transfer and minimizes encrustation. In the draft tube agitated unit (Figure 1b) a high rate of internal recirculation ensures efficient mixing. These crystallizers may be fitted with internal baffles to facilitate fines removal (as in the DTB crystallizer) and an elutriating leg to effect some degree of product classification.

In the Oslo-Krystal crystallizer (Figure 1c) a fluidized bed is maintained by the upflow of supersaturated liquor in the annular region surrounding the central downcomer. These units were originally designed to act as classifying crystallizers, but nowadays they are frequently operated as fully mixed units.

CRYSTALLIZER DESIGN AND OPERATION

The solid and liquid flow patterns in these vessels differ significantly, depending on the type and size, and impose their influence on the crystal product. Sizes can vary enormously from small bench-scale models of one or two litres to large industrial crystallizers of 100 m^3 working capacity and more, with production rates in excess of 500 ton/day.

A high proportion of all industrial crystallizers are of the mixed suspension type. However, even with this basically simple system there are several possible modes of operation, as demonstrated in Figure 2. In the total discharge MSMPR (mixed suspension, mixed product removal) unit the crystal and liquor residence times are identical. If the MSMPR crystallizer is operated with clear liquor over-flow the crystal and liquor residence times become independent. This is a useful method of exerting control over the magma density in the crystallizer. It is possible to provide some product classification by recycling some clear liquor through an elutriating leg as in the MSCPR (mixed suspension, classified product removal) crystallizer, while an MSCPR unit operated with clear liquor overflow provides even greater flexibility of operation.

LABORATORY OR PILOT PLANT

One of the basic problems of crystallizer design is the choice of method for measuring design data. It may not be possible to select the appropriate mode of operation until the data are available. So how should the data be measured and how can we be sure that data measured in one type of apparatus will be applicable to another?

The problem is not only concerned with the type of apparatus; the scale of operation has also to be considered. Basically there are three choices, viz. to measure the data in the laboratory, in a pilot plant or on a full-scale plant. The latter is quite common, because many crystallizers are designed on the basis of past experience, but generally the choice lies between laboratory and pilot units.

In terms of approximate physical dimensions we may have:

	Laboratory	Pilot	Industrial	
Volume	1	100	10,000	litre
Diameter	10	50	250	cm
(Ratio)	(1)	(5)	(25)	

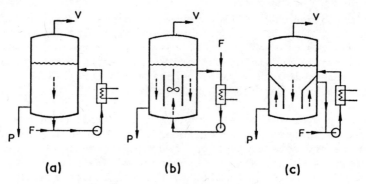

Fig. 1. Three basic types of continuous crystallizer, (a) forced circulation, (b) draft-tube agitated, (c) fluidized-bed (Oslo-Krystal).

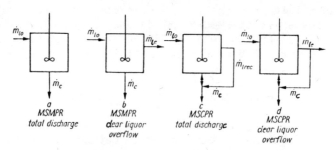

Fig. 2. Basic Types of Mixed Suspension Crystallizer. (a) MSMPR; (b) MSMPR with liquor overflow; (c) MSCPR; (d) MSCPR with liquor overflow and recycle.

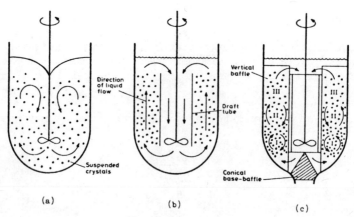

Fig. 3. Simplified flow patterns in agitated vessels, (a) central propeller, (b) draft-tube, (c) draft-tube with wall baggles and conical deflector.

So despite the large volumetric differences the diameter ratios are quite small, and diameter is an important parameter. Laboratory-scale crystallization operations, therefore, may not be greatly different from pilot plant work, and of course it is much easier to work on the smaller scale. In any case the perils of pilot plant operation are well known: they are expensive to design, construct and operate, and they are frequently unstable in operation due to the ease of pipe line blockage and the need for carefully controlled low flow rates. Laboratory scale operations can be increased, sometimes with advantage, to about 5 litres without much trouble, when the above ratios become even more favourable.

It is now generally accepted that much useful information can be obtained in the laboratory so long as the fundamental differences between the laboratory and industrial plant scale of operation are fully appreciated. The two hydrodynamic conditions are vastly different. Liquor paths, or turn-over times, are much shorter in the smaller vessels. The suspension circuit may only be a few centimeters in a beaker, whereas it may be 10 or 20 m in a full-scale plant. If the solution desupersaturates rapidly, very high circulation rates would be needed in the large crystallizer to utilize fully the working volume. In a laboratory beaker there would be no problem.

The exact scale-up of crystallizers is not possible because it would be necessary to preserve identical flow characteristics of both liquid and solid phases and temperatures and supersaturations in all equivalent regions. The scale-up of simple one-phase agitated vessels has long been recognized as a difficult problem, involving such impossibilities as attempting to keep both the dimensionless Reynolds and Froude numbers constant: the former is proportional to the stirrer speed and the square of the stirrer diameter while the latter is proportional to the diameter and the square of the speed.

In a two-phase (solid-liquid) system, scaling-up to maintain exact geometrical (shape), kinematic (velocity) and dynamic (force) similarity is impossible, and in crystallizing systems there are the additional requirements of thermal and chemical potential similarity. So it seems pointless to argue that one experimental model is more 'realistic' than another. The gross deficiencies of all models must be accepted. They are all, to some extent, unrealistic.

This does not mean to say that data obtained from small-scale experimental units are always unreliable, but it does emphasise that caution should always be exercised in the use of such data. The possibility of unreliability always exists and it should never be ignored.

AGITATION AND FLOW RÉGIMES

The processes of crystal growth and nucleation are powerfully influenced by the system hydrodynamics. So the method of agitation in a crystallizer would appear to be very important. However, there is very little reliable design information available for agitated vessels for solid-liquid systems.

We know that different types of agitator produce different flow patterns. For example, a centrally mounted propeller (Figure 3a) induces tangential flow and a well-established pattern of circulation. At high speeds a vortex is generally produced, but the introduction of wall baffles creates a chaotic turbulent motion. Draft tube agitation (Figure 3b) gives a very different type of circulation. The propeller acts as a pump and the liquid may be forced to flow either up or down the draft tube. Tangential flow occurs in the annular zone unless wall baffles are installed.

It is very difficult to ensure near-perfect mixing in suspensions of solid particles. One partially successful arrangement is shown in Figure 3c. Four wall baffles extend down to the base of the vessel to create four equal annular zones. A conical deflector eliminates the 'dead space' at the bottom of the vessel underneath the draft tube which operates with a down-flow.

Despite these precautions the contents of the vessel are rarely perfectly mixed and frequently three distinct flow régimes can be identified. In region 1 crystals circulate vigorously and stay mainly in this zone which is similar to that in a turbine agitated vessel. Region 2 is less vigorous than region 1. Crystals tend to migrate up and down. This zone is rendered unstable by voids which are propagated periodically (a behaviour often seen in liquid fluidized beds) due to the pumping action of the impeller. Region 3 is a low-voidage, gently fluidized zone in which there is a slow circulation of crystals.

The overall system voidage can affect considerably the extents of the three zones, but it is very difficult to quantify these effects.

Despite the popularity of draft tube agitation there are virtually no design guides available. There is frequent reference in the literature to the desirability of arranging that the areas of the draft tube and annulus are roughly equal. For single-phase flow it is easy to demonstrate that this is a feasible solution for either minimum power input for a given circulation rate or maximum circulation rate for a given power input.

However, for two-phase (solid-liquid) flow the situation is by no means clear-cut. There is no unique solution for the optimum draft-tube diameter since it depends on the mass and size distribution of the suspended crystals, which in turn determine the necessary upflow velocities. In a batch crystallization, of course, all these quantities can change significantly over a period of time.

Just by way of illustration, Figure 4 shows how a draft tube was designed for a 25 litre (30 cm diameter) batch-operated laboratory crystallizer for potassium sulphate crystals. In this particular case the flow was down the draft tube and crystals were maintained in the annular zone to prevent damage by the impeller.

The predicted optimum diameter of the draft tube varies over a wide range when, as would happen in a batch crystallizer, crystals grow from small size and mass (low fluidization velocity) to large size and mass (high fluidization velocity). However, in all cases the draft tube diameter is always less than the equal area value of 21 cm. As the solids content is increased the draft tube diameter decreases. In the particular case illustrated in Figure 4 a compromise was made by installing a 15 cm diameter draft tube, giving an annulus:draft tube area ratio of 3:1. Clearly there is scope for much more work on this subject.

CRYSTALLIZATION IN THE PRESENCE OF IMPURITIES

A great deal of attention is being paid nowadays in both academic and industrial circles to the subject of crystallization in the presence of impurities. This is an important topic since no industrial liquor is pure, so in effect all industrial crystallization takes place in the presence of impurities. Furthermore, impurities are often added deliberately to exert some control over the nucleation or the crystal growth processes.

There is nothing new in the use of additives to influence crystallization. There is evidence that primitive salt production has been carried out for hundreds of years under the influence of biological fluids of various kinds. Today surfactants, polyelectrolytes and long-chain polymeric substances are widely used.

The modes of operation of impurities are many and varied, but probably the most difficult to understand are the dramatic effects often induced by small traces of certain ionic substances. The truth of the matter is that in these cases we are often working in the area of non-equilibrium phenomena (the attainment of equilibrium is both a thermodynamic and a kinetic problem) and this is an under-explored branch of science.

The now well-known effects of $K_4Fe(CN)_6$ on the system $NaCl-H_2O$ are now attributed to the non-attainment of equilibrium and a similar behaviour can be observed with K_2SO_4 in the presence of Cr^{3+}.

It is not possible at the present time to explain these phenomena fully, but one thing is clear: if the driving force for crystallization is changed by the presence of an impurity, then yet another variable has to be introduced into this already complicated process.

There is another aspect of crystallization in the presence of impurities that deserves attention, viz. the phenomenon of constitutional supercooling, well-known to single crystal growers and metallurgists, but perhaps not widely appreciated by chemical engineers. Nevertheless, I believe it may be of considerable importance in industrial crystallization.

When a crystal grows in an impure system, impurity is rejected at the solid-liquid interface. If the impurity cannot diffuse away fast enough it will concentrate near the crystal face and affect the equilibrium conditions in the interfacial region. For crystallization from the melt, the melting point of the solid is lowered by the presence of impurities and the supercooling, the driving force for crystal growth, is decreased and the growth rate is retarded. A similar argument may be developed for growth from solution. If the equilibrium saturation concentration of the bulk solution is lower than that at the interface then the driving force increases over a short distance from the crystal face towards the bulk solution.

This condition, called constitutional supercooling, or constitutional supersaturation in the case of crystallization from solution, causes interfacial instability. The growing interface breaks up into finger-like cells in a more or less regular bunched array. In this way the liberated heat of crystallization is more easily dissipated and the tips of the projections advance clear of the concentrated impurity. The regions between the fingers entrap impure solution and a succession of inclusions may be left behind.

Under conditions of low constitutional supercooling (low level of impurity) the advancing interface is cellular. At high constitutional supercooling dendritic branching generally occurs and this sort of behaviour is common, for example, in the casting of metals. It can also be important in industrial crystallization.

Its importance is not in the actual crystal growth process, but afterwards during the solid-liquid separation and drying operations. A crystal with its surface wet with mother liquor has all the requirements for constitutional supercooling - a thin layer of relatively impure solution on a crystal face with no opportunity for the impurity to diffuse away from the interfacial region.

Thus the last stage of the crystallization process can result in the development of an irregular, perhaps even dendritic growth.

CRYSTALLIZER DESIGN AND OPERATION

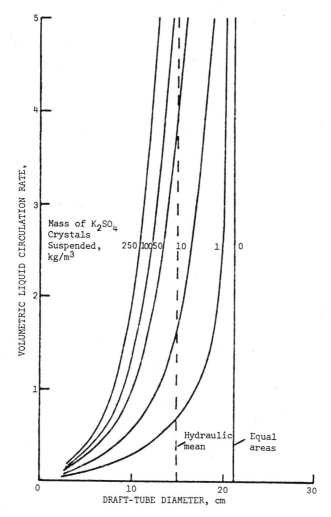

Figure 4. Optimum draft-tube diameters in a 30 cm diameter vessel (251)

This fragile surface layer is easily detached from the parent crystal on subsequent handling and storage, resulting in a 'dusty product'. The dust has not come from excessive nucleation during the crystallization process nor from crystal attrition, it has been caused by 'drying-out growth' under conditions of constitutional supersaturation.

CONCLUSIONS

Progress towards the development of reliable design and operating procedures for industrial crystallization could be hastened by gaining a better understanding of the fundamental processes involved. Industry has an important part to play too: more feed-back is necessary, and it is encouraging to see that some of the larger manufacturing companies appear to be more willing than they were in the past to release information for publication.

It is worth remembering also that there are many research workers in a wide range of academic disciplines, far removed from chemical engineering, such as medicine, biology, geology, and so on, who are studying phenomena which could have some relevance to industrial crystallization. Admittedly, it is very difficult these days to keep up with the technical literature, but contact with these groups of workers could be very rewarding. To appreciate each other's problems might enable us to see our own in clearer focus.

BIBLIOGRAPHY

Jones, A.G. and Mullin, J.W., The design of a draft-tube agitated vessel, Chemistry and Industry (1973) 387.

Larson, M.A. and Garside, J., Crystallizer design techniques using the population balance, Chem. Engr Lond., (1973) 318.

Mullin, J.W., Crystallization, 2nd Ed., 1972, Butterworths, London.

Mullin, J.W. and Garside, J., Crystallization in the presence of impurities, Chem. Engr Lond., (1974) 402.

Mullin, J.W. and Nývlt, J., The design of classifying crystallizers, Trans. Instn Chem. Engrs, 48 (1970) 7.

Nývlt, J., Industrial Crystallization from Solutions, 1971, Butterworths, London.

Nývlt, J. and Mullin, J.W., Design of continuous mixed-suspension crystallizers, Krist. Tech., 9 (1974) 144.

Randolph, A.D. and Larson, M.A., Theory of Particulate Processes, 1971, Academic Press, New York.

THE DESIGN OF A CRYSTALLIZER

K. TOYOKURA, F. MATSUDA, Y. UIKE, H. SONODA AND
Y. WAKABAYASHI

Waseda University

Nishiookubo 4 Chome, Shinjukuku, Tokyo, Japan

INTRODUCTION

The production rate in a crystallizer depends on the reactant feed rates or solvent evaporation rate. The product crystal size is decided from the nucleation rate, growth rate and crystal retention time. The nucleation and growth rates are affected by many operational factors, especially the supersaturation which is decided by the removal rate of solution and the reaction and crystallization rates. The crystallization rate is divided into two independent parts (the nucleation and growth rates) that are correlated independently against supersaturation. However, supersaturation is frequently difficult to measure, so nucleation rate is sometimes correlated against growth rate directly.

In this study, new design methods for a mixed-bed type crystallizer are studied from a correlation between nucleation and growth rates and new charts for design are proposed. Nucleation and growth rates of sulfamic acid crystal are observed by continuous reaction crystallization tests between urea-sulfuric acid and fuming sulfuric acid. Operational conditions for the reaction crystallization of sulfamic acid are also discussed.

DESIGN OF CONTINUOUS MIXED-BED TYPE CRYSTALLIZER FROM NUCLEATION AND GROWTH RATES

Continuous Well-Mixed Bed

The crystallization model of Toyokura (1) gives the following equations:

$$V' = P \ell_m/3 \, (d\ell/d\theta)(1-\varepsilon) \, \rho_c \tag{1}$$

$$P/\rho_c = 2F'_v \ell_m^3 V'/9 \tag{2}$$

V' = crystallizer volume, P = crystal production rate, ℓ_m = dominant size of product crystal, ε = void fraction, ρ_c = density of crystal, F'_v = nucleation rate, θ = time.

The new chart for the correlation of $P/\rho_c V'$, ℓ_m, F'_v, $d\ell/d\theta$ and ε from Eqs 1 and 2 is given in Fig. 1. When the correlation between F'_v and $d\ell/d\theta$ is known, a continuous well-mixed bed crystallizer is easily designed as follows.

When the dominant crystal size and production rate are set as ℓ_m^* and P^*, the design line B which gives the correlation between F'_v and $P/\rho_c V'$, is drawn parallel to the line A from ℓ_m^* on the ℓ_m axis. The crystal growth rate $(d\ell/d\theta)^*$ and suspension density $(1-\varepsilon)^*$ are decided from preliminary tests. Another design line E is also drawn parallel to line D from point E ($(1-\varepsilon)^*$, $(d\ell/d\theta)^*$) on part i Fig. 1. The cross point F between lines B and E is the design point which gives F'^*_v and $(P/\rho_c V')^*$ against ℓ_m^*, $(d\ell/d\theta)^*$, and $(1-\varepsilon)^*$. Then the volume of the crystallizer is easily determined from P^* and $(P/\rho_c V')^*$. When F'^*_v at point F does not agree with the correlation between F'_v and $d\ell/d\theta$ supposed in the crystallizer, the assumed $d\ell/d\theta$ and/or $1-\varepsilon$ should be changed to reset. Alternatively, a fines trap and/or other device might be planned.

Continuous Conveying Bed and Classified Product (DTB Type)

The design equations, obtained from the previous paper (1) are

$$V' = P \ell_p/4 \, \rho_c \, (d\ell/d\theta)_{av} \, (1-\varepsilon) \tag{3}$$

$$P/\rho_c = F'_v \ell_p^3 V' \tag{4}$$

ℓ_p = size of classified product crystal, $(d\ell/d\theta)_{av}$ = log mean crystal growth rate.

The design chart for this type of crystallizer is Fig. 2. It is used for design in the same way as Fig. 1 for the well-mixed type.

DESIGN OF A CRYSTALLIZER

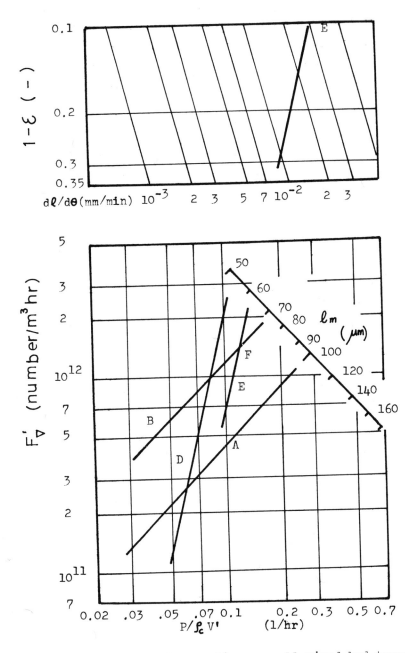

Fig. 1 Design chart for continuous well-mixed bed type

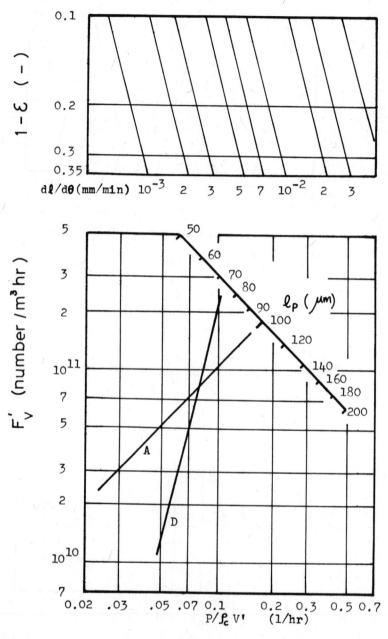

Fig. 2 Design chart for continuous conveying bed type

DESIGN OF A CRYSTALLIZER

EXPERIMENTAL

Sulfamic acid crystals were precipitated in a continuous well-mixed reactor by the reaction between 4 mol/ℓ urea-sulfuric acid solution (98%) and 30% SO_3-fuming sulfuric acid:

$$NH_2CONH_2 + SO_3 + H_2SO_4 = 2NH_2SO_3H + CO_2 \qquad (5)$$

The reactor, a stainless steel agitated tank, 150 mm diam. and 200 mm high, was kept at 90 ± 0.5°C. The paddle type agitator was 100 mm long and 25 mm wide. Raw materials were fed at 25 ml every 5 to 25 minutes. When steady state was obtained after continuous operation for more than five times the solution retention time, the crystallization rate and product size distribution were observed. Reaction rates of urea, r_A, (mol /ℓ min) were obtained from the volume change of carbon dioxide generated (Eq. 5). Reaction rates were correlated against urea concentration C_A (mol /ℓ) and crystal production rate P_o (mol /ℓ min) as

$$r_A = 22 \, C_A^{0.5} \qquad (6)$$

$$P_o = 0.6 \, r_A^{0.84} \qquad (7)$$

Crystal size distributions plotted on a semilog chart gave straight lines and crystal growth rates $dℓ/d\theta$ (mm/min) were obtained from the slopes. These were correlated against the reaction rate as

$$dℓ/d\theta = 14.5 \, r_A^{1.02} \qquad (8)$$

The product of the nucleation rate and shape factor, $F_v'k$, was calculated from reference 1:

$$F_v'k = 9P_v/2\rho_c \, ℓm^3 \qquad (9)$$

P_v = crystallization rate per unit volume (g/ℓ min), k = shape factor.

The correlation between $F_v'k$ and $dℓ/d\theta$ is plotted in Fig. 3. These data were obtained under the following conditions: agitator 160 to 200 r.p.m., dominant crystal size 62 to 80μm, suspension density 0.05 to 0.1.

DESIGN METHOD

The correlation of growth rate against nucleation rate and the chart in Fig. 1 or Fig. 2 are used to establish the product crystal size. For example, using the data in Fig. 3 for discussion, 1-ε is assumed 0.1. Setting the growth rate as 30μm/h, the design

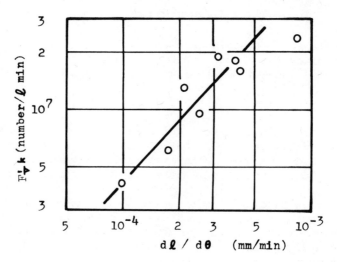

Fig. 3 Correlation of crystal growth and nucleation rates of sulfamic acid

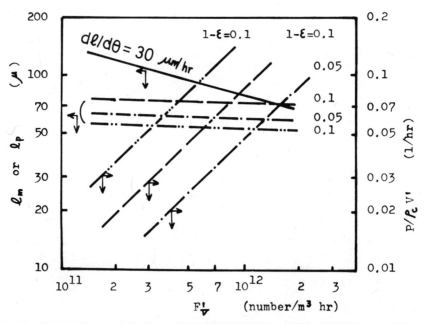

Fig. 4 Correlation of product crystal size and production rate with operational conditions.
Well-mixed bed —— —·—
Conveying bed —··—

DESIGN OF A CRYSTALLIZER

line E is drawn on part i of Fig. 1 as described in the previous section. Since $F_V^! k$ corresponding to $d\ell/d\theta$ is decided from Fig. 3, the operation point F becomes the cross point between $F_V^! k = 1.4 \times 10^{12}$ (number/m^3 hr) and the line E. Then line B is drawn parallel to line A and the dominant size becomes 72μm, and the production rate per unit volume of crystallizer becomes 0.12 (1/hr).

The product crystal size and production rate are also calculated for the cases of $d\ell/d\theta$ = 18 or 6, and shown by the dotted line for ε = 0.1 in Fig. 4. These results show that crystal size does not change over the range of these growth rates. When the effective nucleation rate reduces, compared with those at the same growth rate in Fig. 3, the crystal size gradually increases, parallel to the line $d\ell/d\theta$ = 30 in Fig. 4. The effective nucleation rate is controlled with a fines trap or changing the agitator. When 1-ε is reduced, the product size and production rate also reduce as in Fig. 4.

The same kind of calculation is performed for a continuous conveying bed type crystallizer and the results are also shown in Fig. 4. When the product crystals are classified, the product size ℓ_p decreases compared with that from a well-mixed bed type, although the expression of crystal size is not the same. But the production rate per unit volume increases.

SELECTION OF OPERATING CONDITIONS

Operating conditions are also decided from Eqs 6-8 and Figs 1-3. For design, the production rate P and the product crystal size ℓ_m are generally set. A well-mixed bed type will now be considered as an example. When the crystal growth rate is assumed within the pilot plant test data, the solution retention time θ (equal to that of the crystals for a well-mixed type) is calculated by

$$\theta = \ell_m/3(d\ell/d\theta) \tag{10}$$

The reaction rate is decided from the crystal growth rate by Eq. 8. Therefore, the crystal production rate and the urea concentration of the solution in the crystallizer are obtained from the reaction rate by Eqs 6 and 7. The void fraction in the crystallizer is calculated from the crystal production rate and the retention time.

When the chart in Fig. 1 is used, the design point F is easily decided against the assumed growth rate and the required nucleation rate for the desired product. They are plotted against the assumed growth rate as in Fig. 5 for different product sizes. However, since these calculated nucleation rates should be equal to those corresponding to growth rate in Fig. 3, the cross points between

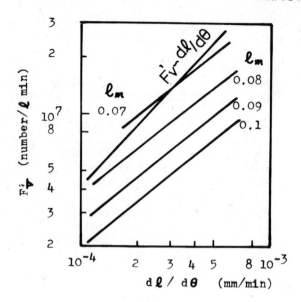

Fig. 5 Correlation of crystal growth and nucleation rates for determination of design conditions

$F'_V \sim d\ell/d\theta$ line and the product size lines in Fig. 5 become the operational points. When the operational point is decided, the operational conditions are easily calculated. These are listed in Table 1 and show that large sulfamic acid crystals cannot be precipitated under the crystallization rate correlations in Fig. 3. If large crystals are desired, some special devices would be required.

Table 1 Calculations to determine operational conditions for different product crystal sizes

product crystal size, μm	70	80	90	100
urea in feed solution, mol/ℓ	1.90	1.97	2.29	2.61
nucleation rate, no./ℓ min x 10^{-6}	15	2.9	0.64	0.19
growth rate, μm/min	0.32	0.073	0.018	0.006
retention time, min	73	365	1700	5650
production rate, g/ℓ min	2.4	0.71	0.22	0.9
suspension density $(1-\varepsilon)$	0.08	0.12	0.18	0.24

REFERENCES

(1) Toyokura, K., Krystal und Technik, <u>8</u> (5) (1973) 567.

DESIGN CRITERIA FOR DTB-VACUUM CRYSTALLIZERS

A.G. TOUSSAINT AND J.M.H. FORTUIN

Central Laboratory

DSM Geleen, The Netherlands

INTRODUCTION

In a draft-tube baffled (DTB) vacuum crystallizer a small portion of the solvent in a thin surface layer is evaporated to create supersaturation and cool the contents (Fig. 1). A propeller stirrer circulates the supersaturated liquid from the surface layer through the annular space and draft-tube. Whilst the liquid is in contact with the crystals, mass is transferred and the crystals grow. The unsaturated hot feed must be carried to the surface layer so that no boiling occurs. The hydrostatic pressure throughout this liquid must therefore be kept higher than the equilibrium pressure at the feed temperature. This temperature and, hence, the equilibrium pressure, can be lowered by mixing the feed internally and/or externally with a suspension flow. For designing a crystallizer the physical properties of the system, as well as the solubility and vapour-liquid equilibria, must be known. Process conditions and crystallizer dimensions must be chosen to comply with a number of criteria, given below, which are based on experience gained in operating a 100 ℓ pilot crystallizer with several binary and ternary organic and inorganic mixtures.

DESIGN CRITERIA FOR DTB-VACUUM CRYSTALLIZERS

1. <u>Cooling water criterion</u> (determines the crystallization temperature).

The minimum crystallization temperature T_c is the sum total of:

a. maximum cooling water temperature, T_{cw}
b. temperature difference across the condenser, ΔT
c. boiling point rise of the mother liquor, ΔT_{BPR}

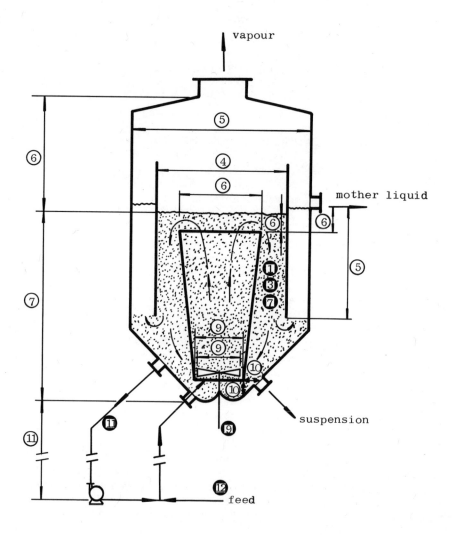

Fig. 1 DTB-vacuum crystallizer.

The numbers in open circles refer to the numbered criteria in the text on the basis of which the dimensions and process conditions can be calculated. The numbers in black circles refer to the following: 1 crystallization temperature, 3 crystal hold-up, 7 suspension volume, 9 stirrer speed, 11 external circulation mass flow, 12 feed condition.

2. **Multiple-effect criterion** (determines the number of stages).

In some cases energy can be saved by arranging two or more crystallizers in series. The heat of condensation of the vapour from crystallizer n-1 can be absorbed by the mother liquor from stage n.

3. **Crystal hold-up criterion** (determines the suspension concentration).

The production per unit volume of a crystallizer is determined inter alia by the total crystal surface per unit volume of the suspension. The concentration of the crystals in the suspension should be:
   a. so high that the required amount of crystals of the required mean size can be obtained.
   b. so low that handling of the suspension remains possible, and crystal breakage is limited.

Generally, the crystal hold-up is in the order of 10 to 30 percent by volume. The concentration of the crystals in the suspension can be controlled by adjusting the ratio between the crystal-free and crystal-carrying mass flows leaving the crystallizer.

4. **Entrainment criterion** (determines the vapour space diameter)

The linear velocity of the vapour u_v is limited by the requirement that no excessive carry-over of liquid droplets will occur. The limit depends on the physical properties of the system, such as densities, surface tension and viscosity. The entrained mass flow $\dot{m}_e$ depends on the value of the entrainment factor ϕ, which can be calculated from

$$\phi = u_v \left[\frac{\rho_v}{\rho_m - \rho_v}\right]^{0.5}$$

We found that an acceptable value for aqueous solutions is $\phi_{max} = 0.017$ m s^{-1}. From ϕ_{max} and the vapour and mother liquid densities ρ_v and ρ_m the value of $u_{v(max)}$ can be calculated.

5. **Settling space criterion** (determines crystallizer diameter and height of the settling space)

If the crystal suspension concentration is too low, the crystal hold-up can be increased by drawing liquid from the annular space between the baffle and the crystallizer wall. The linear velocity of the mother liquid in the annular space is then limited by the settling velocity of the crystals to be retained in the crystallizer. The height of the settling space is determined by the fluid-solid bed expansion.

6. **Liquid level criterion** (determines height of the vapour space, upper diameter of the draft-tube, distance from the upper end of the draft-tube to the liquid level, and the distance from the upper end of the draft-tube to the level of the mother liquor overflow)

The height of the vapour space should be so large as to prevent splashing against the top of the crystallizer and the vapour outlet. We recommend:
Height of vapour space $\geqslant 0.6$ x diameter of vapour space.
The linear velocities in the suspension in the radial and axial directions near the top of the draft-tube should be equal. Therefore it should be arranged that:
Upper diameter of draft-tube = vapour space diameter/$\sqrt{2}$,
Distance from upper end of the draft-tube to liquid level = 0.25 x upper diameter of draft-tube.
The distance from the upper end of the draft-tube to the level of the mother liquid overflow depends on the densities of mother liquid and suspension.

7. **Production rate criterion** (determines the suspension volume)

At a given crystal hold-up the maximum allowable specific production rate of crystals $\dot{m}_c/V$ of the desired mean crystal size depends on the properties of the system. The $\dot{m}_c/V$-value determines the suspension volume of the crystallizer at the desired production rate. Generally, the $\dot{m}_c/V$-value for inorganic compounds is 30 to 50 times the $\dot{m}_c/V$-value for organic compounds.

8. **Supersaturation criterion** (determines the internal circulation mass flow)

The supersaturation in the boiling zone near the surface of the suspension is limited by the requirement that excessive nucleation and, hence, scaling on the wall and the internals of the crystallizer should not occur. Excessive nucleation also leads to a small mean crystal size. Therefore, the internal circulation mass flow should be large enough to prevent excessive supersaturation at a given crystal mass flow. The maximum allowable supersaturation ρ_{max} depends on the properties of the system; generally:

$$0.5 < \Delta\rho_{max} < 5 \text{ kg m}^{-3}$$

9. **Propeller tip velocity criterion** (determines the minimum diameter of the propeller; the maximum propeller speed; the lower diameter of the draft-tube)

The velocity of the stirrer tip must not exceed the limit above which excessive breakage of crystals will occur. This limit depends on the properties of the crystals, and equals the product of the propeller diameter and speed. If a correlation is known for the pumping capacity of the propeller, the minimum diameter and maximum speed of the propeller may be determined. To ensure that the propeller will produce a truly axial flow, the lower diameter of the draft-tube should be made only slightly larger than the propeller diameter.

10. <u>Crystal suspension criterion</u> (determines the dimensions of the crystallizer bottom)

The crystals should remain suspended in the crystallizer. Therefore:
    a. the bottom should be streamlined, and
    b. the velocity of the suspension near the bottom should not drop below a given limit.
The internal circulation mass flow must be large enough to keep the biggest crystals in suspension.

11. <u>Boiling criterion</u> (determines the external circulation mass flow and the feed inlet level)

Boiling should occur only at the surface of the suspension in the crystallizer. In a vacuum crystallizer the feed should be introduced at a level where the hydrostatic pressure is large enough to prevent local boiling. This demands internal (and in many cases also external) recycling of the crystallizer contents. If the vapour pressure of the feed is lower than the static pressure near the propeller stirrer, there is no need for external recirculation. However, external recycling is also employed to supply additional heat, should this be necessary. The vapour pressure of the feed must be lowered so that boiling will occur only at the level of the propeller in the crystallizer. The external circulation mass flow can then be calculated. Otherwise, the internal circulation mass flow must be high enough to ensure that boiling will be restricted to a thin surface layer of the suspension in the crystallizer.

12. <u>Mixing criterion</u> (determines the specific enthalpy and composition of the feed)

In processes where mass flows have to be mixed, and the formation of new phases has to be prevented, it is essential to adjust the relation between composition and specific enthalpy so that, whatever the mixing ratio, the ultimate mixture will, under physical equilibrium conditions (temperature and pressure), consist of one single phase. The requirements imposed on the composition and specific enthalpy are called the mixing criterion. Fig. 2 gives the specific enthalpy-concentration diagram for a binary system. The curve SMS' is the solubility line. Through point M, which denotes the condition of the mother liquid, a pencil of operating lines can be drawn.

It has been shown (1) that points located on the same operating line represent feed conditions with the same ratio R_{vc} between the mass flow rates of the vapour and crystals produced. Feed conditions for which $0 \leq R_{vc} \leq \infty$ lie in the region VMCZ'Z. For points located on MC the ratio $R_{vc} = 0$; for points located on MV the ratio $R_{vc} = \infty$. Point V denotes the specific enthalpy of the vapour leaving the crystallizer, and point C denotes the specific enthalpy of the crystals formed. The practical operating region is enclosed by the solubility curve. This is the region marked VMS'Z'Z. Line WW' is the tangent to the solubility curve SMS' in point M. This tangent is called the nucleation boundary line. The mixing criterion demands that at a given condition of the mother liquid the feed condition h_f and w_f must satisfy the relation:

$$\frac{h_f - h_m}{w_f - w_m} \geq \left[\frac{dh}{dw}\right]_M$$

Feed conditions corresponding to points F or F' in Fig. 2 satisfy this formula. Generally the feed condition for a binary system must lie in the shaded area of Fig. 2 (for practical purposes on the tangent MW').

In the case of a ternary system the requirements imposed by the mixing criterion are more complicated (1). The importance of the mixing criterion appears from the results we obtained in our pilot plant research. In a 100 ℓ pilot DTB-crystallizer $NH_4H_2PO_4$ was crystallized from the ternary system $NH_4H_2PO_4$-NH_4NO_3-H_2O. The feed, as well as the additional heat needed, were supplied to an external recycle flow. Blockage occurring in the inlet region reduced the duration of each run to ∼ 4 h. Moreover, the average crystal size, as evaluated from a Rosin-Rammler distribution (2), was no larger than ∼ 250μm.
Application of the mixing criterion to this ternary system (1) yielded the following improvements:
    a. no more blocking, either in the inlet region or in the supply line to the crystallizer (completely trouble-free operation).
    b. an increase of the average crystal size by a factor 3.5 (from 250μm to 875μm at direction coefficient = 3), implying that the production of nuclei in the system had declined by a factor of ∼ 40.

DESIGN CRITERIA FOR DTB VACUUM-CRYSTALLIZERS

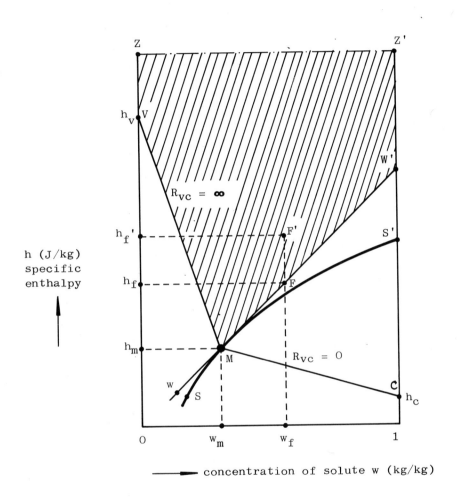

Fig. 2 Feeding conditions satisfying the mixing criterion for a binary system (shaded area).

SYMBOLS

h	specific enthalpy	J kg^{-1}
$\dot{m}$	mass flow-rate	kg s^{-1}
n	number of stages	–
R	ratio of mass flow-rates	–
T	temperature	K
u	velocity	m s^{-1}
V	suspension volume	m^3
w	mass fraction	–
ΔT	temperature difference	K
$\Delta \rho$	supersaturation	kg m^{-3}
ρ	density	kg m^{-3}
ϕ	entrainment factor	m s^{-1}

Subscripts:
BPR boiling point rise
c   crystallizer, crystals
cw cooling water
f   feed
m   mother liquor
max maximum
v   vapour
vc vapour-crystals

REFERENCES

1. Toussaint, A.G. and Donders, A.J.M., Chem. Engng Sci., <u>29</u> (1974) 237.
2. Puffe, E., Erzmetall, <u>I</u> (1948) 97.

INTEGRAL DESIGN OF CRYSTALLIZERS AS ILLUSTRATED BY THE DESIGN OF A CONTINUOUS STIRRED - TANK COOLING CRYSTALLIZER

C.J. ASSELBERGS AND E.J. DE JONG

Laboratory for Chemical Equipment, Delft University of Technology, Mekelweg 2, Delft, The Netherlands

INTRODUCTION

In the past decade a number of crystallizer design procedures based on population balance techniques has been put forward. The key to success of these procedures lies in the proper accumulation of kinetic data for crystal growth and nucleation and in an adequate modelling of the actual industrial system in terms of population balance equations. This may involve a rather extensive investigation of the system in bench-scale equipment, which should closely approach the Mixed Suspension Mixed Product Removal (MSMPR) concept. If the system to be designed does not meet the requirements of a MSMPR-crystallizer, it may moreover be necessary to perform experiments in pilot-plant equipment. This pilot-plant should be carefully scaled down from industrial size in order to reproduce the same regime of crystallization. It will be evident that crystallizer design may still involve much research effort, if the potentials of design via the population balance are to be exploited to their full advantage.

In the present paper a blueprint for design will be given, which given the present state of knowlegde, may serve as a tool for design of well mixed crystallizers. The model will comprise number, material, and energy balance equations. Moreover attention will be paid to those aspects of design, which are often neglected in the literature. For instance the size of a crystallizer does not only depend on requirements of crystallization (e.g. crystal residence time), but also on requirements of heat transmission. The design of the heat exchanger should therefore always be taken into consideration. The intention of this paper is to show the interactions between the conservation of mass, number, and

energy and the transport phenomena, which play a role. The design model will be illustrated for the case of a continuous stirred tank cooling crystallizer. The following aspects of design will be discussed:
- material, energy, and population balances
- crystallization kinetics
- incrustation
- heat transfer
- suspension of crystals
- supersaturation criterion

An important feature of the design model will be the development of two equations for crystallization and heat transfer respectively, which describe the size of the crystallizer as a function of the power input of the impeller of pump or agitator per unit mass of slurry.

MATERIAL BALANCES

In formulating the material balances over a crystallization system the choice of the various process parameters and their dimensions is rather important. For the present development the definition of the symbols is given in Fig. 1.

For a binary system with one feed stream containing no crystals, one product stream and evaporation of the solvent, the general balance equations are:

overall mass balance: $\phi_{mi} = \phi_{mos} + \phi_{mol} + \phi_{mv}$, (kg/s) (1)

solute balance : $\phi_{vi} c_i = \phi_{vol} c_o + \phi_{vol} M_L$, (kg solute/s) (2)

The balance equations can be solved to give the slurry concentration:

$$M_L = \frac{c_i \frac{\rho_{lo}}{\rho_{li}} \left(1 + \frac{\phi_{mv}}{\phi_{mol}}\right) - c_o}{1 - \frac{c_i}{\rho_{li}}}, \text{ (kg crystals/m}^3 \text{ mother liquor)} \quad (3)$$

It should be noted that the slurry concentration M_L is based on unit mother liquor volume and that in the overall mass balance a distinction is made between a crystal mass flow rate ϕ_{mos} and a mother liquor mass flow rate ϕ_{mol}. Thus the contribution of the solid phase to the volumetric flow rate and the mass flow rate has not been neglected in this case. Eq. 3 can be simplified to a great extend, if the crystal content of the product stream is low. This is however seldom permitted for industrial systems.

For the case of a cooling crystallizer ($\phi_{mv} = 0$), it will be evident from eq. 3 that there will be only one possible value of the slurry concentration, when the feed conditions and the temperature of crystallization are chosen. Then, since the rate of production P is given by:

$$P = \phi_{vol} \cdot M_L, \text{ (kg crystals / s)} \quad (4)$$

CONTINUOUS STIRRED-TANK COOLING CRYSTALLIZER

concentration c_i : kg/m³
density ρ_{li} : kg/m³
temperature θ_i : °C
volumetric
flow rate ϕ_{vi} : m³/s
mass flow rate ϕ_{mi} : kg/s

ϕ_{mv} : kg/s vapour

c_e : kg/m³
ρ_{le} : kg/m³
θ_e : °C
ϕ_{vol} : m³ liquor/s
ϕ_{mol} : kg liquor/s
ϕ_{mos} : kg crystals/s
M_L : kg crystals/m³ liq. slurry concentration

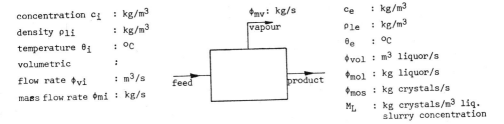

Figure 1. Definition of symbols.

$$\Delta\theta_{ln} = \frac{\theta_{ce} - \theta_{ci}}{\ln\dfrac{\theta_b - \theta_{ci}}{\theta_b - \theta_{ce}}}$$

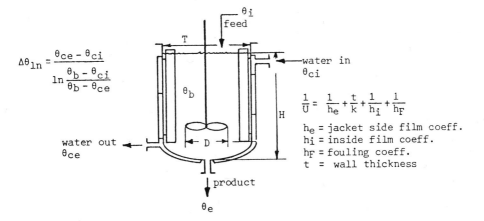

$$\frac{1}{U} = \frac{1}{h_e} + \frac{t}{k} + \frac{1}{h_i} + \frac{1}{h_F}$$

h_e = jacket side film coeff.
h_i = inside film coeff.
h_F = fouling coeff.
t = wall thickness

Figure 2.

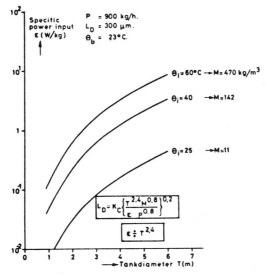

Figure 3. Crystallisation design equation.

it follows that the value of the volumetric liquid withdrawal rate will be specified for a given capacity.

For evaporative systems a greater choice exists for the value of the slurry concentration: the amount to be evaporated will be fixed for a given rate of production and crystallizer temperature; when moreover the withdrawal rate is chosen the value for the slurry concentration will be established.

ENERGY BALANCE

The general energy balance equation for the system described above is given by:

$$\phi_w = \phi_{mi} \, c_{pi} \, (\theta_i - \theta_e) + \phi_{cr} \, P + \phi_{mv} \, r, \, (W) \qquad (5)$$

ϕ_w represents the total heat flow, which has to be transferred to or from the system, for evaporative and cooling conditions respectively. c_{pi} is the specific heat of the feed, ϕ_{cr} is the heat of crystallization and r is the heat of vaporization. θ_i is the feed temperature, and θ_e is the product temperature, which will equal the temperature of crystallization θ_b for a well mixed crystallizer. In eq. 5 the heat produced by agitation which is small compared to the other terms, has been neglected. For a cooling crystallizer it follows from eqs. 3, 4 and 5 that when the feed conditions, the crystallizer temperature, and the production capacity are chosen, the value of the heat flow to be transferred to the cooling water will be fixed.

POPULATION BALANCE

The complete modelling of an industrial crystallizer in terms of population balance equations is at present only possible for relatively simple systems. To start with crystallizers which are substantially well mixed, without internal classification or classification occurring at the product discharge, will be briefly reviewed. Furthermore it will be assumed that the effect of abrasion and breakage of crystals may be neglected, although this may not be permitted for industrial crystallizers. When the feed stream does not contain crystals the basic number balance at steady state is given by the following well known expression:

$$\partial(Gn)/\partial L = n/\tau \qquad (6)$$

For the ideal conditions given above the number balance can be solved if the linear crystal growth rate is independent of crystal size:

$$n/n^o = \exp\,[-L/G\tau] \qquad (7)$$

For industrial design this approach may however be too simple, since many industrial systems do not satisfy the MSMPR-conditions. It is therefore necessary to extend the simple population balance and to take deviations into account. For those instances in which an industrial crystallizer already exists, it is possible to analyse the crystal size distribution (CSD) of the product crystals and of the crystals in the vessel as a function of the various process parameters. The evidence accumulated by such an analysis can for instance indicate size dependent crystal growth. In this case one of the simplest empirical expressions to describe the relationship between crystal size and growth rate is a first order polynomial approximation, as proposed by Canning and Randolph (1):

$$G = G_o \left(1 + z \frac{L}{G_o \tau}\right), \quad z < 1/3 \qquad (8)$$

By substituting this equation in the population balance and performing the integration, the CSD becomes:

$$\frac{n}{n^o} = \left(1 + z \frac{L}{G_o \tau}\right)^{-\frac{1+z}{z}} \qquad (9)$$

From the third moment of this distribution the slurry concentration can be obtained:

$$M = k_v \rho_c \int_o^\infty n L^3 \, dL = \frac{6 k_v \rho_c n^o (G_o \tau)^4}{(1-z)(1-2z)(1-3z)} \qquad (10)$$

k_v = crystal volume shape factor; ρ_c = crystal density (kg/m^3). The dominant particle size based on the maximum of the weight fraction distribution is given by:

$$L_D = \frac{3 G_o \tau}{(1-2z)} \qquad (11)$$

It will be evident that the values of z for an experimentally or industrially obtained CSD, will have to be determined by curve fitting.

If one can reasonably assume that the growth rate is independent of crystal size, a deviation from the straight population density plot can also be caused by classification at the product discharge or by internal classification. In particular the former can be important since it is rather difficult to meet the requirements of isokinetic product withdrawal. When classification occurs the residence time of the crystals will not be equal the liquor residence time and will be a function of crystal size. In its most simple form this function can again be described by a first order polynomial:

$$\tau_c = \tau_o \left(1 + q \frac{L}{G \tau_o}\right), \quad q < 1/4 \qquad (12)$$

τ_o is the residence time of the smallest crystals, which will be identical to the liquor residence time. For this case the CSD of the product becomes:

$$\frac{n}{n^o} = \left(1 + q\frac{L}{G\tau_o}\right)^{-\frac{1+q}{q}} \tag{13}$$

The slurry concentration of the product can be derived from the third moment of the distribution, to give:

$$M = \frac{6 k_v \rho_c n^o (G\tau_o)^4}{(1-q)(1-2q)(1-3q)} \tag{14}$$

The dominant size of the product based on the maximum of the weight fraction distribution becomes:

$$L_D = \frac{3 G\tau_o}{(1-2q)} \tag{15}$$

Again the values of q will have to be determined by trial and error. It can be concluded from the above that the equations for the CSD, the slurry concentration and the dominant size of the product are identical for both size dependent growth and for classification: q and z are completely interchangable. If the models for G and τ_c (eqs. 8 and 12) do indeed apply, one cannot conclude from the shape of the CSD alone, whether the deviations from the MSMPR concept are caused by anomalous growth or by classification.

CRYSTALLIZATION KINETICS

In addition to the description of the CSD, suitable kinetic expressions are needed for design. Since the growth and the nucleation rate cannot be calculated from first principles, the kinetics of crystallization must be determined experimentally. When the CSD of a system conforms to the MSMPR concept (eq. 7) the linear crystal growth rate and the nucleation rate may be readily calculated. When deviations from this concept occur the data can be accumulated by trial and error determination, involving the parameters q and z as illustrated above.

Both growth rate G and nucleation rate B^o can be described by simple power law forms:

$$G = k_G \sigma^p \quad (m/s) \tag{16}$$

$$B^o = k_B \varepsilon^h \sigma^r M^j \quad (\#/m^3 \cdot s) \tag{17}$$

σ = relative supersaturation ($= c_b - c_{eq}/c_{eq}$); ε = viscous energy dissipation rate or power input per unit mass of magma (W/kg); p, h, r, j, k_G, and k_B are empirical constants.

The rate constant k_G is likely to depend on the presence of impurities, the temperature and the degree of agitation. The rate constant k_B can be a function of the presence of impurities, the temperature, the geometry of the crystallizer, the method of mixing (e.g. circulation pump or agitator) and the impeller type. It is assumed here that nucleation is predominantly secondary in nature, hence the dependence on the specific power input (2). From eqs. 16 and 17 the supersaturation can be eliminated to give:

$$B^o = k_N \, \varepsilon^h \, G^i \, M^j \tag{18}$$

$i = r/p$ and $k_N = k_B \, k_G^{-r/p}$

Since the nucleation rate is defined by $B^o = n^o G$, the population density at zero crystal size becomes:

$$n^o = k_N \, \varepsilon^h \, G^{i-1} \, M^j \tag{19}$$

On the basis of eq. 19 and proper expressions for the slurry concentration and the dominant crystal size an equation can be derived which can be directly applied to crystallizer design. Although this derivation is substantially the same as presented by other authors (e.g. (3)) an extension will be given here, which may result in wider applicability.

Considering the case of size dependent growth, it has been shown that the slurry concentration can be described by eq. 10. Combining this equation with the kinetic equation 19, with $G = G_o$, and eliminating n^o gives:

$$M = \frac{6 \, k_v \rho_c \, (G_o \tau)^4 \, k_N \, \varepsilon^h \, G_o^{i-1} \, M^j}{(1-z)(1-2z)(1-3z)} \tag{20}$$

Rewriting this expression gives:

$$(G_o \tau)^{3+i} = M^{1-j} \, \tau^{i-1} \, \varepsilon^{-h} \, \frac{(1-z)(1-2z)(1-3z)}{6 \, k_v \rho_c \, k_N} \tag{21}$$

By combining equation 21 and 11 the dominant crystal size becomes:

$$L_D = \frac{3}{(1-2z)} \left[M^{1-j} \, \tau^{i-1} \, \varepsilon^{-h} \, \frac{(1-z)(1-2z)(1-3z)}{6 \, k_v \rho_c \, k_N} \right]^{1/i+3} \tag{22}$$

The dominant size characterises the product and is an important input variable for crystallizer design. From eq. 22 some important conclusions may be drawn: for instance when $j = 1$ the slurry concentration has no effect on the dominant particle size. The exponent i gives the relative kinetic order of nucleation and growth. When $i > 1$ the dominant size can be increased by increasing the mean retention time of the crystals.

Eq. 22 can be rewritten in terms of the rate of production P and the principal dimension of the crystallizer T, as shown below:

rate of production : $P = \phi_v M$ (23)
mean residence time: $\tau = V/\phi_v$ (24)
crystallizer volume: $V = k_{VT} \cdot T^3$ (25)

k_{VT} is a crystallizer volume shape factor, which describes the geometry of the vessel (e.g. for a flat bottomed tank where the liquid depth equals the tank diameter $k_{VT} = \Pi/4$). By combining eqs. 23, 24 and 25 τ becomes:

$$\tau = k_{VT} T^3 \frac{M}{P} \qquad (26)$$

Substitution in eq. 22 gives the following design expression:

$$L_D = K_c \cdot f(z) \left[\frac{T^{3i-3} M^{i-j}}{\varepsilon^h P^{i-1}} \right]^{1/i+3} \qquad (27)$$

$$K_c = 3 (6 k_v \rho_c k_N k_{VT}^{1-i})^{-1/i+3} \quad ; \quad f(z) = \frac{[(1-z)(1-2z)(1-3z)]^{1/i+3}}{(1-2z)}$$

Completely analogous to the above developement a design equation for the case of classification can be derived. This equation is iden- to eq. 27, wherein L_D and M must represent the dominant size and the slurry concentration <u>of the product</u> and f(z) is to be replaced by f(q):

$$f(q) = \frac{[(1-q)(1-2q)(1-3q)(1-4q)^j]^{1/i+3}}{(1-2q)}$$

It may be concluded that, when the empirical kinetic constants are known and the characteristic product size L_D and the production P are given, eq. 27 gives a relationship between T, ε, M and f(z) (or f(q) when classification occurs). As has been pointed out in the previous pages, the value of the slurry concentration is fixed by the material balances, when the crystallizer feed conditions and the temperature of crystallization have been chosen. Thus eq. 27 reduces to a relationship between the principal vessel dimension T, the power input ε and z (or q). In order to apply this expression to design it is necessary to know the values of z (or q) for various process conditions since z can for example depend on slurry concentration, temperature, power input etc. This implies that an investigation of the behaviour of a system will be more complicated since both kinetics <u>and</u> the functional dependence of z on the process conditions have to be studied. It is evident that when neither size dependent growth nor classification occur, matters are greatly simplified as z and q will equal zero and f(z) = f(q) = 1. That is the crystallizer satisfies the conditions of a steady state MSMPR system, where the ΔL law holds. Thus for such a system eq. 27 represents a function of two choice variables $f_1(\varepsilon,T)$ once the characteristic product size, the production rate and the slurry concentration are fixed. The application to design of eq. 27 for $f(z) = 1$ will be illustrated later on.

INCRUSTATION

Incrustation may occur when the difference between the temperature of crystallization and the inside wall temperature at any point exceeds a certain maximum value $\Delta\theta_{max}$. Of course incrustation can be caused as well by a number of other factors such as the wall material and roughness, the presence of impurities and the velocity of the slurry. For design purposes it is however convenient to describe incrustation by one criterion, $\Delta\theta_{max}$, which obviously has to be determined experimentally under conditions representative for the industrial system. By selecting $\Delta\theta_{max}$ one or two degrees lower than the temperature difference at which incrustation actually occurs, a safety margin may be included.

HEAT TRANSFER

In addition to eq. 27 a second basic design equation is needed, which may follow from heat transfer considerations, as the size of a crystallizer also depends upon the area of the heat transfer surface to be installed. The general equation for heat transfer is given by:

$$\phi_w = U A \Delta\theta_{ln} \qquad (28)$$

wherein ϕ_w is the total heat load (see eq. 5), U is the overall coefficient of heat transfer, A is the heat transfer area, and $\Delta\theta_{ln}$ is the logarithmic mean temperature difference. In order to elaborate eq. 28 it is necessary to make a choice of the type and geometry of the crystallizer to be designed. In the present case the application of eq. 28 will be shown for a well mixed cooling crystallizer fitted with a jacket and a marine propeller agitator. This choice is rather arbitrary since the derivations go along the same lines for other systems. The main features and dimensions of the jacketed vessel are illustrated in Fig. 2. The liquid height H is equal to the tank diameter T.
The inside film heat transfer coefficient can be obtained from correlations given in literature. For a vessel with baffles and a marine propeller h_i is given by (4):

$$Nu = 0.64 \, Re^{2/3} \, Pr^{1/3} \qquad (29)$$

This equation can be used for suspensions if Nu, Re, and Pr are based on suspension properties (5).

Eq. 29 can be rewritten in terms of the power input ε and the tank diameter T to give:

$$h_i = k_w \varepsilon^{2/9} T^{1/9} \tag{30}$$

The derivation of eq. 30 is given in the appendix. k_w is dependent on the physical properties of the suspension, the geometry of the vessel and the Power-number of the agitator. As can be seen, the effect of changing the power input or the tank diameter on h_i is rather small. The jacket size heat transfer coefficient, which also can be obtained from the literature (e.g. (6)), is generally not a function of the vessel diameter and will be of the same order of amgnitude as the inside heat transfer coefficient.

The thermal resistance of the wall is in general small compared with the other resistances and may be neglected. Fouling can occur on both sides of the heat transfer surface. The fouling coefficient can be estimated for a given system or obtained from previous experience. In the present discussion fouling will not be taken into account, though this may not be permitted in industrial practice. The overall coefficient of heat transfer thus becomes:

$$\frac{1}{U} = \frac{1}{h_e} + \frac{1}{k_w \varepsilon^{2/9} T^{-1/9}} \tag{31}$$

When the heat transfer area is expressed in terms of the tank diameter and a tank surface shape factor, according to $A = k_{AT} \cdot T^2$, eq. 28 becomes:

$$\phi_w = \left[\frac{1}{h_e} + \left[\frac{1}{k_w \varepsilon^{2/9} T^{-1/9}} \right]^{-1} \right] \cdot k_{AT} \cdot T^2 \, \Delta\theta_{ln} \tag{32}$$

As has been shown ϕ_w is fixed by the energy balance when the feed conditions, the temperature of crystallization and the production capacity are chosen. When furthermore the geometry of the vessel and the jacket, the cooling water velocity, and the logarithmic mean temperature difference are given, eq. 32 reduces to a function $f_2(\varepsilon, T)$ of the power input and the tank diameter. It is abvious that the effect of the power input in this equation is of minor importance, even if the inside heat transfer coefficient controls the rate of heat transfer.

The logarithmic mean temperature difference can be selected on the basis of the incrustation criterion $\Delta\theta_{max}$. Since the lowest system temperature is the cooling water inlet temperature θ_{ci}, the temperature of the crystallizer θ_b (which is assumed to be uniform throughhout the vessel) can be fixed at $\theta_{ci} + \Delta\theta_{max}$. As the actual inside wall temperature in the vicinity of the cooling water inlet nozzle will be somewhat higher than θ_{ci}, one can safely assume that the incrustation criterion will be satisfied.

By choosing the crystallizer temperature in this manner, the maximum crystal yield may be obtained. The cooling water outlet temperature can be readily chosen considering the cooling water flow rate and the cooling water temperature rise. Thus the logarithmic mean temperature difference is fixed.

SUSPENSION OF CRYSTALS

For well mixed crystallizers the power input of the agitator must exceed a minimum value in order to attain suspension of the crystals. The degree of suspension can be defined in different ways:
- off-bottom suspension, in which no crystal remains on the tank bottom for longer than for instance 1 s.
- homogeneous suspension, in which the CSD (and therefore the slurry concentration) is uniform throughout the tank.

For a given mixing-tank arrangement homogeneous suspension can be obtained at the cost of higher power comsumptions than required for off-bottom suspension. The latter degree of suspension is however sufficient for most industrial systems. The value of the power input required for off-bottom suspension can be calculated from correlations reported in the literature for a variety of tank designs (7, 8, 9, 10). Since the literature data may not be in complete agreement, it may be necessary to use additional data from actual practice. For aqueous inorganic suspensions a value of 1 W/kg will be satisfactory in most cases for off-bottom suspension.

SUPERSATURATION CRITERION

If the supersaturation is too high profuse (primary) nucleation may occur or the product quality (purity, sales appeal) may be poor. For design it is therefore useful to specify a maximum supersaturation permissible, which for convenience sake can be translated into a maximum growth rate. Hence the criterion for design becomes: $G < G_{max}$. The value of G_{max} has to be determined experimentally. The criterion can be written in terms of the principal crystallizer dimension T as follows: Since by definition $L_D = 3\ G\tau$ for the systems considered here, the maximum growth rate can be translated into a minimum residence time for a given dominant size:

$$\tau > \tau_{min} = \frac{L_D}{3\ G_{max}} \tag{33}$$

Combining eq. 26 and 33 gives:

$$\frac{M}{P} k_{VT} T^3 > \frac{L_D}{3\ G_{max}} \quad \text{or:} \quad T > \left[\frac{P\ L_D}{3\ M k_{VT}\ G_{max}} \right]^{1/3} \tag{34}$$

Thus when the quantities of the right-hand term of eq. 34 are specified, a criterion results for the minimum crystallizer diameter.

ILLUSTRATION OF THE DESIGN PROCEDURE

The proposed design procedure will now be illustrated for crystallization of potassium alum, which system substantially satisfies the MSMPR-concept. The kinetic constants for potassium alum are known (2): $h = 1$, $i = 1.8$, $j = 1$, and $k_N = 0.9 \cdot 10^{19}$. The unit employed for crystallization is a simple baffled vessel, fitted with a water-cooled jacket and a marine propeller, conforming to the following specifications:

- liquid height = tank diameter ($H = T$)
- ratio of impeller diameter to tank diameter = 1/3 (D/T)
- agitator Power-number = $Po = 0.6$
- tank volume shape factor = $k_{VT} = 0.73$
- tank surface shape factor = $k_{AT} = 3.55$

The inside heat transfer coefficient for this vessel can be calculated from eq. 30. To simplify matters it will be assumed here, that in all cases the jacket can be designed in such a way that the jacketside heat transfer coefficient $h_e = 3000$ W/m^2 °C. The crystallizer must operate under the following specifications:

- product crystal size = $L_D = 300$ μm
- production: $P = 900$ kg/h
- supersaturation criterion: $G_{max} = 10^{-6}$ m/s
- off-bottom suspension criterion: $\varepsilon_{min} = 0.75$ W/kg
- incrustation criterion: $\Delta\theta_{max} = 3$ °C
- cooling water inlet temperature: $\theta_{ci} = 20$ °C

The crystallization temperature can be determined from the incrustation criterion. Accordingly: $\theta_b = \theta_{ci} + \Delta\theta_{max} = 23$ °C. The calculations will be performed for three different feed temperatures: 25, 40, and 60 °C. By assuming that the input flow is in all cases saturated, the effect of the slurry concentration on the design can be illustrated by varying the feed temperature; Specification of the cooling water outlet temperature now permits calculation of $\Delta\theta_{ln}$. For $\theta_{ce} = 22$ °C the LMTD becomes 1.8 °C. Given the above input data the slurry concentration (and the slurry withdrawal rate) is fixed for a given feed temperature by the material balances (eq. 3 and eq. 4). The rate of heat transfer ϕ_w is then fixed by the energy balance (eq. 5).

Considering that all quantities in the basic crystallization design equation (eq. 27) wherein $f(z) = 1$, are known with the exception of ε and T, the relationship between the principal crystallizer dimension and the specific power input can be depicted graphically. The results of the calculation, obtained by numerical methods, are shown in Fig. 3, for the three feed temperatures mentioned. It can be seen that ε increases with T when the production, the dominant size, and the slurry concentration are held constant; for this example ε is proportional to $T^{2.4}$. The curves in Fig. 3 give the locus of points, which all satisfy the design specifications.

The second basic design expression (eq. 32), derived from heat transfer considerations also gives a relationship between ε and T, since the other quantities in the equation are already known and h_e is held constant. This relationship is illustrated in Fig. 4, and it is evident, as stated before, that the effect of ε on the tank diameter T is of only minor importance.

The solution of the design problem can be obtained by solving the crystallization and the heat transfer design equation for T. This is illustrated graphically in Fig. 5: the solution is given by the point of intersection of the crystallization curve and the heat transfer curve, both for the same feed temperature. In the figure solutions, which all conform to the same specifications, are shown for three different feed temperatures.

Finally it must be checked if the solutions obtained, satisfy the off-bottom suspension criterion and the supersaturation criterion. The suspension criterion is given by: $\varepsilon > \varepsilon_{min} = 0.75$ W/kg. For the present specifications the supersaturation criterion can be calculated from expression (34) to give $T > T_{min} = 0.90$ m. Both criteria are depicted for feed temperatures of 40 and 60°C satisfy both criteria, while for $\theta_i = 25°C$ off-bottom suspension will not be attained. In the latter case the solution therefore does not conform to the design specifications.

The complete (ε, T)-diagram clearly pictures the constraints imposed by the design specifications. The crux of the design procedure presented is that the design variables are fixed in such a sequence that ε and T remain eventually.

SOME DESIGN CONSIDERATIONS FOR EVAPORATIVE CRYSTALLIZERS

The proposed design model can equally well be applied to evaporative systems. For evaporators the vessel diameter is generally fixed by the rate of evaporation in order to avoid entrainment. The liquid height therefore becomes an important design variable. Also the number of degrees of freedom may be more limited for evaporative systems once the geometry of the crystallizer is chosen. For forced circulation machines the specific power input will figure more prominently in the heat transfer design equation, since the inside film coefficient may predominantly control the overall rate of heat transfer and will be a stronger function of ε.

The modelling of evaporative systems in terms of population balance equations may however be more complicated, due for instance to internal fines dissolution (as a result of the temperature rise over the heat exchanger) or due to short circuiting and internal classification when a tangetial inlet is employed. These complexities arising from among other things non-ideal behaviour may necessitate research in pilot plant evaporators.

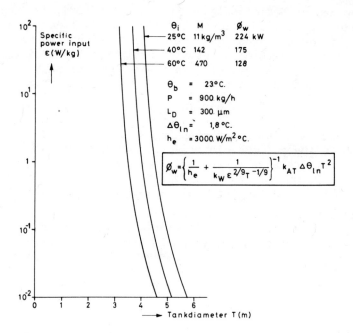

Figure 4. Heat transfer design equation.

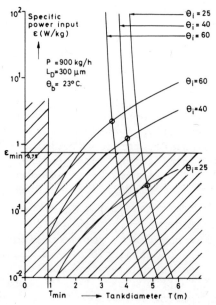

Figure 5. Solution of design problem.

APPENDIX: DERIVATION OF EQUATION 30

$$Nu = 0.64 \, Re^{2/3} Pr^{1/3} \longrightarrow \frac{h_i T}{k} = 0.64 \left(\frac{ND^2}{\nu}\right)^{2/3} \left(\frac{\nu}{a}\right)^{1/3}$$

N = rotational speed of impeller (1/s)
D = impeller diameter (m)
k = thermal conductivity (W/m °C)
ν = kinematic viscosity (m²/s)
a = thermal diffusivity (m²/s)

$$h_i = 0.64 \frac{k}{a^{1/3} \nu^{1/3}} N^{2/3} \frac{D^{4/3}}{T} = 0.64 \frac{k}{a^{1/3} \nu^{1/3}} \left(\frac{D}{T}\right)^{4/3} N^{2/3} T^{1/3} ,(a)$$

Power input = $Po \, \rho_{SL} \, N^3 D^5$ (Watt)
Po = power number
ρ_{SL} = density of slurry (kg/m³)
Tank volume : $V = k_{VT} \cdot T^3$ (m³ slurry)

Power input/kg slurry $\longrightarrow \varepsilon = \frac{Po \, N^3 D^5}{k_{VT} \, T^3} = \frac{Po}{k_{VT}} \left(\frac{D}{T}\right)^5 N^3 T^2$, (b)

Combining equations (a) en (b) and eliminating N gives:

$$h_i = 0.64 \frac{k}{a^{1/3} \nu^{1/3}} \left(\frac{D}{T}\right)^{2/9} \left(\frac{k_{VT}}{Po}\right)^{2/9} \varepsilon^{2/9} T^{-1/9}$$

$$\boxed{h_i = k_w \, \varepsilon^{2/9} \, T^{-1/9}} \qquad (30)$$

REFERENCES

1. Canning, T.F. and Randolph, A.D.; A.I.Ch.E. J. 13 (1), (1967),5
2. Ottens, E.P.K.; Thesis Delf University of Technology (1973)
3. Larson, M.A. and Garside, J.; The Chem. Engr. June (1974), 318
4. Edwards, M.F. and Wilkinson, W.L.; The Chem. Eng. Aug. (1972), 310.
5. Kwasniak, J.; Verfahrenstechnik 7 (1973), 287.
6. Lehrer, I.H.; Ind. Chem. Chem. Proc. Des. Dev. 9 (4),(1970),553
7. Bourne, J.R. and Sharma, R.N.; Chem. Eng. J. 8 (1974), 243
8. Zwietering, T.N.; Chem. Eng. Sci. 8 (1958), 244
9. Kolar, V.; Coll. Czech. Chem. Comm. 26 (1961), 613
10. Weisman, J. and Efferding, L.E.; A.I.Ch.E. J. 6 (3), (1960),419

NOMENCLATURE

A	heat transfer area
B^o	nucleation rate
D	impeller diameter
G	growth rate
H	liquid height
h	film heat transfer coefficient
k_{VT}	tank volume shape factor
k_{AT}	tank surface shape factor
k_w	constant in eq. 30
k_v	crystal volume shape factor
K_c	constant in eq. 27
L	crystal size
L_D	dominant crystal size
M	slurry concentration
N	rotational speed of impeller
n	population density
P	rate of production
Po	agitator power number
T	crystallizer diameter
U	overall heat transfer coefficient
V	crystallizer volume
z, q	parameters
h, i, j, k_N	kinetic constants
ε	power input per kg
ϕ_w	heat flow rate
ϕ_m	mass flow rate
ϕ_v	volumetric flow rate
ρ	density
τ	residence time
θ	temperature

DESIGN OF BATCH CRYSTALLIZERS

JAROSLAV NÝVLT

Research Institute of Inorganic Chemistry

400 60 Usti nad Labem, Revolucni 86, Czechoslavakia

There exists only a very limited number of papers dealing with the design of batch crystallisers. This is mainly due to the complexity of the problems involved: the mass of precipitated crystals is increasing during the run and so also is their surface area, whereas the supersaturation varies in a complex way depending on the cooling or evaporation rate as a function of time. Briefly, nothing remains constant and so the solution of the corresponding integro-differential equations becomes extremely difficult. Nevertheless, batch crystallisers are still used in chemical industries and it would be very desirable to have certain guidelines for the design of such equipment.

Theoretical considerations led us (1) to the conclusion that, under certain simplifying conditions, the supersaturation of a solution may be supposed to reach a maximum during the initial stages of a batch run and then to decrease exponentially to a certain low value at which it remains virtually constant. The result of a mathematical description of such a model shows that, in this case, the size distribution of crystals will be sufficiently approximated (2) by an equation formally identical with that known to hold for continuous crystallisers:

$$M(L) = 100 \left(1 + z + \frac{z^2}{2} + \frac{z^3}{6}\right) \exp(-z) \qquad (1)$$

where z is the dimensionless crystal size:

$$z = 3(L - L_N)/(L_m - L_N) \qquad (2)$$

and L_m corresponds to 64.7 wt% oversize fraction.

When plotting the dimensionless crystal size z (using corresponding oversize fractions as calculated from equation 1 for each particular value of z) against the crystal size L, linear plots should be obtained. And, in fact, this often proves to be true, as shown in Fig.1. In this figure the product crystal size distribution of Glaubers salt is demonstrated as produced in a batch crystalliser (volume about 10 m^3). This seems to corroborate our rather oversimplified considerations.

The next step of our investigations should give an answer to the question under which conditions the above model would be valid. Starting from fundamental equations describing the kinematics of nucleation and growth:

$$\dot{m}_N = k_N \Delta w^n \tag{3}$$

$$\dot{m}_G = k_G A \Delta w^g \tag{4}$$

we obtain the supersaturation balance (3):

$$\frac{d\Delta w}{dt} = s - k_G A \Delta w^g - k_N \Delta w^n \tag{5}$$

which must be solved together with the crystal population balance.

The mathematical model serving as a basis for the simulation has been based on equation 5 and its stepwise solution is shown schematically in Fig.2. The supersaturation, Δw, is repeatedly calculated from equation 5 for each time interval from the known supersaturation rate at the given instant of time and for the instantaneous crystal surface area. The supersaturation is then used for assessing the nucleation and growth rates in the subsequent step. These values are combined for obtaining a new value of crystal surface area, which is used for a new determination of the supersaturation. In this model, no further approximations and simplifications were introduced except for those involved in fundamental kinetic equations, equations 3 and 4. The results were published in references (4) to (6) and the following figures are chosen to serve as an example.

Fig.3 shows the effect of the cooling rate on the course of the supersaturation with time. The shape of these curves corresponds to the considerations leading to the simplified algebraic model discussed above. Fig.4 represents a plot of differential size distributions for various operating conditions of the batch crystalliser with seeding. Increasing numbers denoting the curves correspond to increasing amounts of seeds and there exists a distinct

DESIGN OF BATCH CRYSTALLIZERS

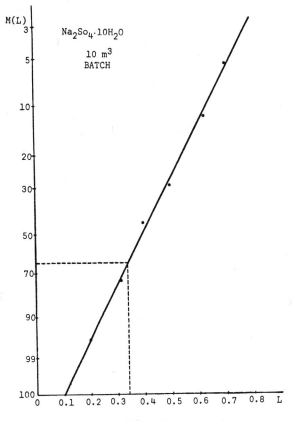

Fig. 1

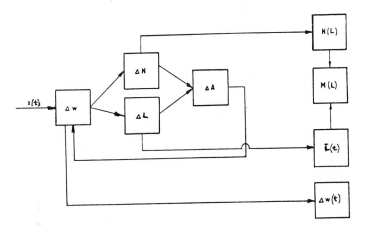

Fig. 2. Batch Crystallizer Simulation

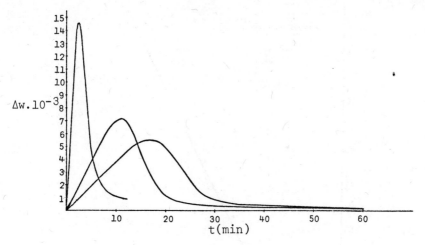

Fig. 3

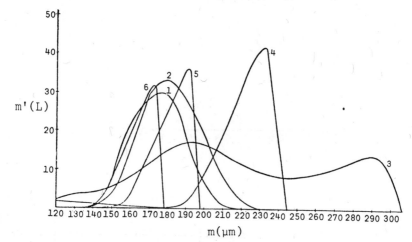

Fig. 4

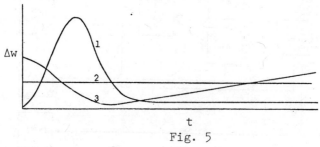

Fig. 5

optimum of seeds that should be used. Finally, for cooling curves calculated so that the solubility decrease corresponds to the increase of surface area of crystals, a quite dramatic effect of the initial supersaturation on the time dependence of supersaturation has been found, which is shown schematically in Fig.5. If the initial supersaturation is lower than the theoretical 'steady state' value, its value grows to a maximum corresponding, in certain cases to the limit of the metastable zone and then it is gradually decreasing again, as it is in the case of curve 1. If the initial value of supersaturation corresponds to the 'steady state' value, it remains constant during the whole batch run (curve 2). For higher initial supersaturations, there occurs a drop in supersaturation at initial stages, the supersaturation passes through a minimum and then it is slowly rising (curve 3). The best results of size distributions correspond to the initial supersaturation equal to the 'steady state' value or possibly slightly higher. An effective linearisation of the CSD curves was possible only for runs effected without a substantial seeding.

Thus, a simplified model of a batch crystalliser can be represented by a rapid nucleation at the beginning of the run, followed by a roughly exponential decrease of supersaturation down to an almost constant low value. Such a model enables us to perform an algebraic solution of a set of equations with the final result (2)(9):

$$[(\bar{L} - L_N)/L_N]^{1 + 3g/n} = 3Bm_c^{g/n} t_e^{(1-g/n)} f(z_N)^{-g/n} \quad (6)$$

where

$$B = \frac{4.5^{g/n} k_G \beta}{3\alpha \rho_c L_N k_N^{g/n}} = \frac{4.5^{g/n} \dot{L}}{\dot{m}_N^{g/n} L_N} \quad (7)$$

$$z_N = L_N/(\dot{L} t_e) \quad (8)$$

$$f(z_N) = 1 + z_N + z_N^2/2 + z_N^3/6 \quad (9)$$

The value of the system constant B can be determined e.g. from the linearised crystal size distribution using equation 7 employing

$$\dot{L} = (\bar{L} - L_N)/3t_e \quad (10)$$

and

$$\dot{m}_N = [z_N^3/6\ f(z_N)]\ m_c/t_e \qquad (11)$$

Thus equation 6 enables us to perform scaling-up of batch crystallisers; starting with laboratory kinetic data or laboratory model experiment and obtaining a product at known CSD, we can calculate the corresponding value of the system constant, B, from equations 7-11 and to use this value for the design equation.

Kinetic parameters of crystallisation of $Na_2SO_4 \cdot 10\ H_2O$ were measured in the laboratory (7)(8) and the value of $B = 1.75$ was calculated using equation 7. A batch crystalliser for the crystallisation of $Na_2SO_4 \cdot 10\ H_2O$ from industrial liquors was then designed. From the product CSD of this crystalliser (about 10 m^3 volume) shown in Fig.1 the value of $B = 1.52$ has been found using equations 7-11. This is certainly in very good agreement with the value calculated previously from laboratory data (the difference in B corresponds to 15% relative only).

On the other hand, the results with sodium thiosulphate were not so satisfactory at all. Kinetic parameters of crystallisation of $Na_2S_2O_3 \cdot 5H_2O$ were measured also in the laboratory (7)(8) and a value of $B = 4.88$ was found from equation 7, whereas $B = 0.45$ resulted from a batch model experiment in a volume of 2 litres and $B = 0.27$ from a batch experiment in 20 litres. This discrepancy might be due to the fact that the experiments were carried out at high magma densities and very intensive agitation, so that secondary nucleation brought about by the particle-particle and/or particle-stirrer collisions might have been of considerable importance. Considering the effect of secondary nucleation

$$\dot{m}_N = k_N \Delta w^n\ m_c^c \qquad (3a)$$

$$\dot{m}_N = [z_N^3/6\ f(z_N)]\ m_c^{1-c}/t_e \qquad (11a)$$

we can modify the design equation to $\qquad (6a)$

$$[(\bar{L} - L_N)/L]^{1 + 3g/n} = 3Bm_c^{(1-c)g/n}\ t_e^{(1-g/n)}\ f(z_N)^{-g/n}$$

and if we now take the value $c = 2$ for the most probable crystal-crystal interactions (10) we obtain closer values of $B = 0.37$ and $B = 0.27$ for the 2 and 20 litres crystallisers, respectively. This corresponds to a roughly 30% difference between the two values leading to maximum error in calculated crystal size of about 19%.

The results of these studies perhaps justify the conclusions that the described design method offers a useful way for estimating the effect of the crystallisation kinetics on the mean crystal size and on dimensioning of the equipment. The results will be the more reliable the closer are the model experiments to the anticipated plant operating conditions.

REFERENCES

1. Nývlt, J. and Skřivánek, J., Coll.Czech.Chem.Commun.(C.C.C.C.) 33 (1968) 1789.
2. Nývlt, H., Kočová, H. and Černý, M., C.C.C.C., 38 (1973) 3199.
3. Nývlt, J. Industrial Crystallisation from Solutions, Butterworths, London, 1971.
4. Nývlt, J., The effect of kinetic parameters on the result of a batch crystallisation process, C.C.C.C. (in press).
5. Nývlt, J., The effect of the cooling way on the size distribution of product crystals from a batch crystalliser, C.C.C.C.(in press).
6. Nývlt, J., The effect of seeding on the CSD of the product from batch crystalliser, C.C.C.C. (in press).
7. Wurzelová, J.and Nývlt, J., Chem.průmysl. 23 (1973) 69.
8. Kočová, H. and Nývlt, J., Chem.průmysl. 22 (1972) 165.
9. Broul, M., Ph.D. Thesis, Technical University of Chemistry, Prague, 1975.
10. Ottens, E.P.K., Janse, A.H. and de Jong, E.J., J.Crystal Growth 13/14 (1972) 500.

NOMENCLATURE

$M(L)$	oversize fraction (wt.%)
z	dimensionless crystal size
z_N	dimensionless original crystal size
L	size of crystals
L_N	size of original crystals
L_m	mean crystal size (64.7% wt. oversize)
L_{max}	maximum crystal size
$\dot{m}_N$	mass nucleation rate
k_N	kinetic constant of nucleation
Δw	supersaturation
n	nucleation exponent
$\dot{m}_G$	mass growth rate
k_G	growth rate constant
A	surface area of crystals
g	growth exponent
s	supersaturation rate
t	time
B	system constant
α and β	volume and surface shape factors
ρ_c	crystal density

APPENDIX

Calculation of B from laboratory data (equation 7 left):

	$Na_2SO_4 \cdot 10\ H_2O$	$Na_2S_2O_2 \cdot 5\ H_2O$
Units:	$kg/m^3\ h$	$kg/kg\ min$
k_G	0.855×10^{-2}	0.1878
β	6	3.8
α	1	0.45
ρ_c	1.460	1.685
$k_N^{1/n}$	0.120	1.004
g	1	1
n	2.55	3.71
L_N	1×10^{-4}	9.6×10^{-5}
B	1.75	4.88

Calculation of B from model experiments (equations 6, 6a and 7 right)

	$Na_2SO_4 \cdot 10\ H_2O$	$Na_2S_2O_3 \cdot 5\ H_2O$	
		20 litres	2 litres
Units:	$kg/m^3\ h$	$kg/kg\ min$	$kg/kg\ min$
t_e	9.5	240	240
m_c	100	0.96	2.06
$\dot{L}t_e$	8×10^{-5}	2.4×10^{-4}	3.7×10^{-4}
$\dot{L}$	8.4×10^{-6}	1×10^{-6}	1.54×10^{-6}
z_N	1.25	0.4	0.26
$f(z_N)$	3.36	1.491	1.296
$\dot{m}_N$	1.02	3.105×10^{-5}	4.55×10^{-6}
B	1.52	0.27	0.37

POPULATION BALANCE MODELS FOR BATCH CRYSTALLIZATION

HUGH M. HULBURT

Department of Chemical Engineering

Northwestern University, Evanston, Illinois 60201, USA

The conservation equation for the mean number of crystals of size L to L + dL, n(L) dL, in a sample of a crystallizing suspension (or magma) is frequently called the "population balance." As formulated by Randolph and Larson (6) and by Hulburt and Katz (3), for a batch crystallizer it takes the form

$$\frac{\partial n}{\partial t} + \frac{\partial}{\partial L}(Gn) = B - D \tag{1}$$

The linear growth rate, G, the nucleation rate, B, and the crystal loss rate, D, are all functions of the supersaturation and possibly crystal size, magma density, temperature and degree of agitation. The mother liquor concentration, C, (kg/m^3 of clear liquor) is related to the magma density, W, (kg crystals/m^3 suspension) through the volume fraction of crystals in the suspension, V_m,

$$V_m = W/\rho_c \tag{2}$$

$$\rho_c V_m + (1 - V_m) C = \rho_c V_{mo} + (1 - V_{mo}) C_o = \text{const.} \tag{3}$$

The magma density, W, is proportional to the third moment of n(L).

$$W/\rho_c = V_m = \Phi_v \int_0^\infty L^3 \, n(L) \, dL \tag{4}$$

Since there is a smallest and a largest crystal in any batch, the size distribution density, n(L), is discontinuous at $L = L_o$ and $L = L_m$. We shall set

$$n(L) = n_1(L) + N_o \, \delta(L - L_o) \tag{5}$$

$$n_1(L) = 0 \quad (L \leq L_o, \; L > L_m) \tag{6}$$

Then
$$V_m = \Phi_v \int_{L_o}^{L_m} L^3 n_1(L') \, dL' + \Phi_v L_o^3 N_o \tag{7}$$

where N_o = number of crystals/m^3 of size L_o, the critical nuclei.

We shall suppose that nucleation produces only crystals of size L_o. This excludes collision breeding and attrition processes. Hence we shall restrict attention to the case $B = D = 0$ for $L > L_o$. Moreover, we shall suppose the growth rate to be expressible as

$$G = G_1(C - C_s) G_2(L) \tag{8}$$

Eq. 1 may be made dimensionless by introducing the variables

$$L = R_o x \tag{9a}$$
$$B = B_o b \tag{9b}$$
$$t = t_o \theta \tag{10a}$$
$$n = B_o f/G_o \tag{10b}$$
$$G = G_o g \tag{11a}$$
$$C = S^* S_1 + C_s \tag{11b}$$
$$C_s = -S^* U + C_o \tag{12}$$

where

$S^* = C_o - C_{so}$ = supersaturation in reference state

B_o = nucleation rate in reference state

G_o = linear growth rate in reference state

The reference quantities are not all independent. Since V_m is dimensionless in Eq. 4,

$$B_o R_o^4 / G_o = 1 \tag{13}$$

By choosing

$$G_o = R_o t_o \tag{14}$$

Eq. 1 becomes

$$\frac{\partial f}{\partial \theta} + \frac{\partial}{\partial x}(gf) = 0 \quad (x > x_o) \tag{15}$$

with boundary conditions

$$f(x,0) = f_o(x) \tag{16}$$

$$f(x_o, \theta) = f_1(\theta) \tag{17}$$

Also
$$g = g_1(S_1) g_2(x) \tag{18}$$

The overall material balance, Eq. 4, becomes

$$U - S_1 = \frac{(1 - C_{ov})(V_m - V_{mo})}{S_v^* (1 - V_m)} \tag{19}$$

where
$$S_v^* = S^*/\rho_c \tag{20}$$

$$C_{ov} = C_o/\rho_c \tag{21}$$

V_{mo} = volume fraction crystals in suspension at reference state (22)

The momentary saturation concentration, U, in a batch crystallizer is a control variable. Hence by appropriate control action, we may make S_1 a function of time, θ, only. In practice this would involve sensing the magma density, W (= $\rho_c V_m$) and adjusting the saturation concentration, U, to satisfy Eq. 19 for a prescribed $S_1(\theta)$. In the more practical case, the saturation is controlled more arbitrarily and S_1 then depends upon the resulting magma density through Eq. 19.

The population balance, Eq. 15, is thus in general a complicated integro-differential equation, but in special instances may become a partial differential equation if $S_1(\theta)$ is a control variable. The mathematical theory in this latter case is well known (1,2)

The functions $f(\theta, x)$ are surfaces in (θ, x, f) space through every point of which there passes a characteristic curve lying in the surface which satisfies the equations

$$\frac{dx}{d\theta} = g_1(\theta) g_2(x) \tag{23}$$

$$\frac{df}{d\theta} = - g_1(\theta) \dot{g}_2(x) f \tag{24}$$

where
$$\dot{g}_2(x) = \frac{dg_2(x)}{dx} \tag{25}$$

and, of course, passes through the boundary curves $f_o(x) = f(0,x)$ and $f_1(\theta) = f(\theta, x_o)$.

Integrating Eq. 23 and 24, we find

$$\theta = \theta_1 + \frac{1}{\bar{g}_1} \int_{x_1}^{x} \frac{dx'}{g_2(x')} \tag{26}$$

where

$$\bar{g}_1(\theta - \theta_1) = \int_{\theta_1}^{\theta} g_1(\theta)\, d\theta \tag{27}$$

and

$$f = f_1\, g_2(x_1)/g_2(x) \tag{28}$$

as parametric equations for the characteristic curve through the point $(\theta, x, f) = (\theta_1, x_1, f_1)$.

The principal base characteristic is given by $x_1 = x_0$ and represents the growth of a critical nucleus. We may define the "age" of any crystal by

$$\tau = \frac{1}{\bar{g}_1} \int_{x_0}^{x} \frac{dx'}{g_2(x')} \tag{29}$$

It represents the time required to grow from a critical embryo to size x. For seeds, $\tau \geq \theta$, the age is greater than the residence time in the batch, since they began in aged condition.

Boundary conditions at $\theta = 0$ specify the initial seed distribution $f(x,0) = f_0(x_1)$. Hence for all points $\theta_1 = 0$, Eq. 28 gives

$$f_1 = f_0(x_1) \tag{30}$$

Moreover, Eq. 26 gives for $\theta_1 = 0$

$$\theta = \frac{1}{\bar{g}_1} \int_{x_1}^{x} \frac{dx'}{g_2(x')} \tag{31}$$

This determines the size, x, of seeds initially having size x, as a function of their elapsed time for growth, θ. If $g_2(x) \equiv 0$ (size independent growth)

$$x - x_1 = \bar{g}_1\, \bar{g}_2\, \theta \tag{32}$$

as would be expected. Eq. 28 then reads

$$f = f_0(x_1) = f_0(x - \bar{g}_1\, \bar{g}_2\, \theta) \qquad (x \geq \bar{g}_1\, \bar{g}_2\, \theta) \tag{33}$$

The initial seeds distribution, $f_o(x)$, thus propagates along the base characteristics, Eq. 31.

However, nuclei born at $\theta = \theta_1$, after the start of the experiment, will have age $\tau = \theta - \theta_1$, less than the batch residence time ($\tau \leq \theta$). For these, the appropriate boundary condition specifies the number of critical nuclei $f(x_o, \theta)$. This will be a function of the supersaturation, $S_1(\theta)$. Thus for $\tau \leq \theta$, in Eq. 28 we take

$$f_1(\theta) = f_1(x_o, \theta) \qquad (\tau \leq \theta) \qquad (34)$$

For size independent growth,

$$f = f_1(x - \bar{g}_1 g_2 \tau, \theta - \tau) \qquad (\tau \leq \theta) \qquad (35)$$

Again the distribution propagates along the characteristics.

In general, the characteristics will not be straight lines if $S_1(\theta)$ and $g_2(x)$ are not constant. Nevertheless, Eqs. 26 and 27 give $x_o(\theta, x, \theta_1)$ in the form

$$\int_{\theta_1}^{\theta} g_1(\theta') d\theta' = \int_{x_o}^{x} \frac{dx'}{g_2(x')} = \bar{g}_1 \tau \qquad (36)$$

We examine first the size dependence for constant supersaturation ($\bar{g}_1 = \text{const.}$)

Nielsen (4) summarizes arguments for expressing the size dependence of the growth rate in the form

$$g_2(x) = a x^\alpha \qquad (37)$$

where

$$\alpha = -1, 0, 2 \qquad (38)$$

according to the mechanisms of diffusion control, constant linear growth, and surface nucleation control. In this case, Eq. 26 becomes

$$\theta_1 = \theta + \frac{1}{a(\alpha - 1)\bar{g}_1} \left[x^{-(\alpha - 1)} - x_o^{-(\alpha - 1)} \right] \qquad (39)$$

Thus when $\alpha = 0$, $g_2 = a$,

$$\theta_1 = \theta - \frac{1}{\bar{g}_1 g_2} (x - x_o) \qquad (40)$$

or

$$x - x_o = \bar{g}_1 g_2 \tau \qquad (41)$$

as expected on more direct grounds.

For $\alpha = -1$, corresponding to rate-controlling diffusion to point-sink crystals,

$$\theta_1 = \theta + \frac{1}{2a\bar{g}_1} [x_o^2 - x^2] \tag{42}$$

and

$$x^2 - x_o^2 = 2a\bar{g}_1 \tau \tag{43}$$

Growth of very small embryos is thus proportional to the square-root of age, as expected for diffusional control.

For $\alpha = 2$, surface nucleation control,

$$\theta_1 = \theta + \frac{1}{a\bar{g}_1} [\frac{1}{x} - \frac{1}{x_o}]$$

or

$$a\bar{g}_1 \tau = \frac{1}{x_o} - \frac{1}{x} \tag{44}$$

Linear growth thus appears to accelerate with age, since

$$x = \frac{x_o}{1 - x_o a\bar{g}_1 \tau} \tag{45}$$

This requires $\tau < 1/x_o a\bar{g}_1$, and hence is limited to very small crystals, since $x \to \infty$ for $\tau = 1/x_o \bar{g}_1 g_2$. Thus diffusion must become limiting before the growth rate reaches this high value.

For these cases, Eq. 30 gives

$$f = f_1 (x_o/x)^\alpha \tag{46}$$

with

$$f_1 = f_o(x_{o\gamma}) \qquad (\tau \geq \theta), \ (\gamma = -1, 0, 2) \tag{47}$$

or

$$f_1 = f_1(x_{o\gamma}, \theta - \tau) \qquad (\tau \leq \theta), \ (\gamma = -1, 0, 2) \tag{48}$$

where

$x_{o\gamma}(x,\theta)$ is found from Eqs. 40, 43, or 45.

With the choice of reference units made here, the dimensionless population balance equation, Eq. 15, contains no arbitrary parameters. The overall material balance, Eq. 19, contains two parameters, the reference volume fraction of crystals, V_{mo}, and the term $(1 - C_{ov})/S_v^*$, the volume of solvent per unit volume of crystal equivalent to the reference supersaturation. If the time course of the supersaturation is identical in two experiments, then the resulting size distributions vs. time should be superposable by appropriate change of scale of time, size and number density. From Eqs. 9 - 14, one finds

$$R_0 = (G_0/B_0)^{1/4} \tag{49}$$

$$t_0 = G_0^{-3/4} B_0^{-1/4} \tag{50}$$

If we adopt the usual power laws for nucleation and growth,

$$B = k_b(C - C_s)^\beta = B_0 S_1^\beta \tag{51}$$

$$G = k_g L^\alpha (C - C_s)^\gamma = G_0 x^\alpha S_1^\gamma \tag{52}$$

we find

$$B_0 = k_b S^{*\beta} \tag{53}$$

$$G_0 = k_g R_0^\alpha S^{*\gamma} = (k_g^4 S^{*(4\gamma - \beta\alpha)}/k_b^\alpha)^{1/(4-\alpha)} \tag{54}$$

$$R_0 = (k_g S^{*(\gamma - \beta)}/k_b)^{1/(4-\alpha)} \tag{55}$$

$$t_0 = [k_g^{-3} k_b^{\alpha - 1} S^{*\beta(\alpha - 1) - 3\gamma}]^{1/(4-\alpha)} \tag{56}$$

$$G_0/B_0 = [k_g^4 S^{*\alpha(\gamma - \beta)}/k_b^\alpha]^{1/(4-\alpha)} \tag{57}$$

If two experiments were carried out with the same $S_1(\theta)$ but different reference states, they should give the same $f(x,\theta)$. Then

$$\ln f = \ln n' - \ln (B_0'/G_0') = \ln n'' - \ln (B_0''/G_0'') \tag{58}$$

$$\ln x = \ln L' - \ln R_0' = \ln L'' - \ln R_0'' \tag{59}$$

$$\ln \theta = \ln t' - \ln t_0' = \ln t'' - \ln t_0'' \tag{60}$$

The plots of $\ln n$ vs. $\ln t$ and $\ln L$ should be superposable if they are translated by the amounts

$$\Delta \ln t = \frac{\beta(\alpha - 1) - 3\gamma}{4 - \alpha} \Delta \ln S^* \tag{61}$$

$$\Delta \ln L = \frac{\gamma - \beta}{4 - \alpha} \Delta \ln S^* \tag{62}$$

$$\Delta \ln n = \frac{4(\beta - \gamma)}{4 - \alpha} \Delta \ln S^* \tag{63}$$

The magma density, W, is a function of dimensionless variables only and hence W(t) for two experiments with different S^* but equal $S_1(\theta)$ should be superposable on a plot of W vs. ln t.

The total number of crystals per unit volume is

$$N_m(\theta) = \int_0^\infty n(L)\, dL = \frac{B_o R_o}{G_o} \int_0^\infty f(x)\, dx = \left(\frac{B_o}{G_o}\right)^{3/4} \int_0^\infty f(x)\, dx \tag{64}$$

Hence

$$\Delta \ln N_m(\theta) = \frac{3\alpha(\beta - \gamma)}{4(4 - \alpha)} \Delta \ln S^* \tag{65}$$

Eqs. 61 and 65 can be solved for β and γ if α is known.

$$\beta = \frac{R_2 - \alpha R_1}{\alpha} \tag{66}$$

$$\gamma = \frac{4R_2(\alpha - 1) - 3\alpha R_1}{3\alpha} \tag{67}$$

where

$$R_1 = \Delta \ln t / \Delta \ln S^* \tag{68}$$

$$R_2 = \Delta \ln N_m / \Delta \ln S^* \tag{69}$$

These relationships hold provided $S_1(\theta)$ is the same for two experiments in which S^* differs. As a practical matter, this requires controlling the supersaturation during the course of the batch and will not often be feasible. However, if the supersaturation holds a constant value, for measurements made during this constant period, the crystal yield, Y, will be a function of dimensionless variables, since

$$Y = \frac{C_o - C}{C_o} = \frac{S^*}{C_o}(U - S_1) = \left(\frac{1 - C_{ov}}{C_{ov}}\right)\left(\frac{V_m - V_{mo}}{1 - V_m}\right) \tag{70}$$

and

$$Y_1 = \left(\frac{C_{ov}}{1 - C_{ov}}\right) Y - \frac{S^* S_1}{C_o} = \frac{S^*}{C_o} U(\theta) \tag{71}$$

Plots of $Y_1(t)$ vs. ln t will now be superposable if both Y_1 and ln t are shifted. Moreover, if the reference concentration, S^*, is taken to be the initial concentration, C_o, (rather than $C_o - C_{so}$, as in Eq. 12) the shift in Y_1 will be equal to the difference in the steady supersaturation values in the two experiments. One then has

$$Y_1(t/t_0) - \left(\frac{C - C_s}{C_0}\right) = 1 - \frac{C_s(t/t_0)}{C_0} \tag{72}$$

Numerical studies of batch crystallization by Nývlt (5) under different cooling regimes suggest that the supersaturation often is nearly constant after an initial transient period. The present analysis suggests that information on nucleation and growth kinetics can be extracted from measurements of crystal number density and crystal yield in such a regime.

REFERENCES

1. Aris, R. and N. R. Amundson, "Mathematical Methods in Chemical Engineering, Vol. 2, First-Order Partial Differential Equations with Applications," Prentice-Hall, New York (1973).

2. Courant, R. and D. Hilbert, "Methods of Mathematical Physics, Vol. II, Partial Differential Equations," Wiley-Interscience, New York (1962).

3. Hulburt, H. M. and S. Katz, Chem. Eng. Sci. _19_, 555-574 (1964).

4. Nielsen, A. E., "Kinetics of Precipitation," Pergamon Press, New York (1964).

5. Nývlt, J., Coll. Czech. Chem. Commun. _39_, 3463 (1974).

6. Randolph, A. D. and M. A. Larson, "Theory of Particulate Processes," Academic Press, New York (1971).

THE CRYSTALLIZATION OF CHLORO-ORGANIC COMPOUNDS IN A BATCH CRYSTALLIZER WITH PROGRAMMED COOLING

I.V. MELIKHOV, E.V. MIKHIN AND A.M. PEKLER

Chair of Radiochemistry, Department of Chemistry

Moscow State University, Moscow 117234, U.S.S.R.

Batch crystallizers are widely used in the chemical industry, but are often operated under conditions which are far from optimum. The optimization of operation, to increase efficiency and reduce costs, is therefore an important problem.

A promising way to optimize the operation of batch crystallizers is by means of programmed cooling (1, 2, 3, 4). Programmed cooling of the crystallizer contents can be carried out either by varying the flow-rate of coolant at constant temperature or by varying the temperature of the coolant at constant flow-rate. The first method has been employed by Nývlt and Skřivánek (5) and the second has been tested by us for the crystallization of hexachloroparaxylene from heptane (6, 7) of paradichlorobenzene from mixed dichlorobenzenes (4) and of hexachlorometaxylene from isopropanol. In all three cases, programmed cooling resulted in higher equipment efficiency and improved quality of product crystals.

The cooling programme was designed using experimental data obtained as follows. Crystallization was performed with different cooling rates, with different levels of agitation of the solution and with different concentrations of impurity. During the crystallization, samples of the crystals and of the mother liquor were taken and analysed for purity. The size distribution of the sample of crystals was found by optical microscopy. The conditions under which incrustations appeared on the walls of the crystallizer were also noted. Meanwhile, in a special laboratory installation, the size and habit of a crystal held in mother liquor on a wire were studied as a function of supersaturation, hydrodynamic conditions and impurity concentration (9). These data, together with a knowledge of both the solubility and of the metastable limit as functions of temperature, enabled us to optimize the operation of the commercial crystallizer.

In analysing the experimental data, we supposed that in ordinary jacketed crystallizers the following processes occur: nucleation both in the bulk solution and at the walls, and subsequently growth, attrition, breakage, and agglomeration. The analysis is complicated by differences between both the temperature and the hydrodynamic conditions in different regions of the crystallizer. If, however, as a result of turbulence, the velocity with which crystals move through regions having different conditions is much greater than the velocity of crystallization, then the behaviour of the crystals at any given time can be related to the average characteristics of the solution. Growth of the crystals can thus be considered as a Markov process, with the assumption that the growth rate fluctuates as a result of the different trajectories of crystals in the crystallizer (8). Growth can then be characterised by only two parameters, a mean growth rate and a dispersion coefficient which reflects any variation of conditions within the crystallizer which may affect the growth.

The instantaneous mass balance and instantaneous crystal population balance can then be written (6):

$$m_w + N\rho \int_0^\infty V\psi(V, t)dV = m_o - V_{soln} \cdot c \qquad (1)$$

$$\frac{\partial \psi(V, t)}{\partial t} = \frac{\partial}{\partial V}\left[D_V \frac{\partial \psi(V, t)}{\partial V} - r_V \psi(V, t) \right] + \omega_V \qquad (2)$$

Here, m_w is the mass of the incrustations on the walls of the crystallizer, N is the total number of crystals, ρ is the density of the solute, V is the volume of a single crystal, t is time, $\psi(V, t)$ is the normalised distribution function of crystal volume, m_o is the mass of solute originally put in the crystallizer, V_{soln} is the volume of solution, c is the mean (supersaturated) solution concentration (g/ℓ), D_V is the dispersion coefficient for the volume of a single crystal, r_V is the average growth rate expressed as a rate of increase of the volume of a crystal and ω_V is the rate of change of the number of crystals having a given volume, due to breakage, agglomeration and/or seeding.

The first term on the right hand side of equation 2 represents possible variations in the number of crystals having any given volume due to fluctuations in their growth rate, the second term represents the effect of steady growth and the third term the effect of other factors such as breakage and agglomeration. Equations 1 and 2 can be used directly if the volume distribution of the product crystals can be found by Coulter counter type methods (11). On the other hand, microscopic methods of determining distributions give the advantage of reliable differentiation between separate crystals and agglomerates. However if microscopic methods are used, a relationship must be established between the crystal volume V and some other size parameter which is more conveniently measured by microscope. We found that for the crystals investigated in the present work, the most suitable microscopic parameter was the length, a, of the longest side of the projected image of the crystal. By observing the variation of a with V for a single crystal on a holder, we found the volume shape factor

$$\alpha_1(t) = V/a^3$$

and, considering the relationship

$$V\psi(V, t) \, dV = \alpha_1(t)a^3 \, \psi(a, t) \, da$$

then from equation 2, we can write

$$\frac{\partial \psi(a, t)}{\partial t} = \frac{\partial}{\partial a} [D_a \frac{\partial \psi(a, t)}{\partial a} - r_a \psi(a, t)] + \omega_a \quad (3)$$

where $\psi(a, t)$ is the normalised distribution function of crystal size (a), D_a is the dispersion coefficient for the size of the crystals, r_a is the average growth rate expressed as a rate of increase of the parameter a, and ω_a is as ω_V but referred to length parameter a rather than to volume V.

It is usually assumed that $D_a = 0$. However, in our experiments on fixed crystals we found wide fluctuations in the growth rates (9, Fig.2), so D_a cannot be zero in our case. Our experiments with fixed crystals showed that the growth rate r_a was practically independent of crystal size, that the dispersion coefficient D_a was proportional to r_a and that the shape factor α_1 was given by $\alpha_1 = 10^3/(0.98 + 1300/a)$ (a in microns) over the whole range of crystal size studied, namely 50 μm < a < 600 μm. Then, when designing the cooling programme for the commercial crystallizer, we used equations 1 and 3, together with the relationships

$$V = \alpha_1 a^3$$

r_a is independent of a

$$D_a = \text{const. } r_a$$

In analysing the behaviour of impurity, we used the concept that impurity could be captured by the crystals both as a result of growth and as a result of independent transfer from the solution to the crystals by a diffusion mechanism (11). Then the total rate of transfer of impurity from the solution to the crystals is

$$\frac{dX_c}{dt} = N \int_0^\infty [Yr_a\lambda + J]\, \alpha_2\, a^2 \psi(a, t)\, da \qquad (4)$$

where X_c is the total quantity of impurity in the crystals, $Y = (X_o - X_c)/V_{soln}$ is the mean concentration of impurity in the solution, X_o is the overall quantity of impurity in the system, λ is the coefficient of capture of impurity as a result of crystal growth, J is the diffusive flux of impurity from solution to crystals and α_2 is the surface shape factor.

Experiments in our laboratory installation showed that $Yr_a\lambda \gg J$ provided the suspension of crystals in mother liquor was not too vigorously agitated. We characterized the level of agitation in both the laboratory and the commercial crystallizer by a Reynolds number in which the length term was the radius of the circle described by the tips of the blades of the agitator and the velocity term was the tangential velocity of those tips. Two different agitation regimes were then recognised, that is, Re below 25000 and Re above 25000. The two regimes are described in more detail below.

CRYSTALLIZATION WITH RE BELOW 25000

With linear cooling, the supersaturation at first inc eased to the metastable limit and then decreased. During the period of increasing supersaturation, which lasted about 17 minutes, nucleation occurred. During the subsequent period of decreasing supersaturation these nucleated crystals grew without breakage or agglomeration ($\omega_a = 0$). For the range of Re investigated (6000 – 25000) with cooling rates, ν, between 1.3×10^{-3} and 6.5×10^{-3} °C/sec, the observed time-dependence of the supersaturation S was expressed by the empirical relationship:

$$S = c - c^* = S_o + \nu K_1 - t\, \exp[-K_2 + (K_3 - K_4 Re)\nu] \qquad (5)$$

where c^* is the solubility, which in the present case is an exponential function of temperature (9), S_o depends on Re but K_1, K_2 K_3 and K_4 are constants with values 104, 6.7 ± 0.2, 520 ± 10 and 6.9×10^{-3} respectively. When ν was more than 8×10^{-3} °C/sec and S was more than 3.1 wt% incrustations occurred on the walls of the crystallizer.

The size distribution function $\psi(a, t)$ satisfied the normal distribution

$$\psi(a, t) = [1/\sigma\sqrt{(2\pi)}]\, \exp[-(a - \bar{a})^2/2\sigma^2] \qquad (6)$$

where $\bar{a}$ is the mean value of a and σ^2 is the variance. The average growth rate r_a was practically constant throughout the crystallization and was decribed by

$$r_a = \frac{d\bar{a}}{dt} = z_1 \exp[z_2 \text{ Re} + z_3 \nu - z_4 \text{ Re } \nu] \qquad (7)$$

where $z_1 = 8.33 \times 10^{-4}$ µm/sec, $z_2 = 3.71 \times 10^{-5}$, $z_3 = 494$ sec/deg C and $z_4 = 0.75 \times 10^{-3}$ sec/deg C. The variance σ^2 was practically independent of the cooling rate, but increased as crystallization continued (see Table I).

Values of the total number of crystals, N, were calculated using equations 1, 5, 6 and 7. These values were constant once supersaturation began to fall, provided ν was less than 8×10^{-3} deg C/sec and S was less than 3.1 wt% (i.e. no incrustations on the walls). This showed that under these conditions the only process occurring after supersaturation started to fall was growth on crystals which had been nucleated during the initial period when supersaturation was rising.

As the crystals increased in size, their growth rate fluctuated considerably, resulting in an increase in the variance σ^2 of the size distribution function $\psi(a, t)$. Under these conditions the system obeys equation 3 with $\omega_a = 0$.

Indeed, if the variable $\bar{a} = \int_\tau^t r_a \cdot dt$ (where τ is the time when supersaturation begins to fall) is introduced into equation 3 and if equation 3 is then integrated with the conditions

$\psi(a_o, \tau) = 1$

$\psi(0, t) = \underset{a \to 0}{\text{Lim}} \psi_1$

$\psi(\infty, t) = 0$

$\bar{a} \gg \bar{a}_o$

$t \gg \tau$

Then, taking into account that r_a is independent of a, and assuming that $D_a/r_a = B = $ const, we can derive equation 6 with $\sigma^2/2a = B$ (see Table I). In the above, ψ_1 is the solution of the diffusion equation for an infinite space and a_o is the crystal size at time τ, when supersaturation begins to fall. Thus fluctuations in the growth of hexachloroparaxylene crystals due to temperature and hydrodynamic fluctuations within the crystallizer (these were much greater than for a fixed crystal (9)) can be described simply by the parameter D_a, while the development of the distribution function $\psi(a, t)$ is via equation 3. Since the coefficient D_a can be related to the real conditions in the crystallizer, the application of equation 3 should be fruitful.

The uptake of the impurity heptachloroparaxylene by product crystals was independent of the crystal size, in other words λ was independent of a. Therefore in the period of decreasing supersaturation, from equations 1 and 4:

$$\frac{dX_c}{dt} = \lambda \frac{(X_o - X_c)}{\rho} \frac{dc}{dt}$$

The coefficient λ was found from equation 8 using experimental data on the uptake of impurity at constant temperature but with various values of ν and Re defining the growth rate r_a. The values of λ were expressed by the empirical formula

$$\lambda = \lambda_P + (\lambda_A - \lambda_P) \exp(-K/r_a) \tag{9}$$

where $\lambda_P = \lambda_P^o \exp(-E_P/RT)$, $\lambda_A \simeq 2.5 \lambda_P$, $\lambda_P^o = 7 \times 10^{-3}$, $E_P = 58$ kJ/mol and $K \simeq 0.033 \pm 0.011$. In order to apply equation 9, a study was made of the equilibrium distribution of heptachloroparaxylene between saturated solution and product crystals at various temperatures. It was found that the coefficient λ_P equalled the coefficient of equilibrium cocrystallization of hepta- and hexachloroparaxylene. Equation 9 agrees with Hall's equation (11) provided the parameter λ_A is assumed to be equal to the impurity adsorption coefficient and K to be equal to the characteristic rate of transfer of impurity molecules from the surface of the crystal to the solution. This agreement indicates that the uptake of impurities by product crystals occurred without equilibrium being reached between the crystal surface and the solution, provided r_a was greater than 0.04 µm/s. The failure to reach equilibrium resulted from the slow interaction of impurity with the crystal surface. At lower growth rates the crystal surface adsorbed a quasi-equilibrium amount of impurity.

CRYSTALLIZATION WITH RE ABOVE 25000

With intense agitation, crystal growth was accompanied by attrition, small fragments being detached from the main crystals. Some of these fragments dissolved but other united to form agglomerates of the same order of size as the attrited crystals. With Re $\simeq$ 25000, growth fully compensated the size reduction due to attrition and the value of a did not decrease during cooling. With Re > 25000, attrition predominated over growth and the value of a decreased while still roughly obeying equation 6, but the value of N increased greatly (see Table I). The occurrence of attrition was also proved by special experiments in the laboratory crystallizer.

It was found that the size distribution function of product crystals held in saturated solution changed with the degree of agitation. The size distribution remained normal and the variance σ^2 and the number N remained constant but the mean size $\bar{a}$ decreased as agitation continued (Table I). It follows that agitation caused fragments to separate from the product crystals and these fragments then completely dissolved. The occurrence of the agglomeration of fragments was proved by the observation of a sharp increase in the rate of change of size of a fixed crystal growing in an agitated solution at constant temperature when finely ground crystals were added to the system. The same phenomenon must occur in industrial equipment. In this case fragments attached to large crystals will be dislodged by high-energy collisions with other large crystals. For agglomerates of smaller crystals the collisions are less intense, so these are able to increase in size. The agglomeration of fragments in supersaturated solutions is also assisted by a reduction in their rate of dissolution. Despite this, dissolution did occur in a supersaturated solution, as can be seen from equation 5, in which the factor $K_4 Re$ (which is considerable for the Re > 25000) reflects the replenishment of the hexachloroparaxylene in the solution as a result of agitation.

The behaviour of the impurity with Re > 25000 was remarkable in that, during the dissolution of fragments, part of the previously-absorbed impurity re-transferred to the solution. Also, intense agitation reduced short-range ordering of the crystals, as a result of their collisions with one another and with the crystallizer internals. Because of this reduction in ordering, the diffusion of the impurity in the surface layer of the crystal was accelerated ($J \simeq Y\, r_a\, \lambda$), equilibrium between the crystal surface and the solution was reached sooner, and hence the product was much purer.

OPTIMIZATION OF THE CRYSTALLIZATION TO ACHIEVE A PURE PRODUCT

We used the results of our studies on crystal growth and impurity capture to optimize the operation of the crystallizer via programmed cooling.

Obviously, optimal crystallizing conditions should permit the maximum rate of production of crystals of good quality, without the occurrence of attrition or of incrustation of the crystallizer walls. The basic requirement was to provide a definite, strictly controlled supersaturation. This can be obtained via programmed cooling in which the temperature of the mother liquor, $T(t)$, and the entry temperature of the coolant, $\theta_{in}(t)$, are varied in conjunction. To determine the functions $T(t)$ and $\theta_{in}(t)$, we used the mass balances for product and for impurity, viz equations 1 and 4, and the population balance, equation 3, substituting the necessary expressions from equations 4 to 8. These equations were considered together with the heat balance

$$G_x c_x (\theta_{in} - \theta_{out}) - q \cdot \frac{dm_c}{dt} = [(m_o - m_c + m_s) c_\ell + c_c m_c] \nu \quad (10)$$

where G_x is the mass flow rate of coolant, c_x is the specific heat of the coolant, θ_{in} and θ_{out} are the coolant entry and exit temperatures, respectively, q is the heat of crystallization, m_c is the mass of crystals and is given by

$$m_c = \rho \int_0^\infty \alpha_1 a^3 \psi(a, t) \, da$$

m_s is the mass of solvent, c_ℓ is the specific heat of the solution and c_c is the specific heat of the crystals.

When applying equations 1 to 8 we bore in mind that the production rate would be a maximum if the supersaturation was held at the maximum which was possible without the formation of incrustations on the walls. Also the Reynolds number could not exceed 25000 without attrition occurring. With these conditions and equation 5, it is easy to find the maximum permissible cooling rate ν_{max}, and hence the function $T(t) = T_o - \nu_{max} \cdot t$ where T_o is the initial solution temperature. The level of impurity in the product for this maximum cooling rate can be calculated from equations 4, 5 and 9, using also the functions $\bar{a}(t)$, $\psi(0, t)$ found from equations 6 and 7, and the derived function $T(t)$. If the calculated level of impurity in the product does not satisfy purity requirements, then the cooling rate, and hence the growth rate (equation 7) should be reduced until the coefficient of cocrystallization (equation 9) falls to such a value that the product is then satisfactory.

Having specified the optimum cooling rate ν_{opt} and calculated m_c and dm_c/dt, one can, with the help of equations 1 and 5, solve equation 10 for $\theta_{in}(t)$, provided θ_{out} is known as a function of θ_{in}. The temperature θ_{out} can be found by calculating the steady heat transfer through the wall with temperature T constant outside the wall and θ_{in} known. The model of steady heat transfer is applicable over a wide range of conditions if the flowrate of coolant is sufficiently great and if its residence time in the cooling jacket is considerably smaller than the duration of the crystallization. In practice it is easy to satisfy these conditions. From equation 10 it then follows that

$$\theta_{out} = T - \frac{(T - \theta_{in})}{\exp\left[\frac{A h}{G_x c_x}\right]}$$

Re	$\nu \times 10^3$ deg C/sec	t sec	$\bar{a}$ μm	σ μm	$N \times 10^{-6}$	B μm
6000	2.7	1480	135 ± 4	62	23 ± 2	17
		8900	161 ± 5	76	31 ± 3	18
15200	2.7	1480	137 ± 5	75	27 ± 3	21
		8900	200 ± 5	72	22 ± 2	13
	4.0	1110	153 ± 6	76	18 ± 2	19
		6000	199 ± 6	99	22 ± 2	25
	1.9	2350	185 ± 5	86	10 ± 1	20
		12600	228 ± 5	81	14 ± 1	14
29300	6.8	1230	126 ± 5	67	43 ± 4	18
		3690	152 ± 5	76	44 ± 4	19
	2.1	1910	173 ± 5	67	16 ± 2	13
		11430	149 ± 5	70	87 ± 9	17
	1.3	2660	155 ± 6	77	22 ± 2	19
		16000	125 ± 5	73	142 ± 12	21

TABLE I Parameters of the crystal size distribution for different levels of agitation (Re), cooling rates (ν) and times (t).

where A is the surface area and h the heat transfer coefficient. These equations enable us to determine the function $\theta_{in}(t)$ for a given ν_{opt} which then gives the necessary variation of coolant entry temperature with time for optimum crystallizer operation.

ACKNOWLEDGEMENT

We thank Dr. V.R. Phillips, of University College London, for help with the translation into English.

REFERENCES

1. Mullin, J.W., Nývlt, J., Chem.Eng.Sci., 26 (3) (1971) 369.
2. Nývlt, J., De Ingenieur (Ned.) 82 (38) (1970) 39.
3. Jones, A.G., Mullin, J.W., Chem.Engng.Sci., 29 (1974) 353.
4. Filatov, L.N., Mikhin, E.V., Pekler, A.M., Khim.Prom., 4 (1973) 265.
5. Nývlt, J., Skřivánek, J., Coll.Czech.Chem.Commun., 33 (6) (1968) 1788.
6. Melikhov, I.V., Mikhin, E.V., Pekler, A.M., Teor.Osn.Khim. Tekhnol., 9 (2) (1975) 219.
7. Kuznetsov, Yu.A., Mikhin, E.V., Pekler, A.M., U.S.S.R. Patent No.465209 (Bulletin No.12, 1975).
8. Nebylitsyn, V.D., Dissertation for the degree of Candidate of Chemical Sciences, Moscow State University (1973).
9. Melikhov, I.V., Mikhin, E.V., Pekler, A.M., Teor.Osn.Khim. Tekhnol., 7 (5) (1973) 670.
10. Khobler, T., 'Teploperedacha i Teploobmenniki', Goskhimizdat, Moskva (1961) ("Heat transfer and heat exchangers, State Chemical Publishers, Moscow, 1961).
11. Melikhov, I.V., Merkulova, M.S., 'Sokristallizatsiya', Khimiya, Moskva (1975) ('Cocrystallization', "Khimiya" Publishing House, Moscow, 1975).

A NEW TECHNIQUE FOR ACCURATE CRYSTAL SIZE DISTRIBUTION ANALYSIS
IN AN MSMPR CRYSTALLIZER

S. JANČIĆ AND J. GARSIDE

Department of Chemical Engineering
University College London, Torrington Place
London WC1E 7JE, England

INTRODUCTION

The design of crystallizers using population balance techniques is critically dependent on a knowledge of the crystallization kinetics of the system under consideration.[1] Accurate experimental determination of nucleation and growth rates from continuous crystallizers presents many problems. Techniques based on the measurement of crystal size distributions (CSD) in a continuous mixed suspension mixed product removed (MSMPR) crystallizer rely on extrapolating the population density plot to zero size.[2] Large errors can result unless the size analysis is extended well into the sub-sieve size range, particularly if the crystal growth rate, and hence the slope of the population density plot, is a function of crystal size.[3]

In this paper we present theoretical results which show how the crystallization kinetics for systems exhibiting size-dependent growth rates may be determined from a population density plot. These results are different from those previously assumed. Experimental crystallization kinetics obtained for the potash alum-water system in a continuous MSMPR crystallizer are also reported. These were obtained using a Coulter counter to extend the size analysis down to about 5 μm and as a result the derived kinetics show substantial differences from those obtained using conventional sieving techniques for crystal size analysis.

THEORY

For an MSMPR crystallizer operating at steady state the general population balance equation reduces to[2]

$$\frac{\partial (nG)}{\partial L} = -\frac{n}{T} \tag{1}$$

For size independent growth this can be integrated to give the well-known equation

$$\ln n = \ln n^o - \frac{L}{GT} \tag{2}$$

A size analysis performed on a sample representative of the crystallizer contents at steady state yields the necessary data for the construction of the ln n vs L graph. Equation 2 indicates that the slope of this graph can be used to determine the crystal growth rate and the intercept provides the value of the nuclei population density, n^o, so enabling the nucleation rate, B^o, to be determined since

$$B^o = n^o G \tag{3}$$

Various growth rate models have been suggested to represent size dependent growth in continuous crystallizers.[4-6] Abegg, Stevens and Larson[6] have proposed the equation

$$G = G^o (1 + \gamma L)^b \quad ; \quad b < 1 \tag{4}$$

Although this equation does not follow directly from known mechanisms of crystal growth it satisfies the following requirements of size dependent growth models:

(i) it is consistent with the population density concept in that all the moments of the population density distributions generated by the model converge

(ii) crystal nuclei grow at a finite rate

(iii) the growth rate expression is a continuous function of L and b in a region which includes the point L = 0, b = 0.

Combining equations 1 and 4 and integrating gives[6]

$$n = n^o (1 + \gamma L)^{-b} \exp \left[\frac{1}{G^o T \gamma (1 - b)} - \frac{(1 + \gamma L)^{1 - b}}{G^o T \gamma (1 - b)} \right] \tag{5}$$

and defining $\gamma = 1/G^o T$, equation 5 can be rewritten as

$$\ln n = \ln n^o + \frac{1}{1 - b} - b \ln (1 + \gamma L) - \frac{(1 + \gamma L)^{1 - b}}{(1 - b)} \tag{6}$$

Equation 6 thus represents the steady state population density distribution in an MSMPR crystallizer when the crystal growth rate is given by equation 4.

Values of the parameters n^o, G^o and b can best be determined from the steady state CSD in an MSMPR crystalliser. In order to evaluate these parameters, however, a rather different procedure from that normally used for the size independent growth case must be followed.

The slope of a ln n vs L plot of equation 6 is given by

$$\frac{d(\ln n)}{dL} = -\gamma\left[\frac{b}{1+\gamma L} + \frac{1}{(1+\gamma L)^b}\right] \quad (7)$$

and the slope at zero size (L = 0) is

$$\left.\frac{d(\ln n)}{dL}\right|_{L=0} = -\gamma(1+b) = -\frac{(1+b)}{G^o T} \quad (8)$$

For size independent growth, b = o and the slope for all values of L is $-1/GT$ where G is the constant growth rate (see equation 2). Equation 8, however, indicates that for size dependent growth the nuclei growth rate, G^o, cannot be calculated from the initial slope of the population density plot unless the value of b is known. Since the value of b is restricted to b < 1, errors of up to 100% can result in deriving the nuclei growth rate if the factor (1 + b) is ignored.

The nucleation rate is given by the expression

$$B^o = n^o G^o \quad (9)$$

the nuclei growth rate G^o replacing the constant growth rate G in equation 3. Determination of the nucleation rate from equation 9 thus also requires a knowledge of the value of b.

The nucleation rate can also be determined from the overall number balance on the crystallizer and is given by integrating equation 1 over the entire size range:

$$\int_{n^o G^o}^{n_\infty G_\infty} d(nG) = -\frac{1}{T}\int_o^\infty n\, dL \quad (10)$$

which, since $n_\infty = o$, gives

$$n^O G^O = B^O = \frac{1}{\tau} \int_0^\infty n \, dL \qquad (11)$$

That is, since on average all the crystals in the crystallizer are removed in one residence time, the nucleation rate is given by the total number of crystals in the crystallizer divided by the residence time. Calculation of the nucleation rate from equation 11 is independent of any growth rate model and relies solely on the determination of the total number of crystals in the crystallizer.

Procedure Used to Calculate Crystallization Kinetics

Based on the above equations, the crystallization kinetics for a size dependent growth system may be calculated as follows:

(i) determine n^O from the intercept at $L = 0$ of the population density plot

(ii) determine $\left[d(\ln n)/dL \right]_{L=0}$, the slope of the population density plot at $L = 0$.

The accuracy of the estimates in steps (i) and (ii) is crucially dependent on being able to extrapolate the size distribution well into the sub-sieve range. Large errors can result if this is not done.[3]

(iii) on the assumption that equation 4 represents the size dependent growth rate, fit equation 6 to the experimental population density plot. Different values of b are tried and a least squares criterion used to determine the value of b which best fits the experimental data. Note that as the value of b is changed, the value of γ used in equation 6 and given by equation 8 also changes

(iv) for this best value of b, determine the nuclei growth rate, G^O, from equation 8

(v) determine the nucleation rate, B^O, from equation 9

(vi) to provide a check on the accuracy with which equation 6 fits the experimental size distribution data, use equation 11 to calculate the value of B^O and compare this with the value determined in step (v).

ACCURATE CRYSTAL SIZE DISTRIBUTION ANALYSIS

EXPERIMENTAL PROCEDURE

Full details of the experimental procedure used in this study are given elsewhere.[3] Briefly, crystallization was performed in a 1.3 litre draft tube baffled continuous MSMPR crystallizer of the same geometry as that described by Jones and Mullin.[7] Agitation was provided by a marine type impeller which was located in the centre of the draft tube. The potash alum-water system was used, crystallization being induced by cooling.

The experiments reported here were performed at 30.0 ± 0.1°C using a stirrer speed of 1350 rpm. The cooling water temperature was 25.0°C. After running until steady state had been achieved (at least 10 residence times) samples for solution concentration measurement and crystal size analysis were taken.

An essential feature of the size analysis technique was that the crystal sample used for the Coulter counter analysis always remained in solution and was not subjected to filtration, washing or drying steps.[3] A combination of sieving and Coulter counter analysis was used for the size analysis. For the former, British Standard sieves[8] in the range 37 to 2800 μm were used, while the model TA Coulter counter fitted with a 280 μm orifice was capable of covering the range between 5 and 112 μm. In order to prevent growth, nucleation or dissolution during the Coulter counter analysis, saturated potash alum solution at the crystallizer operating temperature was used as the electrolyte.

RESULTS

A typical population density plot obtained as described above is shown in Figure 1. The inset shows the small (< 150 μm) size range plotted on a larger scale. The good agreement obtained in the size range where both size analysis techniques were used (i.e. between 37 and 105 μm) was typical of all the results and provides evidence of the absolute accuracy of the method. The rapid increase in population density at sizes less than about 40 μm was also a consistent feature of all results. This increase was not detected when the crystals were filtered and dried before being subjected to a Coulter counter analysis and in this latter case much lower nucleation and nuclei growth rates were obtained.[3]

As noted above, the values of n^o and $\left[d(\ln n)/dL \right]_{L=0}$ obtained from the population density plot are very sensitive to the minimum crystal size down to which the size analysis is performed. Consequently in the present work this minimum size was standardized at 6 μm.

In order to calculate n^o and $\left[d(\ln n)/dL \right]_{L=0}$, the intercepts at L = 0 and the slope of the best straight line through the

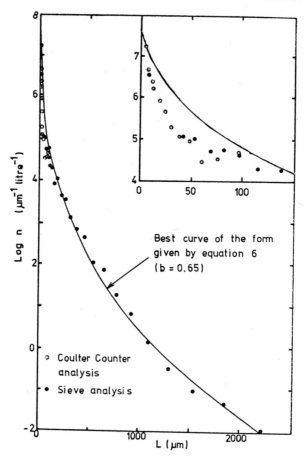

Figure 1: Typical Population Density Plot
(temperature = 30.0°C; Δc = 3.7 x 10^{-3} kg alum/kg solution; magma density = 22 kg alum/m^3 solution; residence time = 40 min)

experimental points corresponding to the five smallest sizes were determined. The advantages of using this method have been discussed elsewhere.[3]

The various parameters evaluated in steps (i) to (vi) (see Section 2) are shown in Table 1. A computer program was used to perform the least squares analysis necessary to calculate b in step (iii), and to evaluate the integrals in equation 11 (step (vi)).

The value of b which best fitted the experimental data was 0.65. When used in equation 6 together with the values of n^o and G^o shown in Table 1, this value gave the curve plotted in Figure 1. This curve fits the experimental points well over most

		Values based on all size analysis data points	Values based on sieve analysis data points
Step (i)	n^o (μm^{-1} litre^{-1})	3.93×10^7	1.94×10^5
(ii)	$[d(\ln n)/dL]_{L=0}$ (μm^{-1})	-0.225	-0.0180
(iii)	b (-)	0.65	0.26
(iv)	G^o ($\mu m\ s^{-1}$)	3.06×10^{-3}	2.92×10^{-2}
(v)	$B^o(=n^o G^o)$ (s^{-1} litre^{-1})	1.20×10^5	5.66×10^3
(vi)	$B^o(=\frac{1}{T}\int_0^\infty ndl)$ (s^{-1} litre^{-1})	1.10×10^5	6.03×10^3

Table 1: Crystallization parameters determined using two different size analysis techniques

of the size range, so confirming that the growth model represented by equation 4 is in general a satisfactory representation of the data. It is only for the size range less than about 100 μm that the theoretical size distribution equation is less satisfactory and does not fully reproduce the rapid change in curvature found for the experimental results.

DISCUSSION

The two values of the nucleation rate shown in Table 1 differ by about 10%. Considering the possible errors associated with the determination of the slope and intercept of the population density plot at L = 0 and with the numerical evaluation of the integral in equation 11, this agreement is extremely satisfactory and provides evidence of the accuracy of the growth rate model (equation 4) in representing the experimental CSD.

It should be noted that the agreement between these two estimates of B^o depends on the inclusion of the factor (1 + b) in the calculation of the nuclei growth rate (see equation 8). The determination of b is very sensitive to the values of n^o and $[d(\ln n)/dL]_{L=0}$ which in turn depend on the accuracy of the crystal size analysis for the smallest crystals in the size range. To illustrate this point the experimental data points determined by sieve analysis were fitted to a 4th order polynomial and the various parameters evaluated in steps (i) to (vi) were recalculated. The minimum crystal size used in the size analysis is now 41 μm. The results are shown in the second column of figures in Table 1 and show significant differences from those obtained when

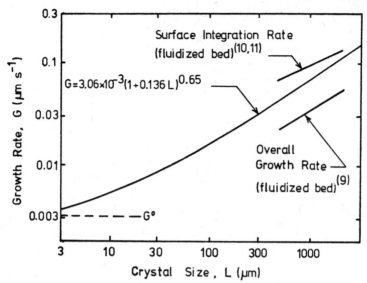

Figure 2: Comparison of Potash Alum Growth Rates

(temperature = 30.0°C; Δc = 3.7 x 10^{-3} kg alum/kg solution)

the size analysis was extended down to 6 μm. In the latter case, the nuclei growth rate is an order of magnitude smaller and the nucleation rate some twenty times larger than the values obtained from the sieve analysis alone. Further, the value of b differs by a factor of 2.5.

The size dependent crystal growth rate evaluated when the size analysis was extended down to 6 μm is represented by equation 4 with the values of G^{o} and b given in Table 1, i.e.

$$G = 3.06 \times 10^{-3} (1 + 0.136 L)^{0.65} \tag{4a}$$

The units of G and L are $\mu m\ s^{-1}$ and μm respectively. This equation is plotted in Figure 2 and represents the overall growth rate of potash alum under the conditions shown in Figure 1.

Growth rates of potash alum have been determined by several previous workers and some of these values are also shown in Figure 2. The results of Mullin and Garside[9] were obtained from batch measurements in a fluidized bed crystallizer and also represent overall growth rates. They cover the size range 530 - 1690 μm. Over this range the growth rates were size dependent and were empirically correlated by an equation of the form

$$G \propto L^{0.63}$$

The curve shown in Figure 2 is that corresponding to the same supersaturation and temperature as the MSMPR results obtained in the present work.

Garside et al.[10,11] have recently re-evaluated the fluidized bed data and derived the surface integration kinetics which were also found to be size dependent. These values are also shown in Figure 2.

The growth rates measured in the present study lie between these two sets of data. Since the degree of turbulence in the agitated vessel was higher than in the fluidized bed, it is reasonable to expect that the diffusional resistance to growth would be less in the agitated vessel and hence the overall growth rate would be higher, though still less than the surface integration rate. The present data are thus fully consistent with this set of previous results over the limited range of crystal sizes shown in Figure 2.

Ottens et al.[12] have reported MSMPR studies using potash alum. They also obtained non-linear population density plots although their size analysis was performed using sieves, the minimum sieve size being about 30 μm. The nuclei growth rates they report were in the range $0.05 - 0.2$ μm s^{-1} although neither the supersaturation nor the exact temperature are recorded. This is between one and two orders of magnitude higher than the values for nuclei growth rates determined from Coulter counter analysis in the present study and corresponds to the growth rates of crystals in the size range $500 - 3000$ μm.

CONCLUSIONS

The difficulties inherent in the determination of crystallization kinetics for systems where the crystal growth rate is size dependent have been demonstrated. It has been shown how these kinetics may be evaluated using the growth rate equation proposed by Abegg, Stevens and Larson and, in this case, the value of the parameter b must be known before the growth rates can be calculated. The good agreement between the growth rates determined in this study and those reported by previous workers provides an indication of the reliability of the present results.

NOMENCLATURE

b	constant in growth rate expression (eqn. 4)	(-)
B^o	nucleation rate	(s^{-1} litre^{-1})
G	crystal growth rate (= dL/dt)	(μm s^{-1})
G^o	nuclei growth rate	(μm s^{-1})
L	crystal size	(μm)

n	population density	(μm^{-1} litre^{-1})
n^o	nuclei population density	(μm^{-1} litre^{-1})
T	residence time	(s)
γ	$1/G^oT$	(μm^{-1})

REFERENCES

1. M.A. Larson and J. Garside, The Chem. Engr., No. 274, June 1973, 318
2. A.D. Randolph and M.A. Larson, "Theory of Particulate Processes", 1971 (New York: Academic Press)
3. S. Jančić and J. Garside, Chem. Engng. Sci., to be published 1975
4. S.H. Bransom, Brit. Chem. Eng. 5, 838 (1960)
5. T.F. Canning and A.D. Randolph, AIChEJ, 13, 5 (1967)
6. C.F. Abegg, J.D. Stevens and M.A. Larson, AIChEJ, 14, 118 (1968)
7. A.G. Jones and J.W. Mullin, Trans. Instn. Chem. Engrs., 51, 302 (1973)
8. British Standards Specification, 410: 1969, "Test Sieves", British Standards Institution
9. J.W. Mullin and J. Garside, Trans. Instn. Chem. Engrs., 45, T291 (1967)
10. J. Garside, J.W. Mullin and S.N. Das., Ind. Eng. Chem. Fundam., 13, 299 (1974)
11. J. Garside, R. Janssen-van Rosmalen and P. Bennema, J. Cryst. Growth, to be published 1975
12. E.P.K. Ottens, A.H. Janse and E.J. de Jong, J. Cryst. Growth, 13/14, 500 (1972)

Crystallizer Operation and Case Studies

TRANSIENT BEHAVIOUR IN CRYSTALLIZATION - DESIGN MODELS RELATED TO PLANT EXPERIENCES

J.P. SHIELDS

ICI Corporate Laboratory

Runcorn, Cheshire, England

INTRODUCTION

At the ICI Corporate Laboratory, the policy for developing a scientific basis for the design and operation of crystallisers involved selecting plant crystallisation processes which highlighted areas of ignorance. The initial exercise with a system under investigation was the derivation of kinetic expressions which related crystal birth and growth rates to the important operating variables such as supersaturation (σ), temperature (T) and solids concentration (m_c/m). These expressions of the form:

$$\dot{d} = k_{GO} \exp\left(-E/R \left[\frac{1}{T} - \frac{1}{T_o}\right]\right) \sigma^g \qquad (1)$$

and $\quad \dot{N}/V = k_N \sigma^n (m_c/m)^c \qquad (2)$

were derived using the well-known population balance approach (1-3). Thereafter these expressions were incorporated in mathematical models of the plant crystalliser. The processes examined initially were continuous and the models assumed steady state. Such a model has been described by Ramshaw and Parker (4) who have shown that this approach is a convenient way of exploring various modes of crystallisation. The need for an unsteady state model to reinforce the steady state models soon became apparent and an encounter with the commissioning stage of a continuous crystallisation process catalysed the development of an unsteady state model. In this paper the experiences of applying the model to three different industrial plants are presented.

THE UNSTEADY STATE MODEL

Theoretical and experimental papers on unsteady state crystallisation differ in the extent to which various simplifying assumptions are made, but each illustrates the complex interaction of growth and nucleation dynamics. Randolph and Larson (1) have made a linearised stability analysis of a fully mixed crystalliser and established the conditions for stability - work which has been supplemented by Sherwin (5). It has been demonstrated (6) that cyclic behaviour can be caused by product classification and can be exaggerated by fines destruction and clear liquor removal. In a recent publication (7) Randolph concluded that the combination of product classification and removal of clear liquor in a continuous crystallisation resulted in cycling of both size distribution and solids concentration. This work will be discussed later.

In the present model of unsteady state crystallisation several common physical assumptions are made, and these together with the solution to the unsteady state equations are presented in the Appendix. Some applications of the model to industrial crystallisation plants will now be considered.

Case 1 - A Highly Soluble System with Mother Liquor Removal

Fig. 1 is a line diagram of the type of plant under consideration. Mother liquor was removed from the crystalliser to maintain an impurity below a fixed level. The separated solids stream was classified in a hydrocyclone and the top flow was returned to the evacuated crystalliser.

<u>Steady state operation.</u> Using the third moment of the population density it can be shown that

$$\sigma^{3g+n} = K \dot{m}_c^4 m^4 / m_c^4 \qquad (3a)$$

where the constant K derives from the empirical relationships 1 and 2. By solving the mass balance equations the changes in steady state resulting from changes in $\dot{m}_f$, $\dot{m}_{ml}$ and $\dot{m}_w$ can be calculated. For this type of operation, Fig. 2 shows the sensitivity of steady state to the flow rates. Lines of different supersaturation are shown in the plot.

In the system investigated the normal production rate was 0.126 kg/s. A supersaturation of 0.1 implied a solids concentration of 0.2. The upper and lower limits of solids concentration were defined as $m_c/m = 0.4$ (i.e. $\sigma = 0.07$) and $m_c/m = 0.1$ (i.e. $\sigma = 0.14$) respectively. For a typical feed of 0.378 kg/s, a 3% reduction in

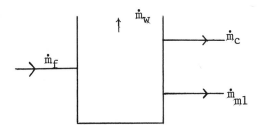

Fig. 1 Line diagram of plant - Case 1

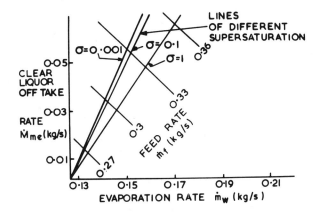

Fig. 2 The steady state situation in Case 1

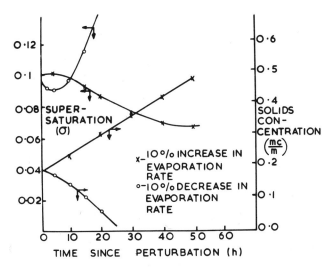

Fig. 3 The effects of changes in evaporation rate on supersaturation and solids concentration

$\dot{m}_{ml}$ coupled to a 1.6% increase in $\dot{m}_w$ causes the steady state to alter from $\sigma = 0.001$ ($m_c/m = 1.0$) to $\sigma = 0.1$ ($m_c/m = 0.2$). It is clear that if the production rate ($\dot{m}_c$) is maintained at a fixed value, the sensitivity of this system is unsatisfactory. At this point a dynamic model of crystalliser behaviour becomes necessary to explore the time constants associated with the changes.

Unsteady state operation. Fig. 3 presents a typical response of this type of crystallisation to 10% alterations in the evaporation rate. It is interesting to note that supersaturation initially moves in the direction opposite to its ultimate movement. This is a result of the alteration in feed rate which is made to maintain a fixed level in the crystalliser. Subsequently, the discharge or accumulation of supersaturation in response to the changing crystal surface area determines the ultimate movement of supersaturation, under the constraint of fixed production rate.

This mode of operation, where the rate of mother liquor removal is proportional to the feed rate, is very sensitive to small changes in process streams. One main advantage of the unsteady state model is that control strategies of all descriptions can be simulated. In this case study one of the control parameters which were investigated was the parameter α:

$$\dot{m}_c = \left[\alpha + (1-\alpha) \frac{m_c/m}{(m_c/m)_o} \right] (\dot{m}_c)_o$$

where $(\dot{m}_c)_o$ was a fixed datum production rate and $(m_c/m)_o$ was a fixed datum solids concentration. The measured values of solids concentration (m_c/m) taken once per working shift determined the production rate (rate of bagging of solids) until the next reading was taken. Fig. 4 shows the effects on supersaturation of this control strategy in response to a 10% increase in evaporation rate at $t = 0$. It can be shown that the penalty for reaching steady conditions earlier is a loss in particle size. The criterion for steady state was defined as the attainment of a production rate close to the desired value. As Fig. 4 shows, there are many ways of attaining this, with various values of supersaturation and size distribution. Many control strategies can be tested in this manner.

Another feature of this mode of operation, as described before, was the return of the overflow from a hydrocyclone to the crystalliser. This overflow contained a significant fraction of sub-100μm crystals. Tables 1 and 2 present the situation when 70% and 100% of the sub-100μm particles are recycled to the crystalliser. Solids concentration rises monotonically from the steady state at $t = 0$ while supersaturation and mean particle size fall in a similar manner. In this particular system no new

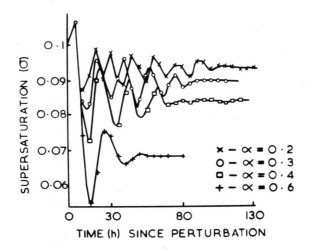

Fig. 4 The effects of control parameter (α)

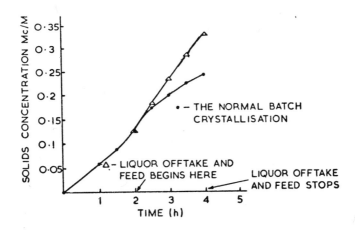

Fig. 5 Simulations of the batch strategy

Time t (h)	Solids Concentration m_c/m	Supersaturation σ	Mean Particle Size $\bar{d}$ (µm)
0	0.0703	0.1	214
2	0.0782	0.087	175
4	0.094	0.063	139
6	0.107	0.041	121
8	0.116	0.029	107
10	0.12	0.02	97
12	0.124	0.017	86
14	0.127	0.014	77
16	0.129	0.011	70

Table 1   The effect of returning all sub-100µm crystals to the crystalliser

Time t (h)	Solids Concentration m_c/m	Supersaturation σ	Mean Particle Size $\bar{d}$ (µm)
0	0.0703	0.1	214
2	0.075	0.092	191
4	0.085	0.076	166
6	0.097	0.058	151
8	0.106	0.044	141
10	0.112	0.035	131
12	0.116	0.029	121
14	0.12	0.024	111
16	0.122	0.021	102

Table 2   The effect of returning 70% of all sub-100µm crystals to the crystalliser

Temperature of Cooling Water	Production	Residual Supersaturation	Mean Particle Size
15°C	100%	0.49	$\bar{d}$
20°C	90.5%	0.44	$1.33\bar{d}$
25°C	81%	0.38	$1.63\bar{d}$
30°C	72.5%	0.31	$1.8\bar{d}$

Table 3   The effects of various cooling water temperatures

steady state occurs nor do limit cycles develop whose amplitude is within the practical range. Clearly in other systems with this mode of operation limit cycles can occur. Such a system has been demonstrated by Randolph (7) in a mode of operation similar to this plant. His observation that the effect of clear liquor offtake appeared to be more pronounced than could be explained in terms of its effect on the product classification function may be very relevant to this case study. The product classification which Randolph induced by altering the orientation of his removal port would itself have been time dependent, since hindered settling is a function of solids concentration which in his system was varying with time. Moreover it is clear from this study that this mode of operation is very sensitive to small changes in flow rates. It would in practice be very difficult to observe small changes in flow streams up to perhaps even 10% of the nominal rate.

Case 2 - Water-cooled Stirred Batch Crystalliser

The intention of this work was to establish the effectiveness of operating strategies which could increase the output from the batch. As a secondary aim the effects of variable temperature cooling water were to be established.

The process. The vessel was charged with hot saturated feed at 70°C and cooling commenced when the vessel was half-filled. The contents were cooled to 30°C then dropped into a slurry holding tank to await centrifugation. The time from initial feeding to dropping the tank contents was about 4 h.

The effects of varying the operating strategy. What was explored in this part of the work were the effects of removing clear liquor from the crystalliser during the batch cycle, and replacing this liquor with fresh feed, thus making the stage partly continuous. Fig. 5 presents a typical simulation of the effects on solids concentration of initiating clear liquor offtake and adding fresh feed at various times. Fig. 6 shows the experimental results obtained in laboratory equipment. The mathematical predictions were that up to about 50% increases in production might be expected to result from the strategy, with about a 10% loss in stage efficiency. Stage efficiency was defined as production rate divided by throughput of solute. The laboratory scale results, as indicated in Fig. 6 typically provided 30% increases in production. In the subsequent plant trial a 45% increase in production was obtained with 16% loss in stage efficiency which was in satisfactory agreement with the predictions. As shown in Table 3, the effects of variable cooling water temperature can also be predicted. At worst a 25% loss in production could be expected for cooling water at 30°C compared with 15°C. The effect on particle size was as expected in the sense that the greater initial rate associated with lower cooling water temperature caused supersaturation to reach a high maximum value at an earlier point in the cycle. Since $g/n \sim 1.5$ in this system, nucleation dominated the size distribution.

Case 3 - Cascade of Well-mixed Crystallisers in Series

The aim of this investigation was to establish the feasibility of a 50% increase in feed to the crystallising system.

The process. The crystallisers consisted of two well-mixed vacuum cooled crystallisers in series. The slurry from the second crystalliser passed to a thickening tank which was subsequently shown to act in a small capacity as a third crystalliser in series. Hot saturated feed solution was introduced continuously to the crystallisers which were operated at 70°C and 36°C respectively. The crystallisers were equipped with a barometric offtake leg for slurry removal and this was shown to be effective in providing representative removal.

Effect of a step change in evaporation. Considering the first crystalliser only, the effects of a 10% increase in evaporation rate on this type of operation can be compared with the effects shown before for Case 1. Table 4 shows the effects of the step change on solids concentration, supersaturation and mean particle size. There is clearly no significant change in crystallising conditions. The imposition of fluctuations on evaporation and feed rates similarly establishes that the crystalliser with representative slurry removal is more resistant to perturbations. It can also be shown that the resistance to perturbation is largely independent of the system solubility.

The predicted effects of a 50% increase in feed are shown in Table 5. The mean residence time for each crystalliser was about 5 h at the increased feed rate. While solids concentration and supersaturation became steady after only a few hours, the mean particle sizes require longer to stabilise. The predicted increase in production resulting from the step change, taking the solids concentration in vessel 2 as 0.208, is 47.5%. In the subsequent plant trial the increase in production was slightly greater due to the crystallisation which occurred in the thickening vessel.

CONCLUSIONS

As a result of these exercises one can have considerable confidence in the applicability of the unsteady state model to plant problems. In Case 1 it has been demonstrated that the mode of operation can have an important effect on the stability of

TRANSIENT BEHAVIOUR IN CRYSTALLIZATION

Solids Concentration m_c/m	Supersat^n σ	Mean Dia $\bar{d}$ (μm)	Time t (h)	Evap^n Rate m_W (kg/h)
0.2	0.05	139	0	1.38×10^2
0.2007	0.0503	140	5	1.52×10^2
0.201	0.0502	140	10	1.52×10^2
0.202	0.0501	139	15	1.52×10^2
0.202	0.05	139	20	1.52×10^2

Table 4 The effects of 10% increase in evaporation on a fully mixed continuous crystallisation with slurry removal

Time t (h)	Vessel No	Super-saturation σ	Solids Concentration m_c/m	Mean Particle Size $\bar{d}$ (μm)
0	1	5.8	0.07	170
0	2	14.3	0.211	117
1	1	7.3	0.067	170
1	2	18.7	0.208	118
2	1	7.5	0.067	169
2	2	18.7	0.208	117
3	1	7.5	0.067	169
3	2	18.6	0.208	115
4	1	7.5	0.067	167
4	2	18.5	0.208	114
5	1	7.5	0.067	166
5	2	18.4	0.208	113
6	1	7.5	0.067	165
6	2	18.4	0.208	113

Table 5 The effects of 50% increase in feed rate on two crystallisers in series

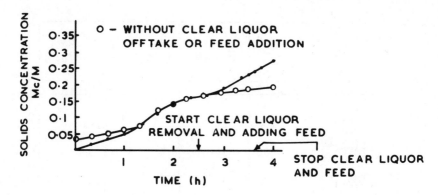

Fig. 6 Laboratory experiments on batch crystallisation

crystallisation. The effects of clear liquor removal and product classifications have been demonstrated at laboratory and plant scale and one type of control parameter has been examined. In Case 2 the model has been applied to batch crystallisation and in Case 3 the stability of a cascade of continuous crystallisers has been explored.

It can be concluded:

1. An unsteady state model is a necessary reinforcement for steady state mathematical models for design, uprating or troubleshooting exercises on crystallisation plants.

2. The mode of operation, particularly one which involves clear liquor removal, can reduce the stability.

3. At plant scale, steady state conditions of crystallisation are not encountered.

ACKNOWLEDGEMENTS

The author wishes to acknowledge the help of his colleagues at the Corporate Laboratory, particularly C. Ramshaw and I.B. Parker, who was responsible for the mathematical development of the model.

NOMENCLATURE

c	index on solids concentration	
d	diameter of crystal	m
E	activation energy for growth of crystals	kJ/kmol
g	index on supersaturation for growth	
k,K	constants	

m	mass	kg
n	population density function or index on supersaturation for nucleation	m^{-1}
N	numbers of crystals	
q	numbers of 'crash nuclei'	
R	gas constant, 8.314	kJ/kmol K
t	time	h or s
$\bar{t}$	mean residence time	
T	temperature	K
U	integration factor	
V	volume of crystalliser contents	m^3
σ	relative supersaturation	
ρ	density	kg/m^3

Subscripts:
- c   referring to crystal
- eq   referring to equilibrium
- f   feed
- G   growth
- i   the ith size interval
- m   maximum
- ml   mother liquor
- N   nucleation
- o   temperature T_o, or zeroth size interval
- sus   suspension or slurry
- v   volume
- w   water or solvent
- x   dissolved solute

REFERENCES

1. Randolph, A.D. and Larson, M.A., Theory of Particulate Processes, Academic Press, New York, 1971.
2. Mullin, J.W., Crystallisation, (2nd Ed.) Butterworths, London, 1972.
3. Larson, M.A. and Randolph, A.D., Crystallisation from Solution and Melts, AIChE Symposium Series, 95 65 (1969) 1.
4. Ramshaw, C. and Parker, I.B., Trans. Instn Chem. Engrs, 51 (1973) 82.
5. Sherwin, M.B., Ph.D. Thesis, The City University of New York (1967).
6. Randolph, A.D., Beer, G.L. and Keener, J.P., AIChEJ, 19 6 (1973) 1140.
7. Randolph, A.D., Ottens, E.P. and Metchis, S.G., Annual AIChE meeting, Washington, D.C., Dec. 1974, paper 100B.

APPENDIX

The Solution to the Unsteady State Equations

The physical assumptions are:

(a) Growth rate of crystals is independent of crystal diameter and is a function of temperature and supersaturation only.

(b) Nucleation rate is a function of supersaturation and is linearly dependent on solids concentration (ie c = 1 in equation 2).

(c) Temperature is a function of time only.

(d) Crystals are assumed to be spherical.

(e) "Crash" or excessive nucleation occurs at some fixed supersaturation (σ_m) in the way described later.

(f) The crystalliser is a completely mixed vessel in which no gradients of any physical quantity exist.

(g) Product from the crystalliser is unclassified.

(h) Crystal size distribution is described by a histogram of numbers against crystal diameter.

Fig. A1 is a line diagram of the crystalliser.

Total mass balance

$$\dot{m} = \dot{m}_f - \dot{m}_w - \dot{m}_{ml} - \dot{m}_c - \dot{m}_{sus} \qquad (3)$$

Solvent balance

$$\dot{w} = \dot{m}_f (1 - c_f) - \dot{m}_w - \dot{m}_{ml} \frac{m_w}{m_x + m_w} - \dot{m}_{sus} \frac{m_w}{m} \qquad (4)$$

where $m_w + m_x + m_c = m$ \qquad (5)

$$\sigma = \frac{m_x/m_w - c_{eq}/1-c_{eq}}{c_{eq}/1-c_{eq}} \qquad (6)$$

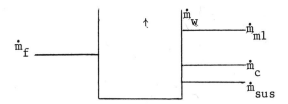

Fig. A1 The Continuous Crystalliser

From equation (1), letting T_o be temperature of crystalliser

$$\dot{d} = k_G \, \sigma^g \tag{7}$$

and, since we are considering that "crash" nucleation occurs above a maximum supersaturation value, equation (2) becomes

$$\dot{N}/V = k_N \, \sigma^n \, (m_c/m)^c + q \tag{8}$$

where q defines the nucleation occurring above σ_m:

$$q = 0, \quad \sigma < \sigma_m$$

$$q = 6m_w \frac{c_{eq}}{1-c_{eq}} \, (\sigma-\sigma_m)/\pi k_v \rho_c \, d_N^3, \quad \sigma > \sigma_m \tag{9}$$

Expression (9) is obtained by dividing the excess of solute over σ_m by the volume of an artificial nucleus, $\pi/6 \, d_N^3 \, k_v$, where k_v is the volume shape factor. This expression is not considered to be other than a mathematical device which allows crash nucleation to be accounted for.

A crystal balance of numbers over those crystals whose diameters lie between d_i and d_{i+1}

$$\dot{N}_i = -N_i/\bar{t} + \dot{d}\,(n_i - n_{i+1}) \quad i = 0, 1 \ldots \nu \tag{10}$$

where $\dfrac{1}{\bar{t}} = \dfrac{\dot{m}_{sus}}{m} + \dfrac{\dot{m}_c}{m_c} \tag{11}$

The population density function is taken to be constant between d_i and d_{i+1} and is equal to n_{i+1}.

Hence, $n_{i+1} = \dfrac{N_i}{\delta_i} \quad i = 0, 1 \ldots \nu \tag{12}$

It follows from this that dn_{i+1} is the number leaving the total N_i per unit time.

Where $d_o = 0$, dn_o is the number of crystals nucleating

$$\therefore dn_o = \dot{N} \tag{13}$$

The total mass of crystals is related to the crystal number by:

$$m_c = \sum_i \frac{\pi}{6} k_v \rho_c \frac{N_i}{\delta_i} \int_{d_i}^{d_i+\delta_i} u^3 du + m_\infty$$

$$\therefore m_c = \frac{\pi}{24} k_v \rho_c \sum_{i=0}^{\nu-1} N_i (d_i + d_{i+1})(d_i^2 + d_i d_{i+1} + d_{i+1}^2) + m_\infty$$

where $d_{i+1} = d_i + \delta_i$ \hfill (14)

The term m_∞ is the contribution to the total mass from the last diameter interval. If the range of particle sizes considered is large enough, a good approximation will be $m_\infty = 0$. The above equations provide a self-consistent mathematical description of a dynamic crystalliser - $12 + 2\nu$ equations and $12 + 2\nu$ unknowns: m, m_w, m_x, m_c, m_∞, σ, $\dot{d}$, $\dot{N}$, q, N_i ($i = 0, 1...\nu$), n_i ($i = 0, 1....\nu+1$). As data the formulation requires the values of: $\dot{m}_f$, $\dot{m}_w$, $\dot{m}_{ml}$, $\dot{m}_c$, $\dot{m}_{sus}$, $\dot{m}_f (1-c_f)$, c_{eq}, k_G, g, k_N, n, σm, ρ_c, d_N, and d_i ($i = 0, 1...\nu$). These data should be either physical constants or known functions of time. The total set consists of a number ($\nu+6$) of differential equations and ($\nu+6$) subsidiary algebraic equations. After rearrangement of the equations a stable algorithm was used to obtain a single parameter iteration at each time step.

The numerical approximations are:

$$m = m^1 + dt\, (\dot{m}_f - \dot{m}_w - \dot{m}_{ml} - \dot{m}_c - \dot{m}_{sus}) \tag{15}$$

$$m_w = \{m_w^1 + dt\, (\dot{m}_f [1-c_f] - \dot{m}_w)\}/\{1 + dt\, (\frac{\dot{m}_{ml}}{m-m_c} + \frac{\dot{m}_{sus}}{m})\} \tag{16}$$

$$N_i = \{N_i^1 + \dot{d}\, dt\, N_{i-1}/\delta_{i-1}\}/\{1 + \dot{d}\, dt + dt/\tau\} \quad i = 0, 1...\nu \tag{17}$$

where the prime denotes the value of that variable at the start of the time step.

All quantities can be found once m_c is known. The Newton Raphson algorithm is used to converge the m_c eq. 14. The cycle of the iteration is as follows:

1) Obtain m from eq. 15.

2) Guess m_c at next time step by linear extrapolation.

3) Obtain m_w from eq. 16.

4) Obtain m_x from eq. 5, σ from eq. 6, $\dot{d}$ from eq. 7, q from eq. 9, $\dot{N}$ from eq. 8, N_i from eq. 17.

Iterate on the guess value of m_c using eq. 14 with $m_\infty = 0$ as a convergence criterion.

STABILITY AND DYNAMIC BEHAVIOUR OF CRYSTALLIZERS

B.G.M. DE LEER, A. KONING, AND E.J. DE JONG

Laboratory for Chemical Equipment, Delft University of Technology, Mekelweg 2, Delft, The Netherlands

INTRODUCTION

Mathematical models can be used to analyse and predict the dynamic behaviour of the crystal size distribution (CSD) in mixed suspension crystallizers of several configurations. In the last ten years much attention has been given to the analysis of absolute stability limits of crystallizer models. These limits can be obtained from the characteristic equation, which is derived from the normalized and linearized dynamic equations of the model. The limits of the stability regions are presented as critical values of the nucleation/growth rate exponent $i = d(\log B^o)/d(\log G)$ versus the dimensionless parameters of the used model.

Randolph and Beer (1) have determined the absolute stability limits of the R-Z model, of which the dimensionless parameters are:

$$R; Z; \quad x_f = \frac{l_f}{\overline{G\tau}}; \quad x_p = \frac{l_p}{\overline{G\tau}}; \quad \beta; \text{ i and j}$$

In the literature it is reported (2) that crystallizers have an oscillatory behaviour, even when the values of the relevant system parameters are within the limits of the stability region. It is therefore important to know what the dynamic behaviour of the model will be within the limits of the stability regions.

In this paper the characteristic equation of the R-Z model derived by Randolph and Beer (1) is used to determine the location of the poles in the complex plane as function of the dimensionless parameters of the model. Each pole has an effect on the stability and the frequency of the state variables of the

model. From the location of all the poles conclusions can be drawn with regard to:
- the stability and the eigen frequency of the uncoupled state variables in the stable regions in relation to the parameter values of the model.
- the question, which poles mainly determine the dynamic behaviour of the state variables.

RESTRICTIONS OF THE R-Z MODEL

Assuming rapid growth kinetics (= class-II) and magma dependent secondary nucleation, the R-Z model gives an idealized description of the CSD in mixed suspension crystallizers with product classification and with fines removal and/or fines destruction. The fines in the vessel, below a certain maximum fines size l_f, are assumed to be removed with (R-1) times the M.S.M.P.R. removal rate. These fines can be dissolved and recycled (fines destruction, $\beta \neq 0$), or permanently removed (fines removal, $\beta = 0$). The parameter β is the ratio of fines mass dissolved and recycled to the external production rate. All the crystals in the vessel, above a certain minimum classification size l_p, are removed with Z times the M.S.M.P.R. removal rate (product classification). The fines and crystals in the vessel with $1 \leqslant l_p$ are removed with the M.S.M.P.R. removal rate. Fig. 1 shows the dimensionless population density plot in a steady state R-Z crystallizer.

Other assumptions are:
1) the growth rate of the crystals does not depend on the crystal size and is proportional to the supersaturation in the vessel.
2) the nucleation rate is proportional to the growth rate of the crystals raised to the ith power and the total mass of the crystals in the vessel raised to the jth power. The magnitude of the parameter i is usually with inorganic salt systems between 0 and 5. The magnitude of the parameter j as found in literature is between 0 and 2. When crystal wall collisions are the dominating source of secondary nucleation the value of j is 1 [Ottens (3)].
3) isothermal crystallization.
4) the crystallizer working volume is constant.
5) the particles are regular solids, which maintain their shape upon growth.
6) the nuclei size is so small it can be taken equal to zero.
7) the feed is a clear solution.
8) the crystal density is taken to be constant.
9) no crystal fracture and attrition.

STABILITY AND DYNAMIC BEHAVIOUR OF CRYSTALLIZERS

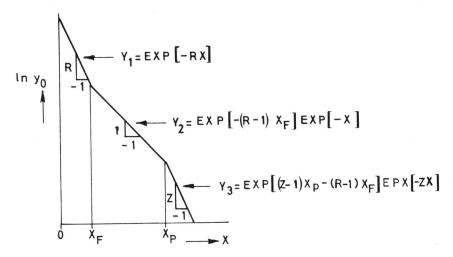

Figure 1: *The dimensionless population density plot in the steady state R-Z crystallizer*

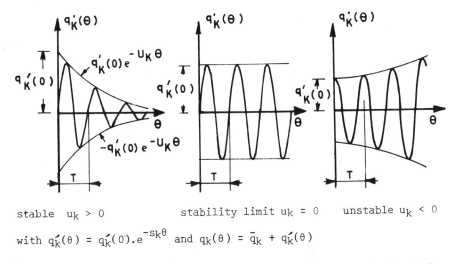

stable $u_k > 0$ stability limit $u_k = 0$ unstable $u_k < 0$

with $q'_k(\theta) = q'_k(0) \cdot e^{-s_k \theta}$ and $q_k(\theta) = \bar{q}_k + q'_k(\theta)$

Figure 2: *The stability (u_k) and eigen frequency (v_k) of the kth uncoupled state variable*

THE LINEARIZED STABILITY ANALYSIS

The characteristic equation of the class-II R-Z model is found by defining the eigen values of the normalized dynamic equations, which are linearized by means of standard perturbation techniques. After substitution of $s = u + \sqrt{-1}\ v$, the real part of the characteristic equation can be separated from the imaginary one. Both parts are functions of u, v, and the dimensionless parameters of the model. The poles of the model (s_k) are that u_k and v_k values for one set of fixed parameter values, for which the real as well as the purely imaginary part of the characteristic equation is equal to zero.

The characteristic equation of the full R-Z model is a ninth degree equation in s, so that nine poles ($s_k = u_k + \sqrt{-1}\ v_k$ with k = 1, 2, .., 9) can be found in the complex plane for one set of fixed parameter values. If one of the poles $s_k = u_k + \sqrt{-1}\ v_k$ for any given k is complex, then also a complex conjugate pole $s_k^* = u_k - \sqrt{-1}\ v_k$ can be found. The model will be stable, if the real part u_k of the kth pole is larger than zero [$u_k = \text{Re}(s_k) > 0$] with k = 1,2,3,..,9 and unstable, if the real part u_k is smaller than zero [$u_k = \text{Re}(s_k) < 0$] for any given k. The transition obviously occurs, when the real part u_k of the kth pole is equal to zero [$u_k = \text{Re}(s_k) = 0$] for any given k.

Fig. 2 shows the response of the kth uncoupled state variable caused by the kth pole after a small disturbance in the initial condition. The stability of the response is determined by the real part u_k of the kth pole and the vibration time T by the imaginary part v_k of the kth pole according to $T = 2\pi/v_k$.

With the aid of a computer program the poles were found by searching the complex plane with small u- and v-steps for one set of fixed parameter values. If u_1 is the smallest real part of all the poles that can be found in the complex plane, the pole s_1 is called 'dominant' if for the real part u_k of all the other poles $u_k > u_1 + 3.22$. This means, that at $\theta = 1$ each value of the enwrapping exponential functions, determined by the real parts of the poles s_k with k = 2,3,..,9, is less than 4% of the mentioned value, which is determined by the real part of the dominant pole s_1.

$$\left[e^{-u_k \theta} \leqslant \frac{4}{100} e^{-u_1 \theta} \text{ at } \theta = 1 \text{ with } k = 2,3,\ldots,9 \right]$$

Our experience is, that the sum of these influences at $\theta = 1$ is less than 10% of the value of the enwrapping exponential function, which is determined by s_1

$$\left[\sum_{K=2}^{9} e^{-u_k \theta} \leqslant \frac{1}{10} e^{-u_1 \theta} \text{ at } \theta = 1 \right].$$

RESULTS OF THE LINEARIZED STABILITY ANALYSIS

Fig. 3 shows the effect of the parameter z, x_p and i on the stability (u) and eigen frequency (v) of the uncoupled state variables (no fines removal, $R = 1$). As can be seen (Fig. 3.1), the real part u_1 of the complex and complex conjugate pole 1 is only influenced by z somewhere in between $z = 1$ and $z = 2$ for the given fixed parameter values. The real parts of the other poles rapidly increase with increasing values of z. This implicates that after a disturbance in the initial conditions the influence of these poles on the dynamic behaviour of the state variables decreases with increasing values of z.

The eigen frequency v however (Fig. 3.2), increases with increasing values of z, so that the vibration time $T = 2\pi/v$ of the uncoupled state variables decreases (faster fluctuations). The real part u_1 of the complex and complex conjugate pole 1 does not change, or increases with decreasing values of x_p for the given fixed parameter values (Fig. 3.3 and 3.5).

Poles with a real part above the 4% limit ($u_k > u_1 + 3.22$) are not given. Their contribution to the dynamic behaviour of the state variables in only of interest between $\theta = 0$ and $\theta = 1$. Rough calculations indicate, that the real part u as well as the imaginary part v of these poles decrease with increasing values of x_p (shaded areas; Fig. 3.3 and 3.4).

The stability and the eigen frequency of the uncoupled state variables decrease with increasing values of i (Fig. 3.7, 3.8, and 3.9). All the results, which are related to the real part u_1 and the imaginary part v_1 of the complex and complex conjugate pole 1, have been worked up in three dimensional figures (Fig. 6 and 8), in which u_1 and v_1 are given as function of the parameters z, x_p, and i for $j = 1$ (and no fines removal, $R = 1$). The absolute stability limits, as given in the figures, have been calculated by Randolph and Beer (1).

Fig. 4 and 5 show the effect of fines destruction ($\beta \neq 0$) and fines removal ($\beta = 0$) on the stability (u) and the eigen frequency (v) of the uncoupled state variables for the given fixed parameter values (no product classification, $z = 1$). As can be seen from these figures, the stability and the eigen frequency are hardly influenced by the parameter R for $\beta \neq 0$ and $\beta = 0$ (Fig. 4.1, 4.2, 5.1, and 5.2). For the given fixed parameter values, they are stronger influenced by x_f for $\beta \neq 0$ and $\beta = 0$ (Fig. 4.3, 4.4, 5.3, and 5.4). In that case however, the fines mass removed ($\beta = 0$), or the fines mass dissolved and recycled ($\beta \neq 0$) per unit of time strongly increase with increasing values of x_f.

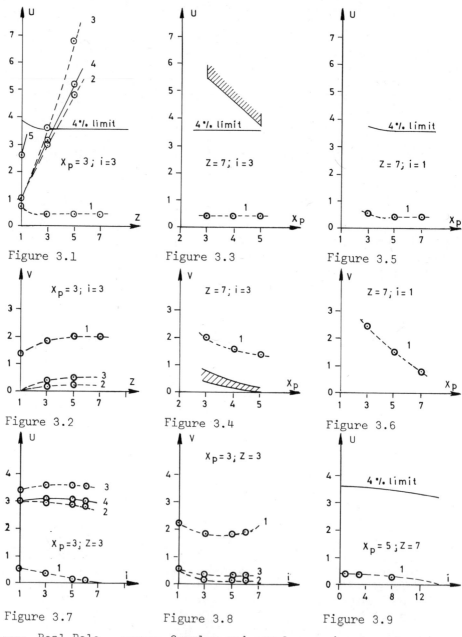

Figure 3: The effect of product classification of the stability (u) and the eigen frequency (v) of the uncoupled state variables.

STABILITY AND DYNAMIC BEHAVIOUR OF CRYSTALLIZERS

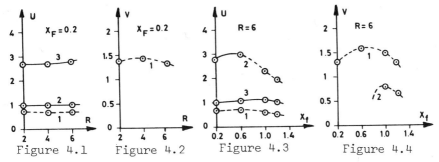

Figure 4.1   Figure 4.2   Figure 4.3   Figure 4.4

Parameters: R and x_f as given in the figures; $\beta \neq 0$; no product classification ($z = 1$); i = 3, j

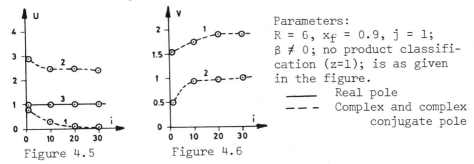

Figure 4.5   Figure 4.6

Parameters:
R = 6, x_f = 0.9, j = 1;
$\beta \neq 0$; no product classification ($z=1$); is as given in the figure.
——— Real pole
- - - Complex and complex conjugate pole

Figure 4: *The effect of fines destruction ($\beta \neq 0$) on the stability (u) and the eigen frequency (v) of the uncoupled state variables.*

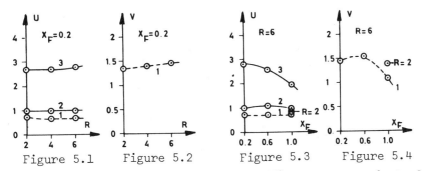

Figure 5.1   Figure 5.2   Figure 5.3   Figure 5.4

Parameters: R and x_f as given in the figures; no product classification ($z = 1$); j = 1, i = 3

Figure 5: *The effect of fines removal ($\beta = 0$) on the stability (u) and the eigen frequency (v) of the uncoupled state variables.*

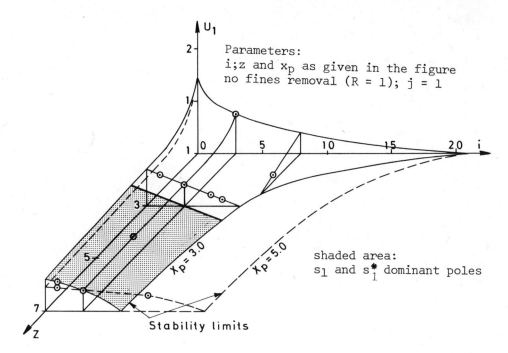

Figure 6: The effect of product classification on u_1

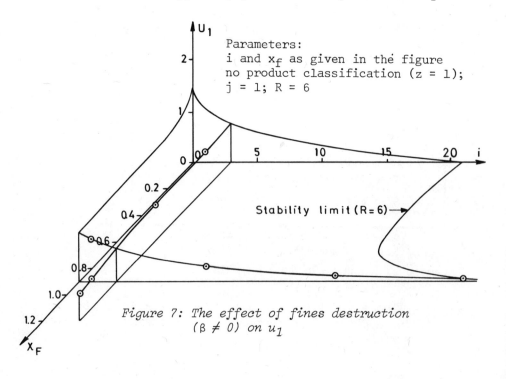

Figure 7: The effect of fines destruction ($\beta \neq 0$) on u_1

STABILITY AND DYNAMIC BEHAVIOUR OF CRYSTALLIZERS

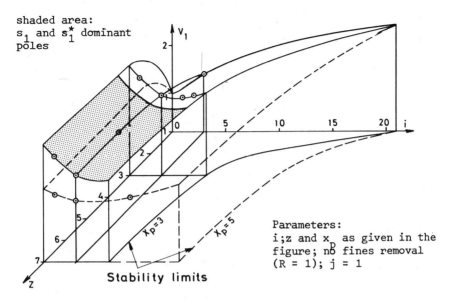

Figure 8 : The effect of product classification on v_1

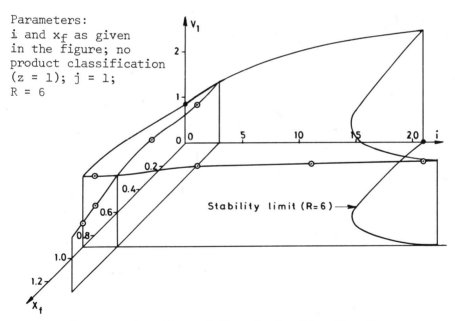

Figure 9: The effect of fines destruction ($\beta \neq 0$) on v_1

Further it can be seen, that the stability and the vibration time, $T = 2\pi/v$ decrease with increasing values of i for the given fixed parameter values (Fig. 4.5 and 4.6). Somewhere in between $x_f = 0.6$ and $x_f = 0.9$ for $R = 6$ and $\beta \neq 0$ the real pole 2 becomes a complex and complex conjugate pole, of which the real part decreases and the imaginary part increases (Fig. 4.3). Again the results for fines destruction ($\beta \neq 0$) related to u_1 and v_1 have been worked up in three dimensional figures (Fig. 7 and 9), in which u_1 and v_1 are given for $j = 1$ as function of the parameters R, x_f, and i (no product classification, $z = 1$).

CONCLUSIONS

Product classification:
1. For certain fixed i values the stability (u) of the uncouped state variables remains constant or even increases with increasing values of z and decreasing values of x_p, although the absolute stability limits in those cases decrease.
2. At the same time however, the vibration time $T = 2\pi/v$ of the uncoupled state variables decreases (faster fluctuations).
3. Using the 4% criterium, the dynamic behaviour of the state variables is mainly influenced by one dominating complex and complex conjugate pole in some parts of the stable region ($z \geq 3.5$; $x_p \leq 5$).
4. For small z values the stability (u_1) of the uncoupled state variable 1 is small and rather insensible to changes of the parameter i at the stability limit.
5. For $x_p \geq 5$ and larger values of z the stability (u_1) of the uncoupled state variable 1 is rather insensible to changes of the parameter i in the stable region and rather sensible at the stability limits.

Fines removal ($\beta = 0$) and fines destruction ($\beta \neq 0$):
1. For certain fixed i values it is possible that the stability (u) of the uncoupled state variables decreases with increasing values of R and x_f, although the absolute stability limits of the model in thoses cases increase.
2. The vibration time of the uncoupled state variables increases (slower fluctuations) with increasing values of the parameters of the fines removal (R, x_f, β) after an initial decrease.
3. At the stability limits, the stability (u_1) of the uncoupled state variable 1 is small and rather insensible to changes of the parameter i.

DISCUSSION

In this work the eigen values (poles) of the model were determined as function of the dimensionless parameters, from which conclusions can be drawn with regard to the stability and eigen frequency of the uncoupled state variables. These conclusions are more relevant and predictive than conclusions based on the absolute stability limits. A state variable however, is a well defined linear combination of the uncoupled state variables, especially when there is not one dominant pole.

With linearized analysis, which are only valid for small changes in the state variables around the equilibrium state, it is not possible to examine the effects of disturbances in the input conditions (feed concentration, residence time of liquid and of crystals) and/or the parameter values of the model. Simulations of the full nonlinear dynamic equations of the R-Z model have to be used to examine these effects and to verify the results of the linearized stability analysis. The nonlinear dynamic equations can also be used to determine the new equilibrium state of the CSD and with that the new equilibrium of the state variables with respect to the M.S.M.P.R. equilibrium state, after permanent changes in the input conditions and/or parameter values of the model. At the same time the dynamic behaviour of the state variables, caused by these permanent disturbances, can be examined.

NOMENCLATURE

B^o	nucleation rate at $l = 0$
G	growth rate of the crystals
$\bar{G}$	growth rate of the crystals in the steady state
i	nucleation/growth rate exponent in the equation $B^o = k_n M_T^j G^i$
j	nucleation/magma density exponent
k	any given constant
l	crystal size
n_o	population density in the steady state
n_o^o	population density at $l = 0$ in the steady state
$q'_k(\theta)$	dynamic behaviour of the kth uncoupled state variable around the equilibrium state $\bar{q}_k$
$\bar{q}_k$	the equilibrium state of the kth uncoupled state variable
$q_k(\theta)$	the dynamic behaviour of the kth uncoupled state variable $= [\bar{q}_k + q'_k(\theta)]$
R	factor describing the decreased residence time of fines for $x < x_f$ (Fig. 1)
s	complex variable
s_k	k^{th} complex pole
s_k^*	k^{th} complex conjugate pole
T	vibration time
u	real part of the complex variable s
u_k	real part of the kth complex and complex conjugate pole
v	imaginary part of the complex variable s
v_k	imaginary part of the kth complex pole
x	dimensionless crystal size ($= l/\bar{G}\bar{\tau}$)
x_f	maximum dimensionless fines size with decreased residence time

x_p	minimum dimensionless crystal size with in- or decreased residence time
y	dimensionless population density ($= n/n_0^o$)
y_0	dimensionless population density in the steady state
y_1	dimensionless population density for $0 \leqslant x \leqslant x_f$
y_2	dimensionless population density for $x_f \leqslant x \leqslant x_p$
y_3	dimensionless population density for $x \geqslant x_p$
z	factor describing the in- or decreased residence time of crystals for $x \geqslant x_p$
Re( )	real part of ( )
Im( )	imaginary part of ( )
β	ratio of fines mass dissolved and recycled to the external production rate
$\bar{\tau}$	mean residence time of the crystals in the M.S.M.P.R. crystallizer in the stady state
θ	dimensionless time ($= t/\bar{\tau}$)

REFERENCES

1. Randolph,A.D., Beer,G.L., and Keener,J.P.; A.I.Ch.E. Journal, 19 (1973), 1140-9.
2. Randolph,A.D., Ottens,E.P.K., and Metchis,S.G.; to be published.
3. Ottens,E.P.K.; Dissertation, Delft University of Technology, May 1973.

THE IMPORTANCE OF CLASSIFICATION IN WELL-MIXED CRYSTALLIZERS

A.H. JANSE AND E.J. DE JONG

Laboratory for Chemical Equipment, Delft University of Technology, Mekelweg 2, Delft, The Netherlands

INTRODUCTION

When a slurry is removed from a crystallizer, classification or non ideal product removal can occur. Classification means that the concentration of particles in the product stream and the average concentration in the crystallizer are different. Recently more information about the classification phenomenon has become available (1-4). It was realized that this phenomenon is very important for the correct interpretation of measured population density curves as well as for the particle size control in industrial crystallizers. Therefore some so-called classification experiments using glass beads and water have been carried out in a 42.5 l crystallizer. In this paper we present some of our results and discuss the importance of the classification phenomenon.

EXPERIMENTAL

From the literature (1,2,4,5) it can be concluded that besides particle and solution properties, the homogenity of the suspension and the velocities in and near the discharge determine the amount of non ideal product removal. Therefore in the experiments we have varied:

- size of the glass beads (70, 230, 710 and 1260μm, the corresponding terminal velocities are 0.005, 0.035, 0.13 and 0.22m/s)

- stirrer speed (1000, 1250 rpm)
- length of insertion of the discharge (10,14--30, 34 mm)
- velocity in the discharge (from 0.57 to 1.05 m/s)

One of our aims was to obtain an ideal product removal. Therefore the fluid velocity at the place of the discharge has been measured (see Fig. 1). With increasing distance from the wall the fluid velocity increases first to a maximum value at 10 mm and then decreases continuously. To obtain an ideal product removal we have located the discharge at the place of the maximum velocity (length of insertion 10mm). From Fig. 1 the average velocity at the place of the discharge opening (at 1000 rpm) can be determined: 0.87 m/s. When the velocity in the discharge is the same as the average fluid velocity the fluid is isokinetically removed. In our experiments we have varied the velocity in the discharge from 0.7 to 1.2 of the average fluid velocity.

The experiments were carried out in a 42.5 l stainless steel crystallizer. The closed crystallizer (inside diameter 395 mm) with a dished bottom was provided with 6 baffles (width 40 mm) and a cooling coil, located just inside the baffles. Agitation was achieved by a three blade propeller stirrer (diameter 151.6mm; power number 0.13) located 170 mm above the bottom of the crystallizer at equal distances from two baffles. The discharge pipe (made of glass) was placed parallel to the direction of the fluid velocity in the vessel. Most experiments were carried out with a discharge with an inside diameter of 6.7 mm.

After filling the crystallizer with water about 2 kg of dry glass beads of known size was added to the crystallizer. At the desired conditions a known flow of water was pumped into the crystallizer for a period of 25 of 40 s depending on the discharge velocity. As the volume was kept constant slurry was discharged from the crystallizer. During 10 s which were used to obtain steady state flow conditions, the removed slurry was collected in beaker no. 1. Then we took a sample of about 500 gram (water plus glass beads) in beaker no. 2. After that we changed to the third beaker and immediately stopped the flow through the crystallizer. This procedure was repeated several times.

After drying the glass beads, the concentration of the particles in the product stream c_e and the average concentration in the crystallizer c_c (expressed in kg/m^3 water) were calculated. The ratio of the two concentrations is the separation coefficient $\lambda(L)$:

$$\lambda(L) = c_c/c_e \tag{1}$$

Preferential withdrawal of the particles occurs when the separation coefficient in smaller than unity. The particles stay longer in the crystallizer when the separation coefficient is larger than unity.

RESULTS

The results of the glass beads experiments are presented in four different ways:
- the separation coefficient versus the discharge velocity, with the particle size as parameter (Fig. 2).
- separation coefficient versus the discharge velocity, with the length of insertion as parameter (Fig. 3).
- separation coefficient versus length of insertion (Fig. 4).
- separation coefficient versus particle size (Fig. 5).

The accuracy of the separation coefficient is about 3%, the accuracy of the discharge velocity about 2%.
From Fig. 2 it can be seen that a representative product removal for the 70 and 320µm particles is achieved at a discharge velocity of 0.83 m/s. This value agrees within 5% with the expected velocity based on the measured fluid velocity. So it can be concluded that (under the stated conditions) isokinetic withdrawal of glass beads smaller than 230µm results in an ideal product removal.

For the larger particles the separation coefficient is smaller than unity, indicating that preferential withdrawal occurs. Although we work isokinetic at 0.83 m/s classification occurs. To transport particles in a vertical pipe the average fluid velocity has to be about 5 times the terminal velocity of the particles. For particles smaller than 710µm the transportation through the discharge is assured.

As the separation coefficient is about 0.7 (L = 710µm and 10 mm length of insertion), near the discharge a concentration of 1.4 times the average concentration in the crystallizer is expected. To obtain more insight in the non homogenity of the suspension, we increased the stirrer speed and the length of insertion.

STIRRER SPEED

The non homogenity of the suspension can be caused by centrifugal forces acting on the particles when the downward flow of the particles changes in an upward flow. According to this mechanism a higher stirrer speed gives larger centrifugal forces and therefore a higher concentration near the wall. This means that for small length of insertion the concentration in the product stream will be higher and the separation coefficient lower.

It is also possible that the high concentration near the wall is caused by the velocity profile. At low velocities the chance that the particles will be transported to the upper part of the crystallizer is small.

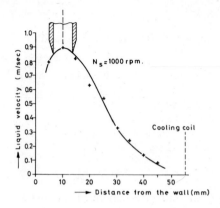

FIG. 1 Velocity profile near the discharge.

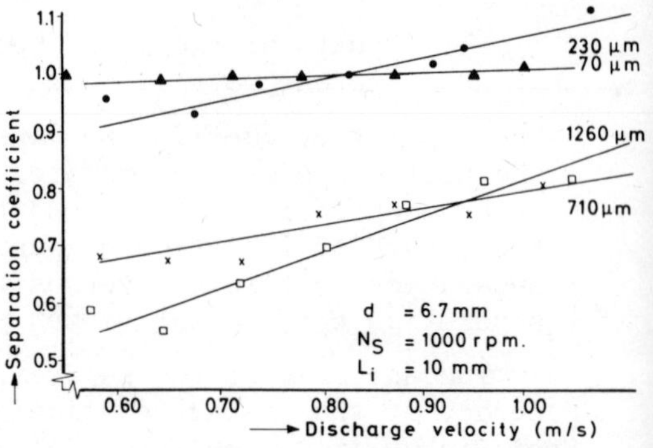

FIG. 2 The influence of particle size and discharge velocity on the separation coefficient.

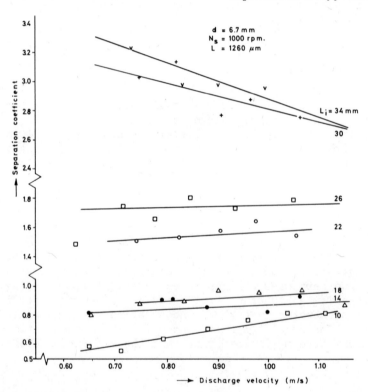

FIG. 3 The influence of the length of insertion on the separation coefficient.

However at high velocities (more close to the wall) this chance is much larger which means that the concentration of particles in the upward flow will be larger close to the wall. In this case a higher stirrer speed gives a lower concentration near the discharge due to the higher velocities and therefore higher separation coefficients. As the separation coefficient (at 10 mm length of insertion) increases with increasing stirrer speed, it can be concluded that the second mechanism predominates.

We have calculated the minimum stirrer speed to suspend 1 mm glass beads. According to Zwietering (6) the minimum stirrer speed (off bottom) is 600 rpm according to Kolar (7) 1200 rpm (complete homogeneity). Although the circumstances are not fully comparable (cooling coil, dished bottom) the calculated values will give a good approximation of the minimum stirrer speeds. So at 1000 and 1250 rpm a more or less homogeneous suspension should be expected. However experiments at 1250 rpm show that the separation coefficient is still considerable smaller than unity. It ranges from 0.60 to 0.85 indicating that the suspension was still not homogeneous.

LENGTH OF INSERTION

As the concentration near the wall is relatively high an increase in length of insertion has to give smaller concentrations near the discharge. So a larger separation coefficient is expected. This effect becomes more pronounced because the fluid velocity decreases with increasing length of insertion (Fig. 1). From Fig. 3 it appears that with increasing length of insertion the separation coefficient of 1260µm glass beads increases indeed. At small length of insertion the separation coefficient increases with the discharge velocity:

$$\lambda(L) \div v_d^{+0.7} \text{ for } L_i = 10 \text{ mm} \qquad (2a)$$

In this situation the fluid velocity and the velocity in the discharge are the same within 25%. The increase in separation coefficient can be explained by the non isokinetic withdrawal of particles with finite inertia.

At larger length of insertion the separation coefficient decreases with increasing discharge velocity:

$$\lambda(L) \div v_d^{-0.34} \text{ for } L_i = 34 \text{ mm} \qquad (2b)$$

Under this condition the ratio of the discharge velocity and the fluid velocity ranges from 3 to 6. Due to the low velocity and high velocity in the discharge liquid will be sucked from all directions. Also liquid which has already passed the plane of the discharge opening. The higher the velocity in the discharge the more suspension will be sucked and the higher the concentration of particles in the product removal stream. Consequently a lower separation coefficient is obtained. We have observed a flow pattern of particles which supports the given explanation.

Elsewhere (3) we have presented results of classification experiments carried out in the same crystallizer but with a different discharge geometry. The reported influence of the discharge velocity (liquid residence time τ_1) on the separation coefficient can be explained by the above mentioned mechanism.

Rushton (5) reported positive values for the exponent from 250μm glass beads in a turbine agitated vessel. We have found positive as well as negative exponents. The value of the exponent is not constant, but depends on the size of the particles, the length of insertion of the discharge, the stirrer speed etc.

In Fig. 4 we have plotted the separation coefficient versus the length of insertion. The influence of the discharge velocity is rather small, therefore only the values for the minimum and maximum velocity have been plotted. The points are marked by "o" and "■". All other values lie between these extremes. This figure shows that the length of insertion has a tremendous effect on the separation coefficient. This means that the location of the discharge is very important. Fig. 4 also shows the change in mechanism for the 1260μm particles (at about 25 mm length of insertion) and for the 230μm particles (at about 33 mm).

CONSEQUENCES FOR THE SIZE DISTRIBUTION

In Fig. 5 the separation coefficient versus size for different lengths of insertion have been plotted. It appears that even the 70μm particles give separation coefficients ranging from 0.98 to 1.16. Through the points we have dotted 3 lines all at the same discharge velocity v_d = 0.79 m/s. They are an estimate of the relation between the separation coefficient and the particle size. For these lines we will show the consequences of classification for the crystal size distribution of a continous crystallizer. As the liquid residence times of all cases are the same (same velocity in the discharge) the production rates are also the smae. Due to different amounts of classification the size distributions will be different. To calculate the size distribution in the product stream the following assumptions were made:

- a steady state continuous crystallizer
- the growth rates of all crystals are equal and size independent
- the nucleation rate can be described by

$$B_o = k_N M_c^j G^i \qquad (3)$$

with i = 1.5 and j = 1

The calculated size distribution in the product stream are shown in Fig. 6. In Fig. 6a the straight line, marked 4, represents the case of an ideal product removal and acts as a reference situation. For case 1 the population density curve decreases faster due to preferential particle withdrawal. In case 3 the crystals stay longer in the crystallizer. Therefore more crystals reach the larger sizes. In this case the nuclei population density is higher due to a high crystal mass concentration in the crystallizer which is only partially compensated by a lower growth rate.

In Fig. 6b the corresponding mass distribution are shown. As can be seen classification changes the median size L_{50} and the coefficient of variation CV. In case 3 the median size is 250μm while in case 4 (ideal product removal) the median size is 175μm. The increase in median size in case of an ideal product removal the liquid residence time τ_1 has according to

$$L_{50} = K \cdot \tau_1^{\frac{i-1}{i+3}} \tag{4}$$

(see ref. 8) to increase with a factor 4.2, which means for a given crystallizer a decrease in production rate by a factor 4.2. In view of this the increase in mean size caused by classification with ($\lambda(L)>1$) is large. It shows that classification can be used to change the product properties.

DISCUSSION

The two factors which mainly determine the separation coefficient are:

- homogenity of the suspension in the crystallizer
- the non isokinetic withdrawal

Both factors are determined by among others the stirrer speed. In practice a lower stirrer speed will be established than in our experiments. So in practice classification can be more pronounced.

On the other hand substances with a much smaller difference between particle and liquid density than glass beads and water will give less classification. For instance a 0.5 mm potassium alum crystal (density 1760 kg/m^3) has a terminal velocity of about 0.035 m/s in a saturated solution of 25°C. Assuming that different solid-liquid systems can be compared on the basis of terminal velocities, the same separation coefficients for 0.23 mm glass beads and 0.5 mm potassium alum crystals can be expected. The crystallizer we have used, was provided with a cooling coil, which influences the velocity profile. In stirred vessels with out a cooling coil the tangential component of the fluid velocity is much larger. As the velocity profile is qualitatively the same similar phenomena can be expected.

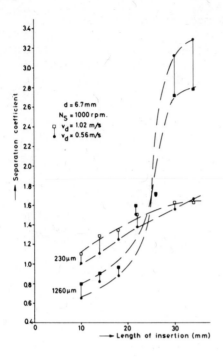
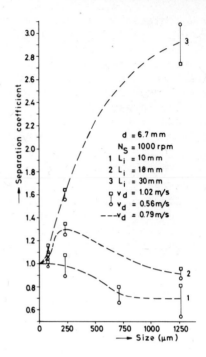

FIG. 4 The separation coefficient as a function of the length of insertion.

FIG. 5 The separation coefficient as a function of the particle size.

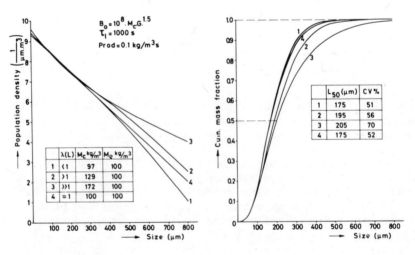

FIG. 6 The calculated size distributions in the product stream.

In a draft-tube agitated vessel the fluid velocity in the wall-tube annular space is nearly constant. If the velocity of the fluid is high enough to transport the crystals, suspension homogeneity is assured and it is possible to effect ideal product removal.

If ideal product removal is not achieved, classification will occur and the population density plots of a steady state continuous crystallizer will be curved. In such a situation accurate kinetic data cannot be obtained. As even the smallest particles (70μm) do not give an ideal product removal under all circumstances it is necessary to check whether classification occurs or not.

The occurrence of classification can be used to change the crystal product, for instance by a change in length of insertion. If only the length of insertion is changed, the liquid residence time, production rate and heat transfer rate remain the same. If there are no constraints (for instance a too high crystal mass concentration) classification can be used to control the crystal size distribution.

NOMENCLATURE

B_o	nucleation rate	$1/m^3 s$
c_c	particle concentration in crystallizer	$kg/m^3 H_2O$
c_e	particle concentration in product stream	$kg/m^3 H_2O$
d	inside diameter of the discharge	mm
G	growth rate	$\mu m/s$
i, j	exponents in eq. 3	-
L	particle size	μm
L_i	length of insertion	mm
M_c	crystal mass concentration in the crystallizer	kg/m^3
M_e	crystal mass concentration in the product stream	kg/m^3
N_s	stirrer speed	rpm
v_d	velocity in the discharge	m/s
$\lambda (L)$	separation coefficient (eq. 1)	-
τ_l	average liquid residence time	s

REFERENCES

1. Aeschbach, S. and Bourne, J.R.; Chem. Eng. J. $\underline{4}$ (1972), 234.
2. Bourne, J.R. and Sharma, R.N.; Paper presented at the first European Conf. on Mixing and Centr. Separation 9-11 Sept. 1974 Cambridge.
3. Janse, A.H. and de Jong, E.J.; Preprint of the G.V.C./A.I.Ch.E. Joint Meeting Munchen 17-20 sept. 1974 no. A6-2.
4. Rehakova, M. and Novsad, N.; Coll.Czech. Chem. Comm. $\underline{36}$ (1971), 3004-3012.
5. Rushton, J.H.; Preprint A.I.Ch.E. Inst. of Chem. Engrs. Joint Meeting London, june 1965.
6. Zwietering, T.N.; Chem. Eng. Sci. $\underline{8}$ (1958), 244.
7. Kolar, V.; Coll. Czech. Chem. Comm. $\underline{26}$ (1961), 613.
8. Asselbergs, C.J. and de Jong, E.J.; Paper in this symposium.

PURIFICATION OF $H_3PO_4 \cdot \tfrac{1}{2}H_2O$ BY CRYSTALLIZATION

YOSHIO AOYAMA AND KEN TOYOKURA

Daidoh Plant Engineering Corp., and Waseda University
1-10 Takami-cho, Konohana-ku, Osaka and
4-170 Nishiokubo, Shinjuku-ku, Tokyo, Japan

PHOSPHORIC ACID HEMIHYDRATE CRYSTAL DATA

Terminal Velocities

The terminal velocity of a falling crystal was measured by tests carried out in the equipment shown in Fig.1. The crystals were classified by JIS standard sieves, and their average sizes are listed with the terminal velocities obtained in the tests. The terminal velocity was correlated with the square of the average crystal size.

Crystal Growth Rate

Crystal growth rates were measured in a fluidized bed 8 cm diam. and 100 cm high. Sieved crystals (~1 mm) were put in the crystallizer as the seed. About 2 m^3 of the feed solution was prepared and stored in a tank in the supersaturated condition. This was fed into the bottom of the crystallizer making a growing crystal bed of up to 40 cm. The 1 mm seed crystals grew to more than twice their size in a 2 hour test. The superficial velocity of the solution was about 0.25 cm/sec.

Examples of plots of cumulative crystal growth against time as a function of supersaturation are given in Fig.2. Crystal growth appeared to be independent of crystal size.

Crystal growth rates obtained from the slope of the lines in Fig.2 were plotted against the supersaturation in Fig.3. They are

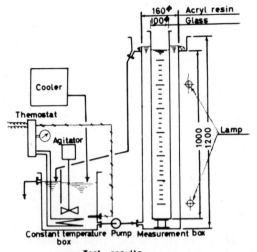

Diameter of crystal (Dp)			Terminal velocity [cm/sec]	
mesh (JIS standard)		average.	temperature of saturated soln.	
pass	on	Dp (mm)	5 (C°)	20 (C°)
8	9.2	2.19	0.81	1.19
9.2	10.5	1.84	0.51	0.71
10.5	12	1.545	0.40	0.53
12	14	1.30	0.26	0.37
14	16	1.095	0.19	0.29
16	20	0.92	0.1	0.18
20	24	0.775	0.09	0.16
24	28	0.65	0.04	0.09
28	32	0.545	0.03	0.08

Fig. 1.   Test equipment for terminal velocity of falling crystal.

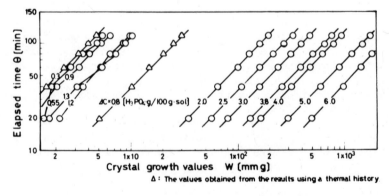

Fig. 2.   Relation between θ and W.

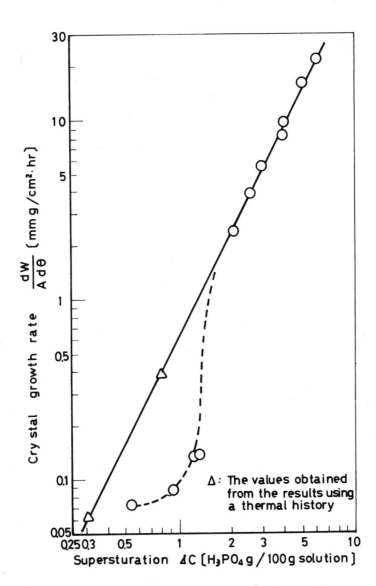

Fig. 3. Relation between dw/Adθ and ΔC.

shown to be proportional to square of the supersaturation over the range ΔC = 2 to 6 g H₃PO₄ /100 g solution and are represented by the equation

$$\frac{dw}{d\theta} = 0.59 A (\Delta C)^2 \qquad (1)$$

where $dw/d\theta$ and A are the crystal growth rate and surface area respectively.

Growth rates at supersaturations 0.3 to 1.3, were much smaller than extrapolated values of equation 1. However, in the tests, feed solution was kept at the same supersaturated condition by precooling for 5 to 6 hours before use. In the later tests, the cooling system was changed to operate as in an industrial crystallizer. In the improved equipment, the solution was kept unsaturated in the feed tank and cooled to the desired temperature, making it supersaturated, just before feeding to the crystallizer. In these tests the crystal growth rates were placed on the straight line of equation 1, even the data for supersaturations less than 1.3. The reason was not discussed in detail, but it was supposed to arise from behaviour of embryos in the supersaturated solution.

The composition of the feed liquid was 29.07 to 32.0 wt% P_2O_5, 0.36 to 0.42 wt% F, 3.23 to 3.42 wt% SO_4, 0.14 to 0.23 wt% CaO, 0.39 to 0.57 wt% Al_2O_3, 0.84 to 0.97 wt% Fe_2O_3, 0.25 wt% MgO, 0.32 to 0.47 wt% SiO_2.

CHOICE OF OPERATIONAL CONDITIONS

Following the measurement of the fundamental crystallization data, a bench scale test plant was set up as shown in Fig.4. Raw liquid was fed into circulation pipe C or C' from the feed tank B. The mixed solution was cooled and supersaturated through the heat exchanger D or D', and fed into the bottom of the crystallizer. The crystallizer was based on the Daidoh classified-bed industrial crystallizer. The outlet solution from the top of the crystallizer, A, which had a residual supersaturation, was returned to the store tank for recirculation. Since the solution was highly viscous, crystals in crystallizer were easily fluidized. But the superficial velocity of the solution was so low that crystals were homogeneously scattered by the slow revolution of the impeller in the crystallizer.

Three kinds of operation were tried. The first was represented by the line A-C-D-A in Fig.5. Concentration of the solution is also shown. This was a typical cooling-operation, but excess fine crystals were born after about 10 hours operation and the solution became milky. This method of operation was unsatisfactory.

PURIFICATION OF $H_3PO_4 \cdot \tfrac{1}{2}H_2O$ BY CRYSTALLIZATION

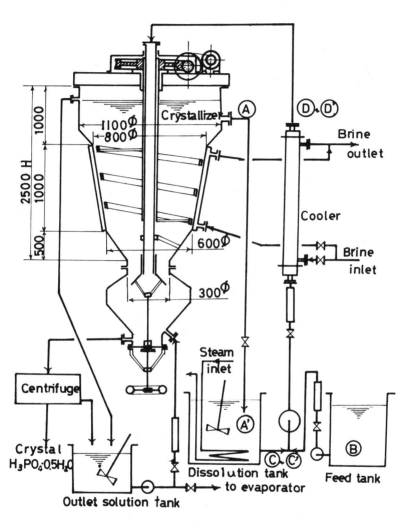

Fig. 4. Test plant

The second operation tried was A-A'-C'-D in Fig.5. A steam pipe was set in the store tank to heat the solution higher than the saturation temperature. After 14 hours operation, excess fines were not found, but scaling was found on the heat exchanger surface. The reason was supposed to be too high supersaturation in the heat exchanger.

The third improved operation was A-A'-C'-D'-A in Fig.5. A jacket-type cooler was additionally fitted to the wall of the crystallizer. Though the temperature of the outlet solution from the heat exchanger was higher than in the second operation, the average supersaturation was almost the same as the previous operation due to the temperature gradient up the crystal bed.

The maximum supersaturation of the solution in the third operation was much smaller than that in the second and the crystallizer was operated satisfactorily for two weeks.

From these test results, an industrial crystallizer (20 times scale-up) was designed.

INDUSTRIAL CRYSTALLIZATION OF PHOSPHORIC ACID HEMIHYDRATE

The process flow sheet is shown in Fig.6. Crude phosphoric acid was fed into a storage tank (1). When the concentration was 25 to 30 wt%, the crude acid was fed to the first evaporator (2), and when about 40 wt%, the acid was fed to the second evaporator (7) in order to concentrate it to about 60 wt%. Impurities precipitated during the evaporating step were separated between the evaporators.

The concentrated acid was cooled to 30 to 35°C at the receiver (9), fed to the circulation line and passed through the heat exchanger to the crystallizer. The crystallizer, which was equipped with a special ribbon-type agitator to make the crystals disperse radially, was separated into two parts. The lower part was for crystal growth and the upper part for fines separation. Each part had a separate circulating line.

The overflow solution from the upper part of the crystallizer came down to the dissolution tank (13) in which the solution was heated 3 to 5°C higher and stayed for 30 to 60 minutes to dissolve all the fines. The solution was then returned to the crystallizer. A portion of the solution was taken from the top of the crystal growing bed and passed through the heat exchanger to increase the supersaturation. This highly supersaturated solution was returned to the bottom of the growing bed. Control of the circulation rates of the overflow solutions from the upper part and lower parts of the crystallizer had an effect on the number of seed crystals in the

PURIFICATION OF $H_3PO_4 \cdot \tfrac{1}{2}H_2O$ BY CRYSTALLIZATION

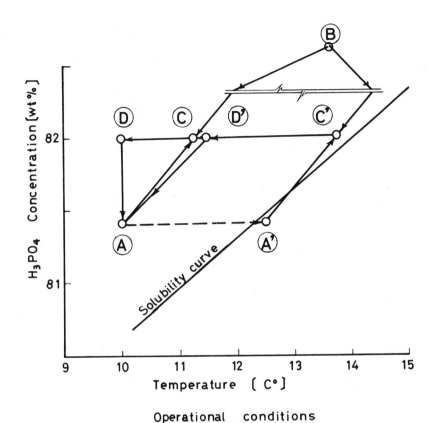

Fig. 5

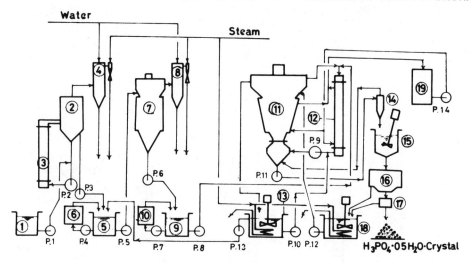

Fig. 6. Flow sheet of purificatory process of $H_3PO_4 \cdot 05H_2O$ by crystallization. 1: Storage tank, 2: 1st evaporator, 3: 1st heater, 4: condenser, 5: 1st storage tank, 6: refining apparatus, 7: 2nd evaporator, 8: condenser, 9: 2nd storage tank, 10: refining apparatus, 11: crystallizer, 12: cooler, 13: dissolution tank, 14: cyclone, 15: crystal storage tank, 16: centrifuge, 17: distributer, 18: seed crystal control tank, 19: refrigerator.

growing bed. This point was the most important one for this process. Other improvements, like temperature control of the dissolution tank (B), were added later.

This industrial crystallizer has been satisfactorily operated at 8 to 11°C and under 1.5 wt% supersaturation. The quality of the crystal passed the JIS (Japanese Industrial Standard). It is necessary, however, to purge some of the mother liquor to prevent accumulation of impurity.

CONCLUSION

The purification of phosphoric acid is not easy on the industrial scale. In this study we have measured the growth rates of phosphoric acid hemihydrate crystals, and some special devices were developed on the basis of the bench scale tests. Following these studies, an industrial crystallizer was designed and good results have been obtained.

THE OPERATION OF A NaOH·$3\frac{1}{2}$H$_2$O CRYSTALLIZER BY DIRECT CONTACT COOLING

T. AKIYA, M. OWA, S. KAWASAKI AND T. GOTO

National Chemical Laboratory for Industry
1-Chome, Honmachi, Shibuya-ku, Tokyo, Japan

K. TOYOKURA

Waseda University
4-Chome, Nishiokubo, Shinjuku-ku, Tokyo, Japan

INTRODUCTION

A serious shortage of fresh water forecast for the future has stimulated the development of sea water desalting plants in Japan. In desalination, a large amount of concentrated sea water (concentrated brine) is produced. Consequently, the national research project of desalination has included by-product recovery processes from concentrated brine.

One of these is the purification of a 39 wt.% caustic soda solution containing small amounts of potassium hydroxide and sodium chloride. In this process, NaOH·$3\frac{1}{2}$H$_2$O crystals precipitate from the impure 39 wt.% caustic soda solution. Purified caustic soda solution is obtained by melting the crystals.

As there is little information available on the industrial crystallization of NaOH·$3\frac{1}{2}$H$_2$O the process was studied using well-mixed crystallizers.

The composition of caustic solution used was as follows: sodium hydroxide 34 to 40 wt.%, potassium hydroxide 2 to 8.5 wt.% and sodium chloride 0 to 1.1 wt.%.

PRELIMINARY TESTS

The growth rate of $NaOH \cdot 3\frac{1}{2}H_2O$ crystals and the generation rate of their seed crystals were studied in a well-mixed crystallizer with an effective volume of two litres. The feed rate was 100 to 200 ml/min. (retention time 10 to 20 min.) and the agitator speed was 300 to 500 rpm. From steady state operation, the crystal size distributions were measured.

When a crystal of $NaOH \cdot 3\frac{1}{2}H_2O$ melts, it turns into a 39 wt.% sodium hydroxide solution. Consequently, the crystallization of $NaOH \cdot 3\frac{1}{2}H_2O$ from a 39 wt.% caustic solution is crystallization of the solvent itself. In this case it is convenient to adopt the degree of supercooling, ΔT (the difference between the saturation temperature and the crystallization temperature) as the driving force for crystallization instead of the supersaturation. The degree of supercooling was affected by the potassium hydroxide concentration and was calculated from the data of Hayano et al. (2).

Crystal growth rates were calculated from the slope of the semi-logarithmic plots of the crystal size distributions (3). The first-order relationship between growth rate and supercooling for the data of the pilot plant tests is shown in Fig. 1 and may be expressed by

$$\frac{dL}{d\theta} = K_o \Delta T \qquad (1)$$

The overall growth rate constant K_o was affected by the crystallization temperature (Fig. 1) and the relationship between the log K_o and the reciprocal temperature is shown in Fig. 2. From the slope in Fig. 2, an apparent activation energy was calculated to be 110 kcal/mol. It seemed that K_o was not affected by the dimensions of the crystallizer and the fluorocarbon-12 from Figs 1 and 2.

The relationship between the generation rates of seed crystals and the degree of supercooling is shown in Fig. 3. The generation rates were calculated from

$$F' = 9P/2L_m^3 \rho_c \qquad (2)$$

and ranged from 10^6 to 10^{10} h^{-1} $litre^{-1}$. But no distinct tendency was observed. For example, the generation rate was proportional to the nth order of the degree of supercooling.

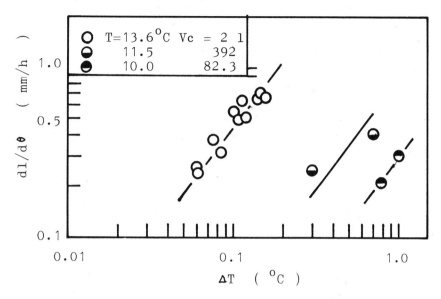

Fig.1 Relationship between $dl/d\theta$ and ΔT

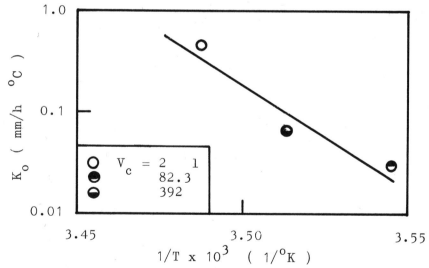

Fig.2 Relationship between K_o and $1/T$

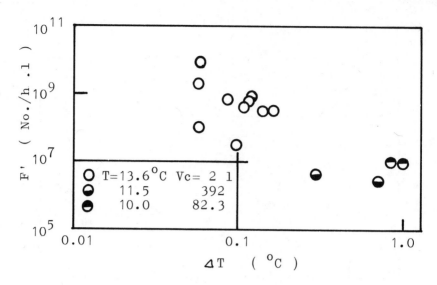

Fig. 3 Relationship between F' and ΔT

PILOT PLANT TESTS

Before pilot plant construction, preliminary tests for scaling were carried out with an annular type model heat exchanger. When a slurry of $NaOH \cdot 3\frac{1}{2}H_2O$ crystals was circulated through the heat exchanger, crystals were spontaneously deposited on the heat transfer surface (4) and stable operation was impossible. Therefore, a direct contact cooling method using a liquefied refrigerant was adopted. A fluorocarbon-12 was chosen as the refrigerant.

Two sizes of crystallizer were used. One was of 82.3 litres effective volume (1.5 ton $NaOH \cdot 3\frac{1}{2}H_2O$/day) and the other was of 392 litres (5 ton/day). These were well-mixed (MSMPR) crystallizers. Fig. 4 shows the flowsheet of the latter crystallizer which was the same as the former. Caustic solution was fed into the crystallizer (1). A slurry was discharged to the centrifugal separator (2) and was sent to the melter (3). The overflowed caustic soda molten solution was stored in receiver (4) and was recycled. The liquefied fluorocarbon-12 was vaporized in the crystallizer and the refrigerant gas was compressed by compressor (5) and condensed in condenser (6).

The crystallizer, constructed from SUS27 stainless steel, is shown in Fig. 5. It has two interesting features in addition to direct contact cooling. One was that the agitator was horizontal

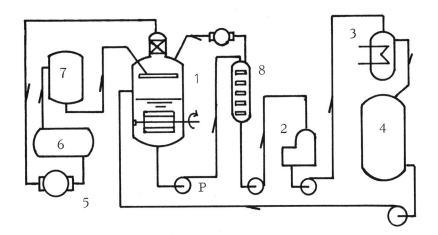

1. Crystallizer 2. Centrifugal Separator 3. Melter
4. Receiver 5. Compressor 6. Condencer
7. Refrigerent Receiver 8. Deaerator P. Pump

Fig.4 Flow Sheet of Pilot Plant

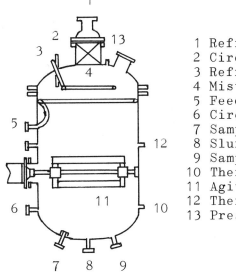

1 Refrigerant outlet
2 Circulating solution inlet
3 Refrigerant inlet
4 Mist separator
5 Feed inlet
6 Circulating solution outlet
7 Sampling nozzle
8 Slurry outlet
9 Sampling nozzle
10 Thermometer nozzle
11 Agitator
12 Thermometer nozzle
13 Pressure gauge

Fig.5 Crystallizer

to prevent leakage of refrigerant gas through the ground seal. The other was that the warm caustic solution was fed as a falling film on the wall of the crystallizer to prevent crystal deposition on the cold wall. The crystallizer was 800 mm diam. and 1090 mm high. Agitation was 45 to 150 rpm and feed rates of caustic solution were 0.8 to 1.0 ton/h (retention time 0.47 to 0.57 h). Feed rates of the liquefied fluorocarbon-12 were 350 to 600 ℓ/h.

After about ten times the mean retention time the caustic soda solution was analysed and the crystal size distributions were measured.

The relationship between the crystal size distributions in the upper and lower parts of the crystallizer are shown in Fig. 6. As these were virtually the same it was assumed that the crystals were distributed uniformly and the operation was well-mixed. The crystal growth rate was independent of crystal size, so McCabe's ΔL law was valid.

A plot of growth rate against time is shown in Fig. 7. The time required for steady state was about eight times the mean retention time. Therefore, ten retention times was considered to be sufficient to obtain steady state growth rates after other operational factors became constant.

The production rate P of caustic soda hydrates was calculated from the KOH mass balance:

$$P = \rho_c F' L_1^3 + (X_{B2} - X_{B1F})/X_{B1} \tag{3}$$

or $\quad P = (X_{B2} - X_{B1F})/X_{B1} \quad$ (no seeding) $\tag{3'}$

From equations 1 and 3 and the Shirotsuka and Toyokura CFC theory (5), the effective volume V_c of a well-mixed, solvent crystallization process and the dominant crystal size were calculated as

$$V_c = \frac{Pl_m}{3\rho_\ell K_o \Delta T} \frac{(X_1^2 + \frac{2}{3}X_1 + \frac{2}{9})}{(X_1^3 + X_1^2 + \frac{2}{3}X_1 + \frac{2}{9})} \frac{1}{(\xi-1)} \frac{1}{(1-\xi)} \tag{4}$$

or $\quad V_c = PL_m/3\rho_\ell K_o \Delta T(\xi-1)(1-\varepsilon) \quad$ (no seeding) $\tag{4'}$

$$L_m = 3K_o \Delta T = 3\tau \cdot dL/d\theta \tag{5}$$

Here ξ is the concentration factor, the ratio of KOH concentration of the outlet solution (X_{B2}) to that of the inlet (X_{B1}). For design purposes ξ is calculated from the products and KOH

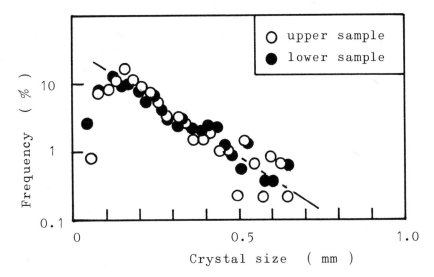

Fig.6 Crystal size distribution
($dl/d\theta$=0.248mm/h , τ=0.568h)

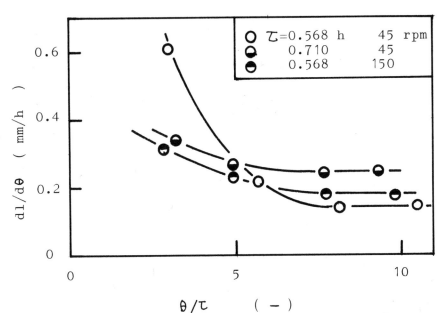

Fig.7 Relationship between $dl/d\theta$ and θ/τ

concentration of the feed, and is used as the dimensionless supersaturation in the CFC theory. When the parameters in the design equations are set, the effective volume of the crystallizer is calculated from equation 4.

Typical crystallization conditions were as follows:

temperature in the crystallizer	11.7 °C
pressure in the crystallizer	3 kg/cm^2 (gauge)
mean retention time	0.568 h
crystal growth rate	0.248 mm/h
generation rate of seed crystal	3.18 x 10^6 h^{-1} ℓ^{-1}
dominant size of crystals	0.455 mm
production rate (NaOH·3½H$_2$O)	5 ton/day

Seed crystal generation rates were adequate for making crystals large enough to be separated by the centrifuge. Therefore, seeding and fines removal were not studied.

The melt of the product crystals contained small amounts of the concentrated mother liquor (3-5 wt.%). Consequently, it was concluded that the crystallization process was successful for the purification of the crude caustic soda solution.

CONCLUSION

Well-mixed MSMPR crystallizers with effective volumes of 82.3 and 392 litres have been operated for the crystallization of NaOH·3½H$_2$O from a 39 wt.% caustic soda and potash mixed solution. Direct contact cooling with fluorocarbon-12 avoided scaling on the heat transfer surfaces and stable operation was obtained. The crystal growth rate of NaOH·3½H$_2$O was directly proportional to the degree of supercooling and the apparent activation energy was 110 kcal/mol.

ACKNOWLEDGEMENT

The authors wish to thank Mr. Hayano and Mr. Takeuchi (National Chemical Laboratory for Industry) for their helpful discussions. This study was done in co-operation with Mitsui Shipbuilding and Construction Co. Ltd.

NOMENCLATURE

F	feed rate	(kg/h)
F'	generation rate of seed crystals	(no./h m^3)
K_o	overall growth rate constant	(mm/h °C)

L	size of crystal	(mm)
L_1	size of seed crystal	(mm)
L_m	dominant size of crystals	(mm)
P	production rate	(kg/h)
T	temperature	(°C)
ΔT	degree of supercooling	(°C)
V	effective volume of crystallizer	(m³)
X_1	dimensionless crystal size (L_1/L_m)	(-)
X_{B1}	concentration of non-crystalline component of inlet solution	(%)
X_{B2}	concentration of non-crystalline component of outlet solution	(%)
ε	void fraction	(-)
θ	time	(h)
ξ	concentration factor (X_{B2}/X_{B1})	(-)
ρ_c	density of crystal	(kg/m³)
ρ_ℓ	density of solution	(kg/m³)
τ	mean retention time	(h)

REFERENCES

1. Goto, T., Hayano, I., Takeuchi, T. and Akiya, T., Proc. 4th Internat. Sympos. on Fresh Water from the Sea, 2 (1973) 517.
2. Hayano, I., Goto, T. and Ishizaka, S., Bull. Soc. Sea Water Science Japan, 24 (1971) 253.
3. Randolph, A.D. and Larson, M.A., Theory of Particulate Process, Academic Press, New York, 1971 p. 68.
4. Akiya, T., Goto, T. and Toyokura, K., 25th Ann. Meeting Soc. Sea Water Science, Japan, 1972, p. 47 (preprints)
5. Shirotsuka, T. and Toyokura, K., C.E.P. Sympos. Ser., No. 110, 67 (1971) 145.

PROGRESS IN CONTINUOUS FRACTIONAL CRYSTALLIZATION

G.J. ARKENBOUT, A. VAN KUIJK AND W.M. SMIT

Institute for Physical Chemistry TNO

P.O. Box 108, Zeist, The Netherlands

INTRODUCTION

The possibilities of separation and purification by crystallization are determined by the solid-liquid phase equilibria. Two well-known types of equilibrium diagrams are those belonging to the eutectic and solid solution systems.

When dealing with a eutectic system it is theoretically possible to obtain a pure compound by one single crystallization. In many cases, however, mother liquor is occluded inside crystal imperfections or is trapped in between crystal agglomerates. When considering a solid solution system, it is impossible to isolate the pure compound by one crystallization stage. So multi-stage separations are often required in order to obtain the purity wanted, even when eutectic systems have to be separated.

Multistage crystallizations can be effected with the simple column for continuous fractional crystallization as developed in the Institute for Physical Chemistry TNO. This column, which was described at the 5th Symposium on Industrial Crystallization (1), shows promising features for technical application: the construction is simple, the operation is easy and reliable. Some favourable features of continuous fractional crystallization, e.g. compared with distillation, are the large separation effect of the single process, the low consumption of energy and the moderate operating temperature.

PRINCIPLE OF THE PROCESS

A schematic flow sheet of continuous fractional crystallization from a melt is given in Fig. 1. The crystals are formed in a cooling scraper crystallizer 3 at one end and melted in the melting section 2 at the other end of the column. In the column 1 the crystals are transported countercurrently to the melt (1-3). As in a continuous distillation process product and/or waste may be withdrawn at both ends of the column.

A good separation efficiency for a continuous fractional crystallization column demands a high rate of recrystallization. The rate of recrystallization may be enhanced by grinding the crystals. On grinding the large crystals are broken into small crystals. The crushed crystals are not stable and will melt or dissolve. This results in the growth of larger and more perfect crystals, especially when the column is held in an adiabatic condition.

When the size of the crystals is reduced continuously by mechanical means, and at the same time the broken and damaged crystals are transported countercurrently to a liquid phase, the crystallization occurs from another (more pure) melt or mother liquor than that from which the original crystal came. In this way concentration differences can be realized in a column in an analogous way to that in distillation and extraction.

The vibration of a column containing proper combinations of balls and sieve discs appears to be a simple method for enhancing recrystallization by grinding. In addition to facilitating grinding, the combination of balls and sieve discs promote efficient countercurrent transport of crystals and liquid and also provide a suitable agitation at the crystal interfaces.

A column for continuous fractional crystallization based on these principles was developed and tested.

EXPERIMENTAL RESULTS

Several systems, including products of technical interest, were separated using columns of 80 mm diam. and with lengths varying from 500 to 1 500 mm (1-3). The mixtures were crystallized both from the melt and from solution.

Mass transport rates of the crystals ranged from 0.03 to 0.15 kg/m^2s at process temperatures ranging from 190 to 370 K. Temperature gradients of more than 50 K per metre and concentration gradients of more that 50 wt.% per metre were established. The column, once set at the proper conditions demanded by the system

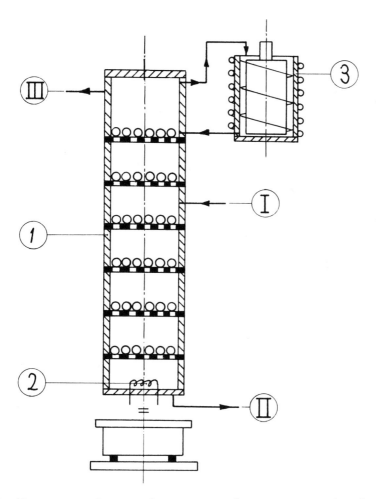

Fig. 1 Flow sheet for continuous fractional crystallization.
1. crystallization column, 2. melting section, 3. crystallizer.
I. feed of the mixture to be purified, II. product withdrawal,
III. waste removal.

to be separated, needed hardly any attention. Full automation of the process appeared possible.

From the multistage separations carried out with the column it was learnt that plate numbers up to 5 per metre could be realized, corresponding to a plate efficiency of about 50%. Even very low concentrations of impurity could be removed, demonstrating that this crystallization column is useful for the preparation of very pure compounds.

The results of separation and purification did not appear to depend strongly on the conditions of crystal formation in the cooling scraper crystallizer. Good settling properties of the crystals facilitate the operation of the column.

SCALING-UP

A plan was devised for scaling-up the column to dimensions of technical interest based on cooperation between a manufacturing company, producers of chemicals and TNO. The manufacturing company, Apparaten- en Ketelfabriek (AKF) at Goes, The Netherlands, constructed a prototype technical column with a length of 500 and a diameter of 500 mm.

The main difference between the designs of the technical and laboratory columns concerns the method of vibrating. The lab-column is vibrated itself in order to move the metal balls. In the technical column the sieve discs are mounted on a common rod connected to an eccentric. By vibrating the central rod the metal balls are agitated.

AKF tested the prototype of the technical column with respect to its mechanical operation. No special difficulties were encountered. No wastage of any importance was observed. The energy consumption of the vibrator was rather low (20 to 30 J/s per sieve disc).

The interest of industrial companies in The Netherlands and abroad for the column has led to sponsored research projects to investigate the possibilities of purification and concentration of distinct products on a technical scale. The results were such, that a few industries are seriously considering participation in the scaling-up project.

The interest of producers of chemicals includes both the purification of fine chemicals and pharmaceuticals on the 100-1000 tonne per year scale and the production of bulk chemicals in quantities of 10 000 to 100 000 tonne per year. In large scale applications this crystallization process may successfully

compete with fractional distillation because it consumes less
energy and often yields a better product. The scaling-up to
dimensions larger than 500 mm is therefore planned.

ACKNOWLEDGEMENT

The authors thank Mr. E. Busink and Mr. W. Voorbergen for
their enthusiastic and skilful cooperation, Dr. H.A. Markens and
his co-workers for carrying out the GLC determinations and Mr.
H.F. Versteeg for solving construction and control problems.

REFERENCES

1. Arkenbout, G.J., van Kuijk, A. and Smit, W.M., 5th Sympos. on Industrial Crystallization (CHISA 72), Prague, 1972 also TNO Nieuws, 27 (1972) 767.
2. Arkenbout, G.J., van Kuijk, A. and Smith, W.M., Chemistry and Industry (1973) 139.
3. Arkenbout, G.J., van Kuijk, A. and Smith, W.M., Dechema-Monographien, 73 (1974) 277.

THE PREVENTION OF DEPOSITION ON CRYSTALLIZER COOLING SURFACES USING ULTRASOUND

M.J. ASHLEY

Ultrasonics Limited

Otley Road, Shipley, West Yorkshire, England

INTRODUCTION

The build up of crystal layers on cooling surfaces is one of the most widespread and costly problems encountered in industrial crystallisation processes. Often the problem is avoided by tight control of operating conditions, or scraped-surface heat exchangers are used to remove deposits soon after formation. Frequently the problem is accepted as a necessary evil and subsequent shutdowns result in high production losses and maintenance costs.

Many crystallisation processes show no tendency to cause fouling problems in laboratory crystallisers, but cause great problems in large industrial units. This can be principally attributed to the lack of effective agitation on the large scale. Conventional agitators can provide good bulk agitation and maintain crystals in suspension, but cannot provide the intense agitation at the surface of the cooling coil or jacket, which is so necessary to prevent deposition.

Ultrasonic agitation, however, has the property of causing intense agitation of liquid at a surface, but without effectively moving, or dissipating energy into the bulk of the liquid.

This principle provides the basis of the work described in the present paper, and to a new practical method of preventing fouling in crystallisation processes using ultrasonics.

DEPOSITION PREVENTION BY ULTRASONICALLY INDUCED CAVITATION IN THE LIQUID PHASE

During the last 20 years Ultrasonic systems for the cleaning of engineering components have become widely used and reliable, and systems are often supplied for the off-line cleaning of heat exchanger surfaces.

Inevitably the question then arises whether heat transfer surfaces can be cleaned "in situ" by incorporating ultrasonic transducers directly into the heat exchangers, thus avoiding costly down-time. In 1950 Loosli patented a method for preventing deposition on metal surfaces, which is based directly on the operating principle of ultrasonic cleaners (1).

High power ultrasonic transducers are immersed in the liquid phase to create intense cavitation in the liquid, and thus erode or remove any deposits that may form on an immersed metal surface. The principle was commercially developed as the "Crustex" method for the prevention of scale in boilers formed by the crystallisation of magnesium and calcium salts, but to the best of our knowledge this system has not achieved much success.

This method of introducing ultrasonics into a heat exchanger requires very high energy inputs (well above economic viability) since intense cavitation is necessary in the liquid close to a surface to have any useful erosive effect. Ultrasonic cleaners use cavitation to accelerate the action of a solvent but considerable experience has shown that erosion of insoluble deposits requires extremely intense cavitation. Also energy losses at the transducer and liquid interfaces are high, and under intense cavitational conditions the attenuation of ultrasonic waves travelling through a liquid, and consequently energy losses also, are very high.

LOW POWER INSONATION OF LIQUIDS IN THE ABSENCE OF CAVITATION

Low power sound pressure waves can be transmitted through liquids in the same way as sound travels through air. Because of the low compressibility of liquids there is a maximum intensity of the pressure wave above which the liquid cannot effectively transmit the sound energy, the liquid structure breaks down and cavitation is induced.

Ultrasonic irradiation of liquids below this cavitation threshold gives rise to some interesting effects and Jackson and Nyborg have shown that certain surface coatings are removed from membranes which are in contact with a liquid and which are ultrasonically vibrated at a power level insufficient to cause cavitation (2). They attribute this effect to high velocity eddies induced close to the moving solid surface - acoustic microstreaming. Although we cannot quantify the size of these eddies or their velocities, it is not suprising that they have such a dramatic effect at the surface. If we consider a surface moving with a maximum amplitude of, say, 0.01 mm at an ultrasonic frequency of 20kHz, the peak acceleration of the surface will be several thousand times that due to gravity. The effect of the microstreaming is to disrupt the boundary layer normally surrounding the surface and this can have significant effects on heat and mass transfer processes.

Clearly if the microstreaming effect can be utilised to prevent crystals adhering to a cooling surface it would represent considerable savings in ultrasonic power consumption and installation cost.

DEPOSITION PREVENTION BY DIRECT SONIFICATION OF HEAT TRANSFER SURFACES

As described previously, the first attempts at deposition prevention were by sonification of the liquid phase, which is extremely inefficient in consumption of ultrasonic power, only a small proportion of the energy reaching the actual heat transfer surface.

A new approach was jointly developed by ICI and Ultrasonics Ltd. which resulted in 1968 in a patent by Williamson & Skrebowski covering a new process for the crystallisation of p-xylene (3). The process involved the direct coupling of magnetostrictive transducers to the heat transfer surface, thus vibrating the surface directly and causing intense agitation of the liquid boundary layer immediately adjacent. This is a direct application of the microstreaming effect reported by Jackson & Nyborg (2) and has several important advantages over the indirect method of sonification of the liquid:

1. Power consumption is far less because energy losses at the transducer/liquid interface and attenuation in the liquid phase are eliminated.

2. The surface can be vibrated sufficiently to prevent deposition by the microstreaming effect without causing cavitation

in the liquid, and power requirement is minimised.

3. Ultrasonic waves can travel uniformly through metallic tubes, but through liquids surfaces can be shielded unless the transducers are immediately opposite.

4. The main dissipation of energy will be very close to the surface/liquid interface, where it can be most beneficial in preventing crystals adhering.

Duncan & West clearly demonstrated the advantages of direct sonification by their own experimental work published in 1972 (4). A two litre continuous crystalliser with a polished copper cooling coil was used to examine fouling in two systems - crystallisation of p-xylene from mixed xylenes, and ice from brine. The magnetostrictive ultrasonic transducer gave a nominal 100 Watt maximum power level at a frequency of 12kHz and was directly attached to the cooling coil. The coil itself was divided into two halves isolated by polythene tube so that only one half was excited by the ultrasonic transducer. At full power, not only was deposition prevented on the surface of the excited coil but because of the cavitation effect, deposition of crystals onto the unexcited coil was also prevented.

When operating at a lower power level only the directly excited coil remained clear while the unexcited coil rapidly became incrusted. Cavitation was absent at this low power level and deposition onto the directly excited coil was prevented by the acoustic microstreaming effect.

EXPERIMENTAL WORK

Figure 1. illustrates the experimental rig used in the present study on the effects of ultrasonics on crystallisation processes. The cooling coil was constructed from 7mm o.d. stainless steel tube, with a heat transfer surface area of $230 cm^2$. The refrigeration unit supplied methanol refrigerant at temperatures down to $-22^{\circ}C$ with an accuracy of $\pm 0.1^{\circ}C$. The magnetostrictive transducer had a maximum power level of 100 Watts, operating at 20kHz, and was directly coupled to the cooling coil by a coupling bar of patented design (5). A variable speed laboratory stirrer was used to provide bulk mixing of the solution.

The first set of experiments confirmed the previous work of Duncan & West in demonstrating incrustation prevention during the crystallisation of ice from a 7% solution of sodium chloride.

Figure 2a shows the ice crystals passing in front of the cooling coil and obscuring the shaft of the agitator, the coil

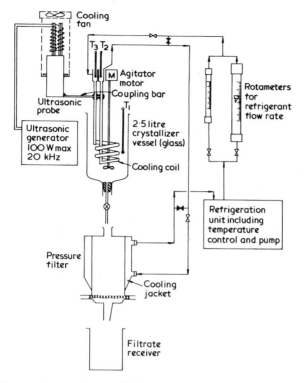

Figure 1. Crystallisation experimental rig

being kept free from encrustation by ultrasonics. Figure 2b shows the unit five minutes after turning the ultrasonic power off - the coil is now totally encrusted with ice and effective crystallisation has ceased.

THE CRYSTALLISATION OF POTASSIUM NITRATE

Previous work has been concerned with severe fouling problems (ice from brine and p-xylene from mixed xylenes). Industrial problems often occur with much less severely fouling systems, typically inorganic salts crystallised from aqueous solutions. A 28% aqueous solution of potassium nitrate was chosen as an example, 2 litres of solution being used for each experiment and the inlet refrigerant temperature maintained at $0°C \pm 0.5°C$. Figure 4 shows cooling curves for crystallisations carried out at varying agitator speeds, but with no ultrasonics.

At the lowest agitation rate (100rpm) the majority of crystals settled freely. It can be seen that heavy encrustation rapidly

Figure 2. The crystallisation of ice from brine (a) with ultrasonics (b) no ultrasonics.

Figure 3. The crystallisation of Potassium Nitrate (a) with ultrasonics (b) without ultrasonics.

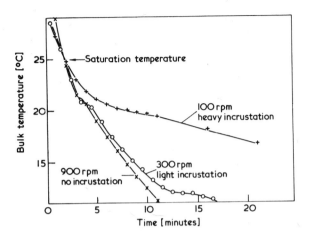

Figure 4. The effect of agitation on the crystallisation of a 28% potassium nitrate solution, without ultrasound.

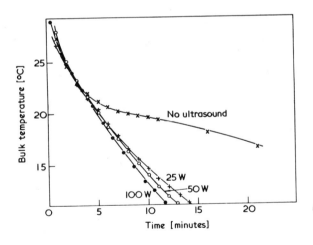

Figure 5. The effect of ultrasonically activating the cooling coil on the crystallisation of potassium nitrate (100rpm agitation).

occurred as soon as the saturation temperature had been passed. The agitator speed of 300rpm was equivalent to a fairly efficient agitation of the bulk solution, allowing only the large crystals to settle. It can be seen that immediate encrustation was prevented by this degree of agitation, but eventually, because of the lower agitation rate, encrustation did take place. The highest agitator speed (900rpm) was equivalent to violet bulk agitation of the solution. Under these conditions all crystals were kept in suspension and no encrustation took place. This degree of agitation would be virtually impossible to achieve on large industrial installations.

All the depositions, shown in figure 3b, were of a dense nature and could not be removed by mechanical agitation once formed.

Figure 5 shows the effect of ultrasonically activating the coil - all these experiments were performed with gentle agitation (100rpm) allowing most crystals to settle out. It can be seen that very little ultrasonic power is required to completely prevent encrustation. Further ultrasonic power from 25 to 50 W also increases the rate of heat transfer because of the microstreaming effect. At 100 W, cavitation was present in the liquid and fine crystals resulted. Clearly the effect of mild ultrasonic vibration of the metal surface prevents crystals sticking or growing there, whereas without ultrasound violent agitation of the bulk liquid is required to achieve the same effect.

OTHER SYSTEMS EXAMINED

During the last two years many other crystallisation systems have been examined, including sodium carbonate, copper sulphate and boric acid all from aqueous solutions, several pharmaceutical materials (including paracetamol derivatives and steroids) and acetic acid/acetic anhydride from organic solvents. Although varying levels of ultrasonic power are required to prevent deposition, the ultrasonic technique has so far been successful with every crystallisation system examined on the laboratory scale.

SCALE UP OF THE TECHNIQUE TO A 1,000 LITRE INDUSTRIAL CRYSTALLISER

About a year ago we were asked to apply the technique to a 1,000 litre copper sulphate crystalliser, which was encountering severe fouling problems on the cooling coil.

In laboratory tests under simulated conditions it was found that 25 watts of ultrasonic power would easily prevent deposition of copper sulphate onto the cooling coil, thus giving a power to

Figure 6. Ultrasonically activated cooling coil.

surface area ratio of 1,000 watts/m^2, which we used as the scale-up parameter.

A new cooling coil was constructed from 250mm o.d. stainless steel tube giving a total heat transfer surface of 0.56m^2. A direct power scale up would require 560 watts for this size, and in fact two 400 Watt ultrasonic transducers were directly attached to the inlet and outlet of the coil, as can be seen in figure 6.

Figure 7. shows the coil following a batch crystallisation without ultrasonics - The coil is obviously severely fouled. Figure 8. shows the coil following a crystallisation in which ultrasonics were used. Fouling has been completely prevented and not only did this alleviate the need to clean the coil after each batch, but also the cooling rate was accelerated by a factor of 5.

A similar transducer/cooling coil assembly has been successfully tested in a 4,000 litre boric acid crystalliser.

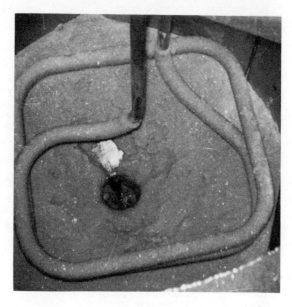

Figure 7. Fouling of the cooling coil by copper sulphate crystals.

Figure 8. Clean cooling coil after crystallising copper sulphate using ultrasonics to prevent fouling.

THE EFFECT OF ULTRASONICS ON CRYSTAL PRODUCT QUALITY

This topic has been covered previously by the author and space does not permit repetition here (6).

Briefly we have found that power levels sufficient to prevent deposition will in general not adversely affect crystal products. However if the ultrasonic power level is increased further, agglomeration of growing crystals can be prevented and nucleation rates increased. Consequently smaller, more regular sized crystals are produced using ultrasonics at higher power levels.

CONCLUSIONS

It has been shown that the ultrasonic technique can effectively prevent deposition of crystals on crystallisation cooling coils, and the work has been successfully followed through to the first full size industrial application.

Although the technique is still being developed, it can already be considered seriously as a viable solution to many fouling problems encountered in small and medium size crystallisers.

At present the main applications are in the crystallisation of pharmaceuticals and fine chemicals where the crystalliser sizes are relatively small and the value of the materials processed is high.

ACKNOWLEDGEMENT

The work described in this paper was jointly sponsored by the National Research Development Corporation and Ultrasonics Ltd.

REFERENCES

1. Loosli, H.,U.K. Patent No. 646882 (1950)
2. Jackson,F.J. & Nyborg, W.L., J Acoustical Society of America $\underline{30}$ (1958) 614.
3. Williamson, J. Skrebowski, J.K., U.K.Patent No. 1104508 (1968).
4. Duncan,A.G. & West, C.D., Trans. Inst. Chem. Engrs. $\underline{50}$ (1972) 109.
5. Goodman, J.E. & Grange, A., U.K. Patent No.1190917 (1970)
6. Ashley, M.J., Ultrasonics, September, 1974, 215.

CASE STUDY OF INCRUSTATION IN AN INDUSTRIAL SALT CRYSTALLIZER

C.M. VAN 'T LAND AND K.J.A. DE WAAL

Akzo Engineering bv

Arnhem, The Netherlands

INTRODUCTION

Common salt (NaCl) can be produced in multiple effect evaporation plants. In such a salt plant there may be the problem of incrustation in the crystallizers. The plant under discussion concentrates brine in 5 effects (see Fig.1). Brine is fed in parallel to the first four effects. The salt discharge is in series from effect to effect. Low pressure steam is fed to the first effect. The vapour from the last effect is condensed in a surface condenser. The levels in the crystallizers are controlled by overflows. Steam consumption is in the range of 30-45 tonnes/h. The tubes are $3/3\frac{1}{4}$" diam. and 6.1 m long. The diameter of the vapour/liquid separator is 5 m.

Forced circulation is maintained by means of a top-driven screw pump. The fluid flows down the central tube. To prevent boiling in the tubes the draft tube is longer than the other tubes. As the salt produced will settle inside the conical bottom, only little salt (2 to 5% by weight, calculated on the slurry) will remain in circulation.

The incrustation inside the crystallizers necessitates daily addition of water. About 6% of the steam consumption of the plant is used to evaporate this water.

We were asked to study the possibility of reducing water consumption with a view to saving steam.

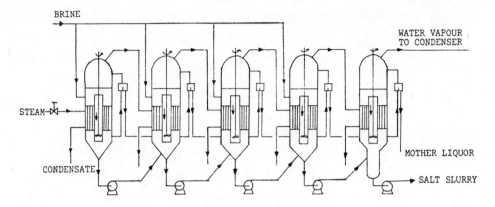

Fig. 1   Multiple effect evaporation plant, design Escher Wyss (1949)

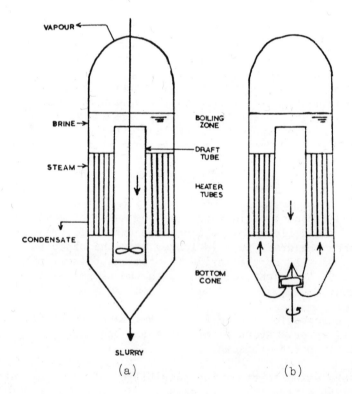

Fig. 4   (a) zones in a crystallizer, (b) crystallizer with a modified bottom

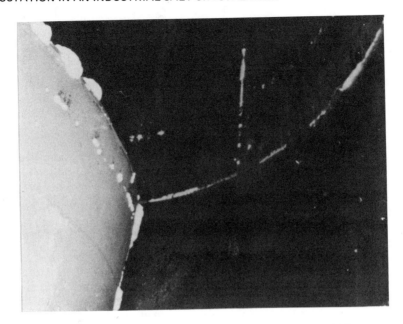

Fig. 2 Crystallizer wall and draft tube in the boiling zone, viewed upward from the upper tube sheet

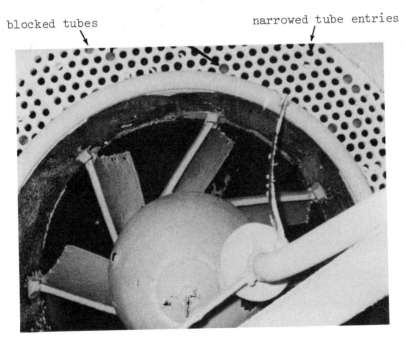

Fig. 3 Crystallizer circulation pump and lower tube sheet

ANALYSIS OF THE PROBLEM

The addition of water was said to be necessary to re-dissolve crusts that had formed on the wall in the boiling zone and had fallen on the upper tube sheet. It was decided that the cause of the necessity of adding water should be studied in more detail. Our study has been concentrated on the first effect for the following two reasons.

First, the evaporation temperature is high here (about $120^\circ C$), so that the crusts will have a high rate of growth. Second, the other crystallizers receive lumps of salt from the preceding ones. We kept a record of the operational conditions of the first crystalliser during a test run. No water was added to the plant.

Whereas the steam pressure remained constant, steam consumption of the plant gradually decreased and energy consumption of the main circulating pump gradually increased. Both phenomena point to a slow growth of crusts impeding the flow of the salt slurry. After 24 hours of operation the crystallizer was taken out of service, drained and opened and the following observations were made.

There were no lumps of salt on the upper tube sheet. Nor was the wall of the boiling zone covered with salt crusts (see Fig.2). To the left one will notice the draft tube and to the right the wall of the vessel. There are only some small salt deposits on the welds in the vapour/liquid separator. The cause of the trouble was not, as was originally suggested, the formation of crusts on the upper tube sheet but gradually narrowing tube entries in the lower tube sheet due to salt growth. Fig.3 is a view from below of the pump with the fixed outlet vanes. The majority of tubes have narrowed to about 50% of their original cross-sectional area. We found incrustations in the tube entries over a distance of only 5 to 10 cm. The remaining tube parts were perfectly clean. Some tubes have wider openings. As these tubes are completely clogged, there will be no flow of brine through them and hence no growth of crusts at their entries.

After some months of operation any salt plant will have a usually small percentage of tubes clogged by salt growth. These tubes are manually cleaned every now and then, which practically never offers any serious problems.

Visual inspection has now enabled us to reformulate the problem, viz. prevention of incrustation at the tube entries. Furthermore, it was reasoned that the sudden addition of water (a remedy practised for many years) decreases the temperature of the contents of the vessel. Due to its coefficient of expansion being different from that of the tube material the salt crusts will be detached and fall into the cone rather than dissolve in the added water. We decided

INCRUSTATION IN AN INDUSTRIAL SALT CRYSTALLIZER

to tackle the problem first by a literature study.

LITERATURE SURVEY

In our company salt has been produced since 1918. We went through the records of former and present plants and found that the phenomenon of narrowing tube entries had been observed before. The crystallizer described also had forced circulation and was used as a second effect (85°C). Its salt output was about 500 kg/h. The slurry density varied irregularly between 2 and 10% by weight.

In the literature one reference was found describing salt incrustations at the tube entries (1). The crystallizer was used as a first effect with forced circulation. The salt plant operating in quadruple effect produced several hundreds of tons of salt per day. The heater contained 1220 1" O.D. tubes. No details were given about the slurry density.

The problem had obviously been encountered in the past. However, no solution had been offered other than continual addition of water.

CONSIDERATIONS

In forced circulation evaporative crystallizers we can distinguish several zones with different flow characteristics (see Fig.4). First the boiling zone: this zone can be considered a continuous-stirred tank reactor (CSTR) with uniform supersaturation. The flow entering this zone is formed by superheated, but undersaturated, brine containing salt particles. The brine flashes, solvent is removed and the brine cools down. These two factors cause supersaturation. The flow leaving this zone is supersaturated.

After the boiling zone the slurry enters the draft tube. The flow in this tube resembles plug flow with supersaturation decreasing with the time spent in the tube.

Third, the direction of the flow is reversed and the slurry enters the bottom cone. The flow in this zone is rather difficult to define. Superficial velocities are too small to keep the salt in suspension. The crystals settle in the lower part of the cone. This is the cause of low slurry densities in the other parts of the crystallizer. Experimentally, slurry densities between 2 and 5% by weight were found in the boiling zone.

Fourth, the slurry enters the heater tubes. There is also plug flow in the heater. Due to heating up, supersaturation is transformed into undersaturation. Through the concentric channel between the draft tube and the vessel wall the slurry re-enters the boiling zone.

The time required to complete a loop is called the cycle time τ. The variation in supersaturation during a cycle is indicated in Fig.5.

We conclude, therefore, that:
1. The flow characteristics of the bottom cone are rather difficult to define. Re-design may have to be considered.
2. Lowering the supersaturation at the tube entries may prevent incrustation.

SUPERSATURATION AT THE TUBE ENTRIES

Incrustation in the entries of the tubes can only take place as a result of the passing brine being supersaturated. Decreasing this supersaturation may eventually solve the problem. The highest supersaturation can be found in the boiling zone. In the plug flow zone between the boiling zone and the tube entries (the draft tube) the supersaturation is to some degree removed.

Rumford and Bain (2) mention that at temperatures higher than 50°C, mass transfer from the liquid to the solid phase in the case of the crystallization of sodium chloride is directly proportional to supersaturation:

$$\phi_m'' = K \cdot \Delta C \text{ kg/m}^2 \cdot s \tag{1}$$

Formally we define

$$C_{av} = \frac{1}{\tau} \int_{t=0}^{t=\tau} \Delta C \cdot dt \text{ kg/m}^3 \tag{2}$$

See the list of symbols.

Per m^3 of slurry in the draft tube:

$$\frac{d\Delta C}{dt} = K \cdot A \cdot \Delta C \text{ kg/m}^3 \cdot s \tag{3}$$

Integration:

$$C(L) = C(L_o) \cdot \exp[-K \cdot A \cdot t] \text{ kg/m}^3 \tag{4}$$

This equation offers the clue to the solution. The slurry density varies from 2 to 5% by weight. If we increase the slurry density from 2.5% by weight to 25% by weight (a factor of 10), A increases

INCRUSTATION IN AN INDUSTRIAL SALT CRYSTALLIZER

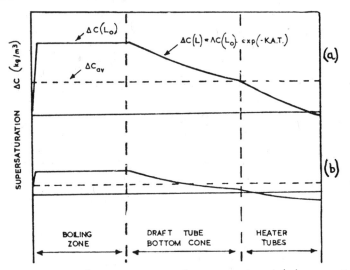

Fig. 5 Supersaturation curves during a cycle: (a) case 1, (b) case 2.

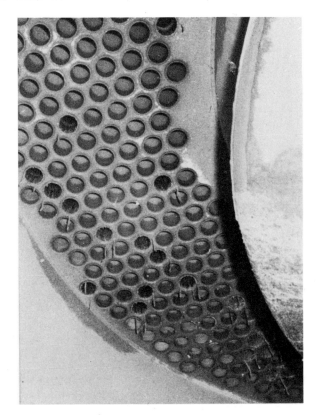

Fig. 6 Lower tube sheet of the modified crystallizer

by a factor of about 8 (not 10 because the salt is coarser). This means an increase of the exponent by a factor of about 8. Furthermore, this leads to a drastic decrease of the supersaturation at the tube entries.

For the two cases a numerical example will be given. It is assumed that, except for the slurry density, all independent variables (liquid level, production rate, circulation rate, direction of circulation, pump speed) are kept constant. The pump speed may be made independent of the circulation rate by adjusting the vanes.

General data:
ρ_s = 2160 kg/m³
ρ_ℓ = 1210 kg/m³
Q_1 = 45 t/h (steam flow)
Q_2 = 14 t/h (salt production)
V = 160 m³ (slurry volume)

		Case I	Case II
Slurry density	% by weight	2.5	25
Salt content	kg/m³	30	338
Average crystal size, 50% by weight	microns	400	600
Uniformity number, RRS-system	-	3	3
Specific area of crystals	m²/kg	7.5	5
Area of crystals (A)	m²/m³	225	1690
Area of crystals in the vessel (A_{tot})	m²	26000	286000

Appendix 1 deals with the estimation of mass transfer coefficients in m/s for the mass transfer of sodium chloride from brine to the solid crystals. It appears that at 120°C the mass transfer coefficient may be assumed to be 5×10^{-4} m/s. The average supersaturation may be calculated using:

$$C_{av} = \frac{\phi'm}{K \cdot A_{tot}} \text{ kg/m}^3$$

	Case I	Case II
Salt production (ϕ'm) kg/s	3.89	3.89
ΔC_{av}	0.22	0.029

From Figs 5a and b the decay of the supersaturation in the draft tube is qualitatively apparent for the Cases Nos. I and II. In case the average supersaturation of Case No. I must be divided by 7.6 (see above) to obtain the average supersaturation of Case

No.II the boiling zone supersaturations will have a ratio of about 5:1 (guestimate).

Dividing the supersaturations for the two cases:

$$\frac{C(L)_I}{C(L)_{II}} = \frac{\Delta C(L_o)_I \cdot \exp[-K A_I t]}{\Delta C(L_o)_{II} \cdot \exp[-K A_{II} t]}$$

$$= \frac{5 \cdot \exp[-5 \cdot 10^{-4} \cdot 225 \cdot 11]}{1 \cdot \exp[-5 \cdot 10^{-4} \cdot 1690 \cdot 11]} = 1.58 \times 10^4$$

Although simplifying assumptions have been made, it is clear that increasing the slurry density by a factor of 10 will decrease the supersaturation at the tube entries by at least 10^4.

SOLUTION OF THE PROBLEM

The bottom of the crystallizer was changed so as to prevent the salt from settling. A bottom-driven pump was installed in order to check the performance of this alternative type of pump (see Fig. 4b). The modifications proved to be a success. With slurry densities between 10 and 25% by weight there was no incrustation at the tube entries. Below 10% by weight the phenomena reappeared. Fig.6 shows the clean tube entries. Lumps of salt were deliberately placed on the upper tube sheet. They were re-dissolved by the undersaturated brine leaving the heat exchanger, thereby proving that the crystallizer can manage incrustations in the boiling zone if slurry densities higher than 10% by weight are maintained. The crystallizer was operated for a month without any water addition at all. After this period, the crystallizer was placed in normal service. A proposal was made to change the bottoms of the other four crystallizers as well.

SYMBOLS

A	crystal area	m^2/m^3
A_{tot}	crystal area in the crystallizer	m^2
ΔC	supersaturation	kg/m^3
ΔC_{av}	average supersaturation	kg/m^3
$\Delta C(L)$	supersaturation with distance L covered in the draft tube	kg/m^3
d	particle size	m
D	diffusion coefficient	m^2/s
K	mass transfer coefficient	m/s
m	mass	kg
Q_1	steam flow	t/h

Q_2	salt production	t/h
Re	Reynolds number ($\rho_\ell vd/\eta_\ell$)	-
Sc	Schmidt number ($\eta_\ell/\rho_\ell D$)	-
t	time	s
T	absolute temperature	K
V	crystallizer slurry volume	m^3
v	particle settling rate	m/s
θ'_m	mass transfer rate	kg/s
θ''_m	mass deposition rate	kg/m^2.s
η_ℓ	brine viscosity	N s/m^2
ρ_ℓ	liquid specific mass	kg/m^3
ρ	solids specific mass	kg/m^3
τ	cycle time	kg/m^3
I and II	numerals referring to cases Nos.I and II	

REFERENCES

1. Richards, R.B., Chem. Eng. Progr., 52 (1956) 6, 58.
2. Rumford, F. and Bain, J., Trans. Instn. Chem. Engrs., 38 (1960) 10.
3. Hughmark, G.A., AIChE. J., 13 (1967) 1219.
4. Perry's Chemical Engineers' Handbook, 5-59, Fourth Edition, McGraw-Hill.

APPENDIX

Estimation of Mass Transfer Coefficients

Approach No.1: This approach extrapolates the experimental results of Rumford and Bain (2), who measured the rate of growth of salt particles. They took 4 g of crystals between 850 and 1400 microns (12 to 18 mesh). The particles were rounded off by abrasion. The crystals were supported by a 20 mesh monel screen in a 1" fluidization tube. The granules were fluidized by supersaturated brine.

Experimental conditions:
Fluidization velocities: 3 and 5 cm/s in the empty tube.
Temperatures: 26 to 73°C (6 values)
Supersaturations: 0.5 to 2.5 g/l (various values).

Their results suggest diffusion control above 50°C, with an activation energy growth of 5.4 kcal/mol. Extrapolation of their Arrhenius-plot to 120°C suggests a mass transfer coefficient of 5.8×10^{-4} m/s.

There is a close analogy between the physical situations in the case of the Rumford and Bain experiment and in the case of the

INCRUSTATION IN AN INDUSTRIAL SALT CRYSTALLIZER

industrial crystallizer. In the experiments of Rumford and Bain the particles were somewhat coarser. Therefore Approach No.2. was made also.

Approach No.2: We will now consider mass transfer coefficients for salt particles falling through a stagnant supersaturated brine. This will be done for spherical particles of 100, 500 and 1000 microns. The particles fall at their terminal velocities. Use may be made of the Froessling equation

$$Sh = 2 + a \cdot Re^{0.50} Sc^{0.33} \qquad [1]$$

Hughmark (3) has analysed the data of many authors. For our case $1 < Re < 450$ and $Sc < 250$ and hence we may use:

$$Sh = 2 + 0.60 \, Re^{0.50} \, Sc^{0.33} \qquad [2]$$

The method given by (4) has been used for determining the settling velocities. The following physical data were already available:

Densities   salt 2.60 kg/m^3, brine at 120°C 1210 kg/m^3

Viscosities   brine at 20°C 2.2×10^{-3} N s/m^2
                  brine at 120°C 0.5×10^{-3} N s/m^2

Diffusivities   Rumford and Bain (2) give 1.35×10^{-9} m^2/s for the diffusion rate of sodium chloride in water at 20°C. On the assumption that $D \eta_\ell / T$ is constant, a value for 120°C can be calculated. In this way 8×10^{-9} N s/m^2 is found.

The results of the calculation are given:

Particle diameter, m	Terminal velocity, m/s	Mass transfer coefficient, m/s
10^{-4}	0.8×10^{-2}	4.1×10^{-4}
5×10^{-4}	6×10^{-2}	3.4×10^{-4}
10^{-3}	12×10^{-2}	3.2×10^{-4}

The K-values according to theoretical predictions clearly vary only little with size.

Taking into account the degree of turbulence in an industrial crystallizer, we may take the mass transfer coefficient to be 5×10^{-4} m/s.

The accuracy is fair enough for this problem.

OPERATION OF A LARGE-SCALE KCl CRYSTALLIZATION PLANT

W. WOHLK AND G. HOFMANN

Standard-Messo Duisburg, 41 Duisburg 1, W. Germany

INTRODUCTION

The crystallization of potassium chloride from mined sylvinite involves a particular problem resulting from the accompanying component, sodium chloride, in the feed solution. To overcome this problem requires very careful planning and the present example of industrial crystallization is ideally suited to describe the many typical problems and their solutions.

Today, the conditions required from a potassium chloride crystallization plant include high capacities (> 100 t/h KCl) most economical operation by reheating the mother liquor with recuperated heat to 80°C, working periods of three weeks without any incrustations and a product with at least 60 wt% of potassium oxide and a mean crystal size of at least 0.6 mm.

The basic process is determined by the composition of the feed solution. For the chosen example, this is given in Fig.1. A product of the specified quality can only be produced by particular measures taken with a view to the solution-flow and the crystallization process. Such measures must be simple and economic to keep the product marketable.

DESCRIPTION OF THE CRYSTALLIZATION PLANT

The operating diagram of the chosen, seven-stage vacuum cooling crystallization plant is presented in Fig.2. This plant produces 140 t/h of potassium chloride from 1200 t/h of feed solution in two parallel seven-stage units. Each stage consists of a turbulence

19,9	wt.-%	Potassium chloride
15,65	wt.-%	Sodium chloride
1,02	wt.-%	Magnesium chloride
1,71	wt.-%	Calcium chloride
0,1	wt.-%	Calcium sulphate
0,25	wt.-%	Potassium bromide
Rest		Water

Temperature of Feed: 102 °C

Fig. 1. Raw Solution

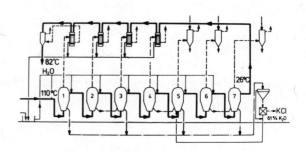

Fig. 2. KCl-Crystallisation Plant (2x1670t KCl/d) Operating Diagram

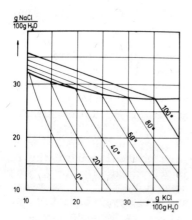

Fig. 3. System KCl—NaCl—H_2O

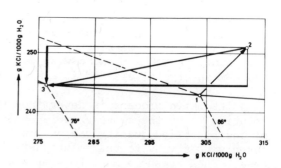

Fig. 4. Crystallisation without Water Addition Vacuum Cooling from 86°C to 76°C

crystallizer. All crystallizers work with the same crystal and suspension residence time, as well as with the same suspension density and supersaturation. The number of crystallization stages resulted from the required reheating of the mother liquor to 80°C and from the final cooling temperature of 26°C which was determined by the available cooling water. Mother liquor reheating is done with the recuperation of heat withdrawn during the preceding vacuum cooling. Of course, only those vapours are used which have a sufficiently high temperature compared with that of the liquor to be reheated. The vapours from the 1st stage, the last heat supplier for the reheating to 80°C, show a temperature of about 74°C. Therefore, indirect heat transfer is not sufficient at this point. As a result of the important boiling point elevation, the required 80°C is finally achieved by condensing vapours direct in the mother liquor. The importance of heat recuperation for the process economy is given by the fact that each additionally recovered degree of heat saves 320000 DM/yr of heating steam costs.

The suspension density of 20 wt% required for the crystallization can only be met on the precondition of providing the separate flow of suspension and mother liquor. This separation of mother liquor and solids is done by the intentionally affected sedimentation in the crystallizers. The overflow of clear liquor from the last crystallization stage is recycled back through the heat recuperation system to the dissolving station. Crystals are withdrawn at four points. This high number of salt discharge points effects an obvious reduction in crystal attrition compared with normal process operation. The suspensions are collected in one prethickener and from there fed into continuously working centrifuges. After separation, the product is dried.

In order to have a long working period free of incrustations, the following factors have been considered: supersaturation only at points where there is sufficient crystal surface available; prevention of heat losses by insulation; prevention of stagnating streams by maintaining permanent flow in all pipes; maintaining the stage temperatures constant by control systems.

The dimensions and arrangements of the erected plant are given on the enclosed photographs. Photograph 1 shows the two parallel, seven-stage units. At the top you can see the three direct contact condensers of the last stages. The photograph was taken at a time when the first set of these two units was already producing. Photographs 2 and 3 allow estimation of the dimensions: they show one of the crystallizers of about 200 m^3 volume.

Process Technique

The process technique is always based on the solubility diagram

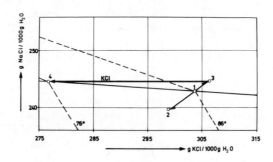

Fig. 5. Crystallisation with Water Addition Vacuum Cooling from 86°C to 76°C

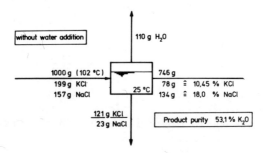

Fig. 6. Product Purity

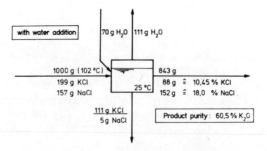

Fig. 7. Product Purity

of the system. In the present case (see Fig.3), potassium chloride, unlike sodium chloride, shows an obvious decrease in solubility by cooling. This explains the choice of vacuum cooling crystallization for the crystallization of potassium chloride from sylvinite. A special feature of this process being used in this system is the necessity of recycling a pre-determined amount of condensate to hinder precipitation of sodium chloride which otherwise is inevitable.

For the second crystallization stage, where the solution is cooled from 86 to 76°C, the concentration relations are given in Fig.4. The solution from the 1st stage has the composition as per point 1. During vacuum cooling, evaporation and cooling happen simultaneously, but for better understanding, the water withdrawal and cooling are shown as two separate operations effected one after the other.

Concentration as per point 2 is reached by evaporation. With regard to the equilibrium temperature of 76°C a supersaturation for both substances, sodium chloride and potassium chloride, results. The supersaturation of sodium chloride, however, has to be hindered because otherwise it would be impossible to obtain the required purity of KCl. This can be done, by adding so much water to the solution from stage 1 to reach point 2 in the solubility diagram (Fig.5). It follows evaporation to point 3 which lies on the horizontal line of point 4. The only resulting supersaturation is that of potassium chloride. Without any water addition, 121 g potassium chloride and 23 g sodium chloride crystallize from 1000 g feed solution. This corresponds to a product purity of only 53.1 wt% (Fig.6).

However, with water addition, every required degree of purity can be achieved. To obtain a product purity of at least 60 wt% potassium oxide, a quantity of added water lower than the quantity of water eveporated by vacuum cooling (Fig.7) will be sufficient.

Among the crystallizers allowing for clear liquor discharge, the classifying and the turbulence crystallizers are the available types. In the present case, the choice between these 2 types was determined, amongst other things, by the investment cost required for equal production capacity. Fig.8 shows a roughly-estimated comparison in size between both types. The classifying crystallizer needs a clearing surface diameter of about 12 to 15 m, while for the magma-type crystallizer a surface diameter of 4.2 m is sufficient. So the type of crystallizer used in this process was that shown in Fig.9. Based on the maximum admissible supersaturation, the internal suspension circulation was determined with 15000 m^3/h in each stage. The determination of the largest diameter of 5.4 m resulted from the requirement for a clear liquor overflow.

Parameter	CSCPR		MSMPR	
Temp. difference between stages	10	degC	10	degC
$\Delta T_{max.}$	1,2 ÷ 0,8	degC	0,5	degC
Feed	500	$m^3 h^{-1}$	500	$m^3 h^{-1}$
Circulation	4160 ÷ 6250	$m^3 h^{-1}$	10 000	$m^3 h^{-1}$
Upstream velocity	10	$mm s^{-1}$	10	$mm s^{-1}$
Diameter of clarifying surface	12,1 ÷ 14,9	m	4,2	m
Production	10	$t h^{-1}$	10	$t h^{-1}$

Fig. 8. Comparison of Size: CSCPR/MSMPR for Estimation of Investment

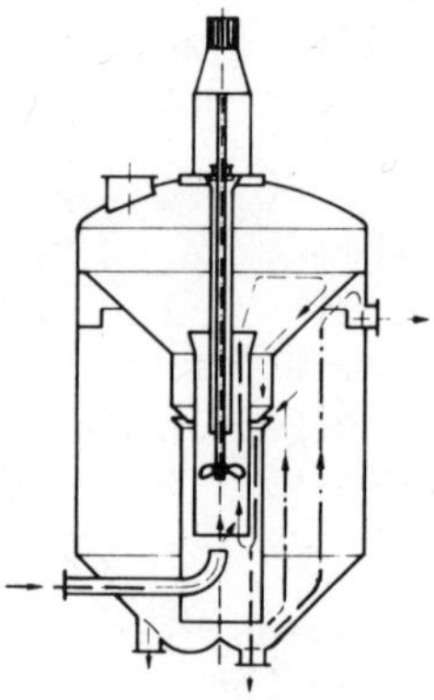

Fig. 9. Standard- Messo Turbulence Crystalliser

98,1 wt. -% Potassium chloride
≙ 62,05 wt. -% Potassium oxide

1,45 wt. -% Sodium chloride
0,084 wt. -% Calcium chloride
0,082 wt. -% Magnesium chloride
0,006 wt. -% Calcium sulphate

Fig. 10. Analysis - Plant Product

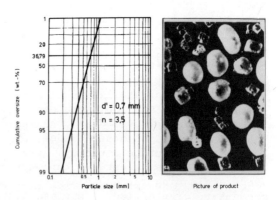

Fig. 11. Product Coming from a Large-Scale KCl-plant

OPERATION OF A KCl CRYSTALLIZATION PLANT

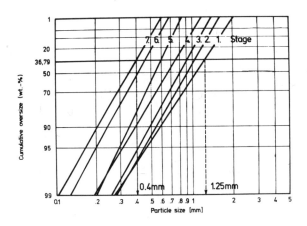

Fig. 12. Crystal Size Distributions in the Stages (RRS).

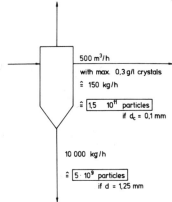

Fig. 13. Calculation of Number of Discharged Particles.

1

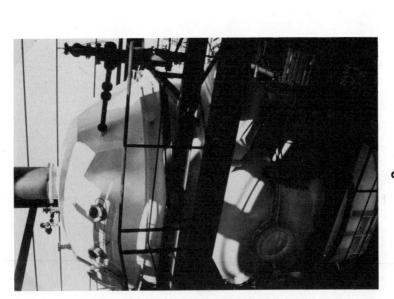

RESULTS

Due to the imperfection of the parameters, the design of a plant can be evaluated only on the basis of the operating results. The main criteria for such a judgement are, in particular: purity, and crystal size, and also the way of meeting the particular requirements for process engineering such as process technical heat recuperation and long working periods without incrustations.

The above described heat recuperation system provides reheating of the mother liquor to 82°C and thereby exceeds the required value by 2°C. Likewise, the measures which were taken to get a long working period without incrustations proved successful; the working period exceeds three weeks.

The composition of the product obtained from the plant is shown in Fig.10. The product purity of 60 w % potassium oxide is absolutely obtained. It is possible to obtain higher degrees by increasing the amount of added water with reduced production capacity, and vice versa. For a purity reduced to 59 wt% potassium oxide, for example, the capacity of the plant was 130%. Fig.11 shows the crystal size distribution and also a photograph of the plant product.

Parameters Influencing on Crystal Size

The decrease in mean crystal size from 1.25 mm in the 1st stage to 0.4 mm in the 7th stage (Fig.12) is remarkable. However, this is well known in potassium chloride crystallization. Examination showed that one of the reasons was the amount of fines increasing from stage to stage. The number of particles withdrawn in the 'clear liquor' overflow exceeded the number in the removed product by several orders of magnitude, e.g. by 10^2 according to the calculations in Fig.13. In the following stage, these small crystals act as nuclei, and their large number effects a decrease in the mean product crystal size.

As a matter of interest, an analysis was made of those factors which could effect an enlargement of crystal size. The result was that variations in suspension density, residence time and supersaturation appeared to be less important than the dissolution of fines contained in the clear liquor overflow.

Fines dissolution could be made with water added to the system to hinder precipitation of sodium chloride. This would only require the provision of a tank of sufficient volume to meet the time required for dissolution. However, the additional investment cost involved would not justify this possible improvement of the crystal product which was already of marketable quality without that measure.

SUBJECT INDEX

Ammonium
  bromide,6
  chloride,6,9,206,210,217
  dihydrogenphosphate,106,123,
      206,208,209,245
  sulphate,253
Attrition of crystals,9,33,59,
      67,77

Barium nitrate,217
Batch crystallizers,335,343,
      353,379
BCF model,98,224
Boric acid,444

Calcium
  carbonate,204
  phosphate,277
  sulphate,173,263
Carbon tetrabromide,124
Chlorobenzenes,353
Citric acid,135
Cobalt carbonate,211
Collision nucleation,10,20,23,
      33,59,77
Constitutional supercooling
      300
Copper sulphate,20,81,444
Coulter counter,24,367
Crystal growth
  computer simulation,91,113
  dispersion,103,145,173,192,
      204

  effects of impurities,8,
      106,167,203,215,239,245,
      253,285,299
  layers,4,123,245
  single crystals,91,123,187,
      223,285
  small crystals,29,363
  spiral,103,130,250
  theories of,3,61,91,203
Crystallization,fractional,
      431
Crystallizers
  batch,335,343
  cascades of,380
  classification in,403
  design of,44,289,291,
      303,311,319,335,375
  direct contact cooling,
      421
  dynamic behaviour of,389
  industrial,51,294,312,373,
      420,449,462,465
  laboratory units,43,58,125,
      138,177,189,246,256,296,
      417,433,441
  MSMPR,28,296,303,319,363,
      389,403
  operation of,373
  pilot plant,295,417,425
  programmed cooling of,353
  removal of crystals from,
      403
  stability of,389
  transient behaviour of,
      375

De Laval nozzle, 67
Dendritic growth, 6, 106, 300
Design of crystallizers, 44, 289, 291, 303, 311, 319, 335, 375
Direct contact cooling, 421
Dispersion, crystal size, 103, 145, 173, 192, 204

Entrainment of evaporators, 313
Equilibrium habit, 203

Fluid energy mill, 67
Fractional crystallization 431

Gel, titanium dioxide, 155
Gibbs-Thomson equation, 34, 173
Grinding of crystals, 67
Gypsum, 263

Habit modification, 201, 203, 215, 239, 245, 253, 285
Hartman-Perdok model, 91, 224
Hexamethylene tetramine, 107, 223
Hydroxyapatite, 277

Ice, nucleation and growth of, 13, 65, 440
Impurities, effect of,
  on growth, 8, 106, 167, 203, 215, 239, 245, 253, 285, 299
  on nucleations, 12, 20
  on Ostwald ripening, 163
Inclusions, liquid, 106
Incrustation, 314, 327, 437, 449

Isomorphous crystals, 269

Kinematic wares, 100, 132
Kossel model, 92

Layer growth, 4, 123, 245
Lithium formate, 105

Magnesium sulphate, 9, 36, 63, 81, 92, 239
Melts,
  fractional crystallization of, 431
  growth in, 113
  nucleation in, 33
Mixed crystals, 269
MSMPR crystallizers, 28, 296, 303, 319, 363, 389, 403
Multiple-effect evaporators, 51, 449, 461

Nickel ammonium sulphate, 81
Nucleation, 1, 3, 23, 33, 51, 61, 67, 77
  effect of impurities on, 12, 20

Optical isomers, 7
Ostwald ripening, 163, 173
Oxalic acid, 205

Paraffin crystals, 106, 205
PBC theory, 203
Pentaerythritol, 23
Phosphate acid hemihydrate, 413
Population balance, 50, 146, 322, 343, 363, 389

INDEX

Potassium
  bromide, 204
  chloride, 105, 123, 163, 204, 209, 210, 461
  dihydrogenphosphate, 106, 145, 285
  iodide, 204
  nitrate, 217, 441
Precipitation, 155
Programmed cooling, 353

Radioactive isotopes, 163
Recrystallization in suspensions, 173
Rochelle salt, 204

Sampling from a crystallizer, 403
Secondary nucleation, 1, 3, 23, 33, 41, 51, 61, 67, 77
Single crystal studies, 91, 123, 187, 223, 285
Size dispersion, 173
Sodium,
  carbonate, 444
  chlorate, 7, 208
  chloride, 51, 105, 107, 163, 187, 207, 209, 210, 449
  hydroxide hydrate, 421
  nitrate, 163
  perchlorate, 208
  sulphate, 342
  thiosulphate, 342
  triphosphate, 205, 210
Solvents, effects of, 205, 223
Spiral growth, 103, 130, 250
Sucrose, 106, 145, 209
Sulfamic acid, 307
Surface energy, 39, 86, 223
Surface integration rates, 370
Surfactants, 263
Sylvinite, 461

Temperature cycling, 168, 173
Tempkin model, 92, 223
Terminal velocities, 414
Titanium dioxide, 155
Triglycine sulphate, 102, 210

Ultrasound, effect on crystallization, 437

Xylene(-p), 441

Zinc sulphate, 239